Polymer Extrusion

Polymer Extrusion

Pierre G. Lafleur
Bruno Vergnes

ISTE

WILEY

The authors gratefully acknowledge Zara Reveley for the English translation of the initial French text.

First published 2014 in Great Britain and the United States by ISTE Ltd and John Wiley & Sons, Inc.

ISTE Ltd
27-37 St George's Road
London SW19 4EU
UK

www.iste.co.uk

John Wiley & Sons, Inc.
111 River Street
Hoboken, NJ 07030
USA

www.wiley.com

Library of Congress Control Number: 2014934410

British Library Cataloguing-in-Publication Data
A CIP record for this book is available from the British Library
ISBN 978-1-84821-650-1

FSC
www.fsc.org
MIX
Paper from responsible sources
FSC® C013604

Printed and bound in Great Britain by CPI Group (UK) Ltd., Croydon, Surrey CR0 4YY

Table of Contents

Introduction

Extrusion is by far the most important and probably the oldest transformation and shaping process of thermoplastic polymers. After synthesis (polymerization) up until the formulation and production of the finished or semi-finished product, this process is almost entirely concerned with synthetic polymers as well as elastomers or food products. However, it is important to specify from the beginning that different applications and processes exist.

The development of the single-screw extrusion, which is directly derived from the Archimedean screw principle, began during the 1880s, initially with rubber and then with polymers around 1940. Nowadays, it is primarily used to make finished or semi-finished products that will undergo a second transformation (e.g. the extrusion of sheets that will then be thermoformed for use in packaging). This process essentially involves the melting and homogenization of the raw materials before pumping the molten polymer through a die (continuous process) or into a mold (in the case of injection, a cyclic process). Globally, more than 90 million tons of thermoplastics are transformed in this way each year.

Table I.1 provides an insight into the main processes involving a single-screw extruder, as well as examples of manufactured products. This process will be clarified and explained in more detail in Chapter 3.

Process	Examples of products	
Profile extrusion	Tubes, pipes, window frames and exterior cladding	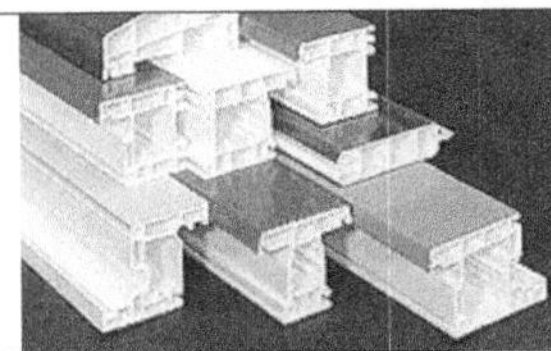
Extrusion – calendering	Wall and floor coverings and tablecloths	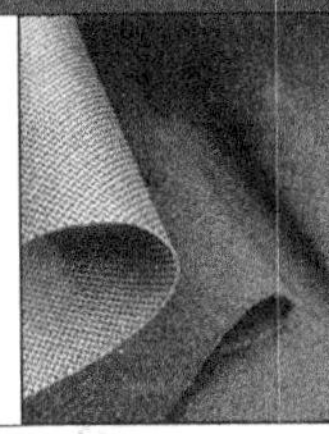
Film blowing	Films for packaging, bags, agricultural mulch, etc.	
Extrusion – blow molding (of hollow bodies)	Water and soda bottles, gasoline containers and tanks.	
Injection molding	Buckets, buttons, containers, syringes, electronic and car parts and pens.	

Table I.1. *Examples of manufactured products from injection and extrusion processes*

Twin-screw extrusion, as the name suggests, is a process involving two screws rotating inside a barrel whose cross section has a figure-eight shape. Also, counter-rotating systems are distinguished from co-rotating systems according to the direction in which the screws rotate with respect to each other.

Counter-rotating extruders are primarily used in polyvinyl chloride (PVC) extrusion, for the manufacture of pipes or profiles. They are not described in this work. A significant development has been seen in co-rotating extruders, due to their great versatility and diverse uses. First and foremost, by being used as a mixing tool or as continuous chemical reactors, they allow specific materials to be produced. It is with this type of machine that polymer alloys, thermoplastic elastomers, filled polymers and nanocomposites are produced as well as all other materials made by reactive extrusion (chemical modifications, grafting, controlled degradation, etc.). This process, shown in Figure I.1, will be reviewed in Chapter 4.

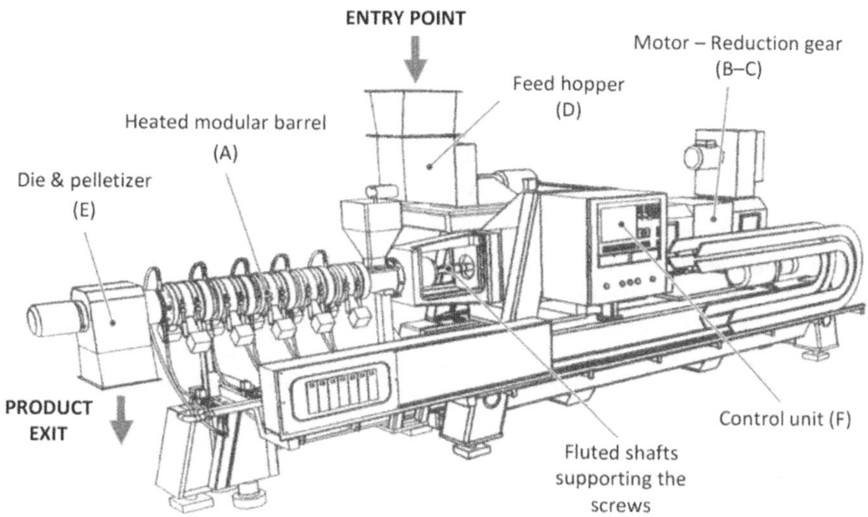

Figure I.1. *Diagram of a co-rotating twin screw extruder (Source: Clextral)*

Manufacturing of a finished product using extrusion is carried out by pushing the molten polymer through a tool called a die, which will give the final shape to the product.

Figure I.2 shows the conventional setup of an extrusion line, with the extruder, the die and the cooling system required to generate an end product.

The die must be designed in a way that allows a product of controlled dimensions to be obtained at a uniform rate and temperature. This poses serious challenges that will be discussed in Chapters 5 (profiles and pipes), 6 (films and sheets) and 7 (cables).

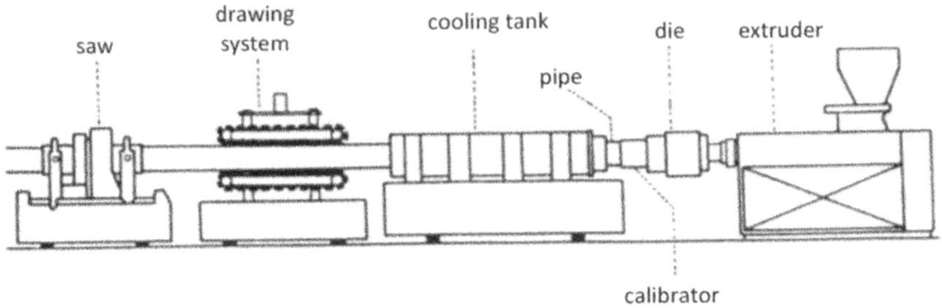

Figure I.2. *Example of a pipe extrusion line*

For each of the points mentioned above, this book first presents the physics of the mechanisms involved and then suggests some models of various sophistication to describe these mechanisms, interpret the results and gain an in-depth understanding of the flow conditions. To help readers with this, some essential reminders about continuum mechanics, rheology and heat transfer are presented in Chapter 1. Chapter 2 includes a short description of the most commonly used calculation methods.

Chapter 1

Continuum Mechanics, Rheology and Heat Transfer Overview

The aim of this chapter is to provide the readers with the basics of continuum mechanics, rheology and heat transfer, which will be of fundamental importance throughout the remainder of this book. We will keep this presentation as concise as possible by avoiding unnecessarily detailed mathematical manipulations, and referring interested readers to other pertinent references in the literature.

1.1. Continuum mechanics

1.1.1. *Strain*

Let us consider the deformation of a continuous medium defined by the displacement vector field of components: $U(x,y,z)$, $V(x,y,z)$, $W(x,y,z)$. The corresponding strain (provided that it is small) can be described by a symmetric tensor $\underline{\underline{\varepsilon}}$ as follows [SAL 88, AGA 14]:

$$\underline{\underline{\varepsilon}} = \begin{bmatrix} \dfrac{\partial U}{\partial x} & \dfrac{1}{2}\left(\dfrac{\partial U}{\partial y}+\dfrac{\partial V}{\partial x}\right) & \dfrac{1}{2}\left(\dfrac{\partial U}{\partial z}+\dfrac{\partial W}{\partial x}\right) \\ \dfrac{1}{2}\left(\dfrac{\partial U}{\partial y}+\dfrac{\partial V}{\partial x}\right) & \dfrac{\partial V}{\partial y} & \dfrac{1}{2}\left(\dfrac{\partial V}{\partial z}+\dfrac{\partial W}{\partial y}\right) \\ \dfrac{1}{2}\left(\dfrac{\partial U}{\partial z}+\dfrac{\partial W}{\partial x}\right) & \dfrac{1}{2}\left(\dfrac{\partial V}{\partial z}+\dfrac{\partial W}{\partial y}\right) & \dfrac{\partial W}{\partial z} \end{bmatrix} \qquad [1.1]$$

Hereafter, a symmetric tensor will be considered as a square matrix, involving six independent terms:

– the diagonal terms $(\varepsilon_{xx}, \varepsilon_{yy}, \varepsilon_{zz})$ correspond to uniaxial deformations of traction or compression;

– the symmetric terms $(\varepsilon_{xy} = \varepsilon_{yx}, \varepsilon_{yz} = \varepsilon_{zy}, \varepsilon_{xz} = \varepsilon_{zx})$ correspond to shear deformations.

1.1.2. *Strain rate*

We will now consider the velocity field $u(x,y,z)$, $v(x,y,z)$, $w(x,y,z)$ which is associated with the aforementioned strain. Just like the strain tensor, the strain rate tensor can be defined by:

$$\underline{\underline{\dot{\varepsilon}}} = \begin{bmatrix} \dfrac{\partial u}{\partial x} & \dfrac{1}{2}\left(\dfrac{\partial u}{\partial y}+\dfrac{\partial v}{\partial x}\right) & \dfrac{1}{2}\left(\dfrac{\partial u}{\partial z}+\dfrac{\partial w}{\partial x}\right) \\ \dfrac{1}{2}\left(\dfrac{\partial u}{\partial y}+\dfrac{\partial v}{\partial x}\right) & \dfrac{\partial v}{\partial y} & \dfrac{1}{2}\left(\dfrac{\partial v}{\partial z}+\dfrac{\partial w}{\partial y}\right) \\ \dfrac{1}{2}\left(\dfrac{\partial u}{\partial z}+\dfrac{\partial w}{\partial x}\right) & \dfrac{1}{2}\left(\dfrac{\partial v}{\partial z}+\dfrac{\partial w}{\partial y}\right) & \dfrac{\partial w}{\partial z} \end{bmatrix} \qquad [1.2]$$

Unlike the strain tensor, defined by relation [1.1] for small deformations, the strain rate tensor is defined in a general manner. It is, therefore, well adapted to the description of fluid flows, for which the deformations are always very large. Just like $\underline{\underline{\varepsilon}}$, the terms for the strain-rate tensor have a particular meaning:

– the diagonal terms are elongation rates, often referred to, below, as $\dot{\alpha}$;

– the symmetric terms are shear rates, often referred to, below, as $\dot{\gamma}$.

EXAMPLE 1.1.– Let us consider two elementary flows, which we will come across often in the remainder of this book. The first flow is a planar shear flow between two plates (see Figure 1.1.(a)). The bottom plate is immobile, whereas the top plate is mobile with a velocity V. The velocity field is, *a priori*, in the following form: $u = u(y)$, $v = 0$ and $w = 0$. The strain rate tensor is reduced to:

$$\dot{\underline{\underline{\varepsilon}}} = \begin{bmatrix} 0 & \frac{1}{2}\frac{du}{dy} & 0 \\ \frac{1}{2}\frac{du}{dy} & 0 & 0 \\ 0 & 0 & 0 \end{bmatrix} \qquad [1.3]$$

The second flow occurs between two immobile plates under a pressure drop (Figure 1.1(b)). This is called a Poiseuille flow. The velocity field has the same form as above, i.e. $u = u(y)$, $v = 0$ and $w = 0$.

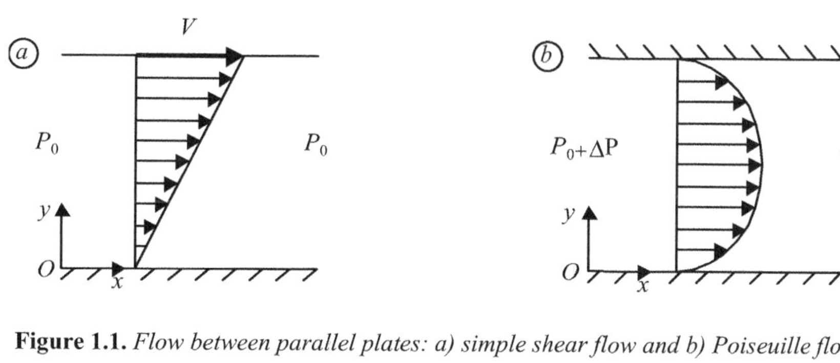

Figure 1.1. *Flow between parallel plates: a) simple shear flow and b) Poiseuille flow*

Consequently, the strain rate tensor will also have the same expression. All flows which give rise to this type of tensor (a single symmetric non-zero term) are called simple shear flows. In the corresponding Cartesian coordinate system, *Ox* is the shear direction, *Oxy* is the shear plane and du/dy is the shear rate.

1.1.3. *Stress*

Let us consider a small surface ds upon which a force dF is exerted (Figure 1.2). By definition, the stress vector $\underline{T}$ at point O of this surface is the limit dF / ds as ds tends toward zero. This vector is projected:

– on the normal $\underline{n}$ to the surface at point O: $\sigma_n = \underline{T}.\underline{n}$ is the normal stress, traction or compression;

– on the tangent plane: σ_t is the shear stress.

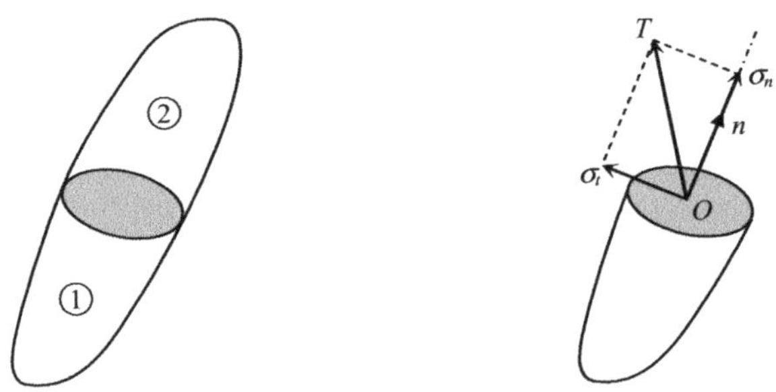

Figure 1.2. *Stress applied on a surface*

The stress vector cannot define a general state of stress, since it is associated with the orientation of the surface on which it is exerted. It is the stress tensor which will define this general state. As with the tensors mentioned above, it is a symmetric tensor written as:

$$\underline{\underline{\sigma}} = \begin{bmatrix} \sigma_{xx} & \sigma_{xy} & \sigma_{xz} \\ \sigma_{xy} & \sigma_{yy} & \sigma_{yz} \\ \sigma_{xz} & \sigma_{yz} & \sigma_{zz} \end{bmatrix} \qquad [1.4]$$

As with the previous tensors, there are six terms involved:

– the diagonal terms, corresponding to normal stresses (along axes x, y and z), of traction or compression;

– the symmetric terms, corresponding to the shear stresses (in the planes xy, xz and yz).

From the stress tensor, the stress vector at any normal point $\underline{n}$ can be calculated using the following equation:

$$\underline{T} = \underline{\underline{\sigma}}.\underline{n} \qquad [1.5]$$

For any stress state, the hydrostatic pressure can be defined by the following equation:

$$p = -\frac{1}{3}\mathrm{tr}(\underline{\underline{\sigma}}) \qquad [1.6]$$

where $\mathrm{tr}(\underline{\underline{\sigma}})$ is the trace of the stress tensor, i.e. the sum of the diagonal terms: $\mathrm{tr}(\underline{\underline{\sigma}}) = \sigma_{xx} + \sigma_{yy} + \sigma_{zz}$.

This allows the stress tensor to be broken down into the sum of the hydrostatic pressure and the traceless tensor, called the deviator:

$$\underline{\underline{\sigma}} = -p\underline{\underline{I}} + \underline{\underline{s}} \qquad [1.7]$$

where $\underline{\underline{I}}$ is the identity tensor. It is the deviator $\underline{\underline{s}}$ that allows us, in section 1.2.2, to define the constitutive laws.

1.1.4. *General equations in continuum mechanics*

From the above concepts, we can now establish the basic equations that will be used to solve a flow problem.

1.1.4.1. *Continuity equation*

The continuity equation expresses the conservation of mass in a flow. It is generally written as:

$$\frac{d\rho}{dt} + \mathrm{div}(\rho\underline{u}) = 0 \qquad [1.8]$$

where ρ is the density, $\underline{u}$ is the velocity vector and "div" is the divergence operator.

In the case of an incompressible medium, as in the case of molten polymers, this equation is simplified to:

$$\mathrm{div}(\underline{u}) = 0 \qquad [1.9]$$

i.e. in Cartesian coordinates: $\frac{\partial u}{\partial x} + \frac{\partial v}{\partial y} + \frac{\partial w}{\partial z} = 0$.

Using the definition of the trace of a tensor, this can also be written as:

$$\mathrm{tr}(\underline{\underline{\dot{\varepsilon}}}) = 0 \qquad [1.10]$$

1.1.4.2. *Equilibrium equation*

The equilibrium equation expresses the equilibrium of forces being exerted upon an element of material. If F represents the forces of mass (e.g. gravity) and $\rho\gamma$ the forces of inertia, the equilibrium equation is written as:

$$\mathrm{div}(\underline{\underline{\sigma}}) + \underline{F} - \rho\underline{\gamma} = 0 \qquad [1.11]$$

The divergence of a tensor is a vector whose components are the divergence of each row of the tensor.

$$\mathrm{div}(\underline{\underline{\sigma}}) = \begin{vmatrix} \frac{\partial \sigma_{xx}}{\partial x} + \frac{\partial \sigma_{xy}}{\partial y} + \frac{\partial \sigma_{xz}}{\partial z} \\ \frac{\partial \sigma_{xy}}{\partial y} + \frac{\partial \sigma_{yy}}{\partial y} + \frac{\partial \sigma_{yz}}{\partial z} \\ \frac{\partial \sigma_{xz}}{\partial z} + \frac{\partial \sigma_{yz}}{\partial z} + \frac{\partial \sigma_{zz}}{\partial z} \end{vmatrix} \qquad [1.12]$$

In the case of an incompressible medium, the unknowns of the problem are the three components of the velocity vector and the six components of the stress tensor. The continuity equation [1.8] and the equilibrium equation [1.11] give us four equations. The five missing equations will be supplied by the constitutive equation, i.e. a relation between the stress tensor (or its deviator) and the strain rate tensor. The definition of these laws is the purpose of rheology, which is presented in the next section.

1.2. Rheology

The molten polymers discussed in this book are naturally viscoelastic. In fact, in the majority of flows encountered in extrusion, it is the viscous characteristic that will predominate and we will limit ourselves solely to the discussion of this behavior. We refer readers who are interested in viscoelasticity to more specialized sources [FER 70, BIR 77, DEA 90].

1.2.1. *Newtonian behavior*

Let us consider the flow between parallel plates presented in Figure 1.1(a). Viscosity is, by definition, the ratio of the stress exerted upon the top plate to the shear rate:

$$\eta = \frac{\tau}{\dot{\gamma}} \qquad [1.13]$$

This viscosity is constant and does not depend on pressure or temperature. The overall equation of Newtonian behavior can be written in terms of the stress and strain rate tensors:

$$\underline{\underline{\sigma}} = -p\underline{\underline{I}} + 2\eta\underline{\underline{\dot{\varepsilon}}} \qquad [1.14]$$

By applying this ratio to the case of shearing between plates, we obtain a shear stress equal to $\eta\dot{\gamma}$ and three identical normal stresses:

$$\underline{\underline{\sigma}} = \begin{bmatrix} -p & \eta\dot{\gamma} & 0 \\ \eta\dot{\gamma} & -p & 0 \\ 0 & 0 & -p \end{bmatrix} \qquad [1.15]$$

The latter is characteristic of inelastic behavior.

The average viscosity of a molten polymer is approximately 10^2 to 10^4 Pa.s; this is very significant compared to the viscosity of conventional fluids such as water (10^{-3} Pa.s) and oil (10^{-2} to 1 Pa.s).

This has many important consequences for flow, especially since, in the majority of cases, the mass and inertia forces are negligible in comparison to viscous forces. Therefore, in extrusion, even at high rates, the flow will always be laminar, well below turbulent regimes. Mixing operations will certainly become much more difficult.

By introducing the equation for Newtonian behavior [1.14] into the equilibrium equation [1.11], and after deeming the body forces and inertia forces to be negligible, we obtain the Stokes equations:

$$\mathrm{grad}\, p = \eta \, \Delta \underline{u} \qquad [1.16]$$

linking the pressure gradient to the Laplacian of the velocity field. For simple shear flows between plates, which have already been mentioned, this is reduced to:

$$\frac{\partial p}{\partial x} = \eta \frac{d^2 u}{dy^2}, \; \frac{\partial p}{\partial y} = 0, \; \frac{\partial p}{\partial z} = 0 \qquad [1.17]$$

From this, we deduce that the pressure gradient is constant. For the shear flow (Figure 1.1(a)), this means that the pressure between the two planes is uniform. From this, we can deduce that the velocity profile is linear and independent of viscosity:

$$u(y) = \frac{V}{h} y \qquad [1.18]$$

For the Poiseuille flow (Figure 1.1(b)), by considering the velocities to be zero on the upper and lower planes, we obtain:

$$u(y) = \frac{1}{2\eta} \frac{\Delta p}{L} (h - y) y \qquad [1.19]$$

where L is the flow length. The velocity profile is parabolic. The average velocity is obtained from equation [1.20]. It is proportional to the pressure gradient and inversely proportional to the viscosity.

$$\bar{V} = \frac{h^2}{12\eta} \frac{\Delta p}{L} \qquad [1.20]$$

1.2.2. *General viscous behavior*

It is well known that molten polymers are not Newtonian, and their viscosity depends on shear rate: $\eta = \eta(\dot{\gamma})$. For the most part, we are dealing with shear thinning behavior, which means that the viscosity decreases when the shear rate increases (Figure 1.3).

A certain number of semi-empirical laws have been proposed to describe this behavior. The laws most commonly used are:

– the power law (or Ostwald-de Waele law):

$$\eta = K\dot{\gamma}^{m-1} \qquad [1.21]$$

where K is the consistency of the material (expressed in Pa.s ^{m}) and m is the flow index. The index m has a value of 1 for a Newtonian fluid (thus, we find that $K = \eta$ = constant) and 0 for a rigid plastic body. The index m is generally between 0.2 and 0.5 for the majority of conventional polymers.

– Carreau's law:

$$\eta - \eta_\infty = (\eta_0 - \eta_\infty)\left[1 + (\lambda\dot{\gamma})^2\right]^{\frac{m-1}{2}} \qquad [1.22]$$

where η_0 is the viscosity at low shear rates (Newtonian plateau), η_∞ is the viscosity at very high (or theoretically "infinite") shear rates and λ is the characteristic time of the material. In practice, for molten polymers, $\eta_\infty = 0$. For high shear rates, this law is equivalent to the power law. By contrast, it allows the Newtonian plateau observed at low shear rates to be introduced.

A more general form is the Carreau–Yasuda law:

$$\eta = \eta_0\left[1 + (\lambda\dot{\gamma})^a\right]^{\frac{m-1}{a}} \qquad [1.23]$$

The parameter a allows the transition between the Newtonian plateau and the power-law zone to be accurately adjusted.

All of these one-dimensional (1D) laws can be generalized as a tensorial equation:

$$\underline{\underline{\sigma}} = -p\underline{\underline{I}} + 2f(\dot{\bar{\gamma}})\underline{\underline{\dot{\varepsilon}}} \quad [1.24]$$

where $f(\dot{\bar{\gamma}})$ is the 1D law, as defined above, and $\dot{\bar{\gamma}}$ is the generalized shear rate, defined by:

$$\dot{\bar{\gamma}} = \sqrt{2\sum_{i,j}\dot{\varepsilon}_{ij}^2} \quad [1.25]$$

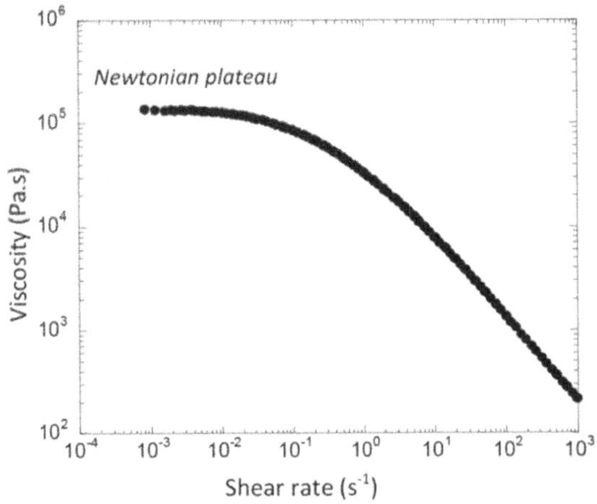

Figure 1.3. *Example of the viscosity curve of a molten polymer*

1.2.3. *Effects on pressure and temperature*

Temperature has a great impact on viscosity. Thermorheologically simple polymers follow the time–temperature superposition principle [FER 70], i.e. we can express the viscosity at temperature T as a function of both its value at a reference temperature T_0 and the shift factor a_T:

$$\eta(\dot{\gamma},T) = a_T\,\eta(\dot{\gamma}a_T,T_0) \quad [1.26]$$

The latter is often expressed by the Arrhenius equation:

$$a_T = \exp\left[\frac{E_a}{R}\left(\frac{1}{T}-\frac{1}{T_0}\right)\right] \quad [1.27]$$

where E_a is the activation energy and R is the ideal gas constant. E_a varies between 20 and 100 kJ/mole for the majority of polymers, which means that a change of 10°C will cause the viscosity to change by 10–70%.

The viscosity also depends on pressure and is expressed as:

$$\eta(p) = \eta_0 \exp(\chi p) \qquad [1.28]$$

The compressibility coefficient χ is approximately 10^{-8} Pa^{-1}, which means that the effect of pressure becomes apparent beyond several dozen MPa only. This effect will often, therefore, be negligible in extrusion.

1.3. Heat transfer [CAR 59, BIR 60, AGA 14]

1.3.1. *The thermal balance equation*

The variation in temperature over time is controlled by the thermal balance equation, which – in the simplest thermodynamic example, used frequently hereafter – is written as follows:

$$\rho C_p \frac{dT}{dt} = -\text{div}\underline{q} + \dot{W} \qquad [1.29]$$

where:

- C_p is the specific heat capacity;
- $\underline{q}$ is the heat flux;
- $\dot{W}$ is the power dissipated per unit volume;
- dT / dt is the material derivative, which is expressed as:

$$\frac{dT}{dt} = \frac{\partial T}{\partial t} + \underline{u}\,\text{grad}\,T \qquad [1.30]$$

where $\partial T / \partial t$ is the variation in temperature over time at a fixed point and $\underline{u}\,\text{grad}\,T$ is the amount of heat "induced" by movement, i.e. convection. Obviously, this term is zero if the material is at rest.

The heat flux $\underline{q}$ is most commonly described by Fourier's law:

$$\underline{q} = -k \, \mathrm{grad} \, T \quad [1.31]$$

where k is the thermal conductivity. Therefore, conduction is written as:

$$-\mathrm{div}\underline{q} = k \Delta T \quad [1.32]$$

where Δ is the Laplace operator.

The power dissipated by deformation is expressed by:

$$\dot{W} = \underline{\underline{\sigma}} \, \underline{\underline{\dot{\varepsilon}}} = \sum_{i,j} \sigma_{ij} \, \dot{\varepsilon}_{ij} \quad [1.33]$$

From the constitutive law used, we can deduce different expressions, for example:

– $\dot{W} = \eta \dot{\bar{\gamma}}^2$ for a Newtonian fluid;

– $\dot{W} = K \dot{\bar{\gamma}}^{m+1}$ for a power-law fluid.

Thus, for a stationary flow of a Newtonian fluid (see Figure 1.1(a)), the thermal balance equation is written as:

$$\rho C_p u \frac{\partial T}{\partial x} = k \left(\frac{\partial^2 T}{\partial x^2} + \frac{\partial^2 T}{\partial y^2} \right) + \eta \left(\frac{V}{h} \right)^2 \quad [1.34]$$

To calculate the temperature field, it is necessary to add the boundary conditions to this equation, both at the entry and at the wall. At the entry, we generally use an isothermal condition: $T(0) = T_0$. The boundary conditions can be:

– either an imposed temperature: $T(0) = T(h) = T_p$;

– or an imposed heat flux: $\underline{q} = -k \dfrac{\partial T}{\partial \underline{n}} = h_T (T_p - T_\infty)$.

where T_p is the wall temperature, h_T is the heat transfer coefficient and T_∞ is an infinite temperature of the surrounding environment.

In the remainder of this book, heat transfer is mainly considered in two cases:

– the case of a flow inside an extruder or a die;

– the case where a material is cooled at the die exit.

1.3.2. *Heat transfer during flow*

When a molten polymer flows through a piece of equipment, its temperature increases according to the power dissipated by deformation and heat transfer by conduction through the walls. Depending on the respective weights of these different terms, different thermal situations can be observed, which we will review using Poiseuille's flow between parallel plates (Figure 1.1(b)). In most cases, for a simple shear flow, the thermal balance equation in a stationary regime for a Newtonian fluid will be written as follows:

$$\rho C_p u \frac{\partial T}{\partial x} = k\left(\frac{\partial^2 T}{\partial x^2} + \frac{\partial^2 T}{\partial y^2}\right) + \eta\left(\frac{du}{dy}\right)^2 \quad [1.35]$$

We can show that in the x-direction, the conduction (in terms of $k\partial^2 T / \partial x^2$) is negligible compared to the convection (in terms of $\rho C_p u \partial T / \partial x$). Equation [1.35] can thus be simplified into:

$$\rho C_p u \frac{\partial T}{\partial x} = k \frac{\partial^2 T}{\partial y^2} + \eta\left(\frac{du}{dy}\right)^2 \quad [1.36]$$

1.3.2.1. *The equilibrium regime*

Temperature is, *a priori,* a function of both x and y: there is a temperature profile along the thickness of the flow which will evolve with the latter. After a certain amount of time, equilibrium should be reached, which then does not alter throughout the rest of the flow. An ideal scenario of thermal equilibrium would be when the dissipated power is entirely compensated for by heat transfer through the walls. Using the velocity field equation of this flow (equation [1.19]), we can, therefore, write:

$$k\frac{\partial^2 T}{\partial y^2} = -\eta\frac{36\bar{V}^2}{h^4}(h-2y)^2 \qquad [1.37]$$

By supposing that the temperature imposed on the walls is equal to T_p, the integration of equation [1.37] gives:

$$T(y) = T_p + \frac{3}{4}\frac{\eta\bar{V}^2}{k}\left[1-\left(1-\frac{2y}{h}\right)^4\right] \qquad [1.38]$$

We obtain a quadratic temperature profile, with a maximum at the center of the flow, characterized by:

$$T_{max} = T_p + \frac{3}{4}\frac{\eta\bar{V}^2}{k} \qquad [1.39]$$

In practice, this equilibrium state is hardly ever found in molten polymer flows. In fact, given the weak thermal diffusivity of these products, the time or distance needed to achieve equilibrium would be very high compared to actual processing times or distances.

1.3.2.2. *The adiabatic flow*

Let us now consider the case where conduction is negligible compared to the dissipated power. The thermal balance equation is, therefore, reduced to:

$$\rho C_p u\frac{dT}{dx} = \eta\frac{36\bar{V}^2}{h^4}(h-2y)^2 \qquad [1.40]$$

In this case, the temperature profile will evolve with the flow, without any limitations, since heat exchange with the exterior is excluded. Equation [1.40] can no longer be solved analytically. We can, however, propose an approximate solution in the form of an average temperature. For this, we will consider a slab with thickness dx, whose average temperature is defined by:

$$\bar{T}(x) = \frac{1}{h\bar{V}}\int_0^h u(y)T(y)\,dy \qquad [1.41]$$

The thermal balance equation in terms of average temperature is therefore written as:

$$\rho C_p h \bar{V} \frac{d\bar{T}}{dx} = 36\eta \frac{\bar{V}^2}{h^4} \int_0^h (h-2y)^2 \, dy \qquad [1.42]$$

that is: $$\frac{d\bar{T}}{dx} = 12\eta \frac{\bar{V}}{\rho C_p h^2} \qquad [1.43]$$

Supposing that the viscosity remains constant and using the average velocity [1.20], by integrating equation [1.43] we obtain:

$$\bar{T}(x) = T_0 + \frac{\Delta p}{\rho C_p} \frac{x}{L} \qquad [1.44]$$

where T_0 is the initial temperature and L is the flow length. We found that the average temperature increases linearly. After a given length L, the increase in temperature is:

$$\Delta T = \bar{T}(L) - T_0 = \frac{\Delta p}{\rho C_p} \qquad [1.45]$$

In this form, we found that the increase in temperature in an adiabatic system is directly proportional to the pressure drop and is independent of viscosity. This value, which is easy to calculate, is overestimated, since heat exchange limits the increase in temperature. We can show that equation [1.45] is valid for any Poiseuille flow, independent of the rheological behavior and the geometry.

1.3.2.3. *Transitory regime*

In fact, in most cases, the terms "conduction" and "dissipation" will be taken into account. We are, therefore, left with the most general form of the thermal balance equation [1.36]. Also, with regard to the numerical solutions of this equation, it is often useful to propose an approximate solution for the average temperature. As in the previous section, by taking into account a slab of material with a thickness of *dx*, we can write:

$$\rho C_p h \bar{V} \frac{dT}{dx} = -2q + 36\eta \frac{\bar{V}^2}{h^4} \int_0^h (h-2y)^2 \, dy \qquad [1.46]$$

where q is the heat flux exchanged via the surface. This is generally expressed using heat transfer coefficient:

$$q = h_T(\bar{T} - T_p) \qquad [1.47]$$

h_T can also be defined as an internal Nusselt number:

$$h_T = \frac{k\,Nu}{h} \qquad [1.48]$$

The thermal balance equation for average temperature is written as follows:

$$\frac{d\bar{T}}{dx} = 2\frac{h_T}{\rho C_p \bar{V}}\frac{T_p - \bar{T}}{h} + 12\eta\frac{\bar{V}}{\rho C_p h^2} \qquad [1.49]$$

which allows the local evolution of temperature in the section of material dx to be calculated. Later on, we will see many examples where these calculation methods are used to estimate temperature fields.

1.3.3. *Cooling temperature*

Let us consider a film with a thickness of $2e$ and a uniform velocity u exiting a flat die (Figure 1.4).

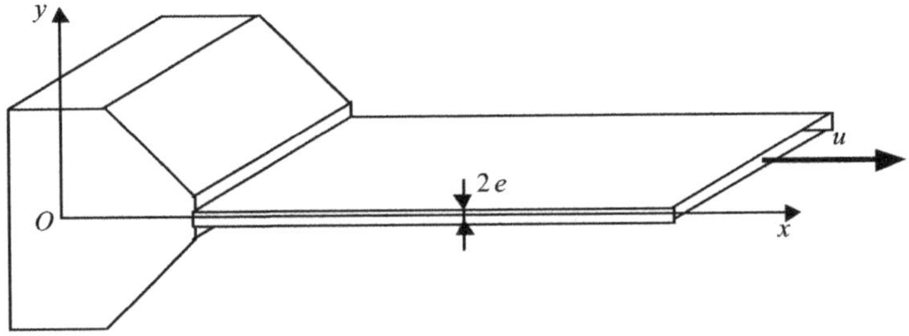

Figure 1.4. *Cooling of a film exiting the die*

If we assume that it is not subjected to stretching, the power dissipated by deformation is negligible. The thermal balance equation is written as:

$$\rho C_p u \frac{\partial T}{\partial x} = k\left(\frac{\partial^2 T}{\partial x^2} + \frac{\partial^2 T}{\partial y^2}\right) \quad [1.50]$$

As discussed in section 1.3.2, the conduction according to x is negligible compared to the corresponding term "convection". This is then reduced to:

$$\rho C_p u \frac{\partial T}{\partial x} = k \frac{\partial^2 T}{\partial y^2} \quad [1.51]$$

By introducing the thermal diffusivity a ($a = k/(\rho\ C_p)$), this equation is written as:

$$u \frac{\partial T}{\partial x} = a \frac{\partial^2 T}{\partial y^2} \quad [1.52]$$

Solving this equation will give us the temperature evolution of the film on exiting the die. This will be strongly dependent on the boundary conditions determining it. Here, we will assume that the cooling occurs on contact with air, with a heat transfer coefficient h_T. The boundary conditions are:

– a uniform temperature exiting the die: $T = T_0$ at $x = 0$;

– a heat flux on the upper and bottom surfaces:

$$k \frac{\partial T}{\partial y} = \pm h_T \left[T(x) - T_f \right] \text{ in } y = \pm e$$

where T_f is the room temperature.

Equation [1.52] can be solved by using numerical methods or series expansion [CAR 59]. We can show that if the thickness e is small and the amount of transferred heat is low, the temperature is uniform and its evolution in the heat direction is, therefore, expressed by:

$$\frac{\bar{T}(x) - T_f}{T_0 - T_f} = \exp\left(-\frac{h_T}{\rho C_p} \frac{x}{eu} \right) \quad [1.53]$$

The major challenge here is to accurately determine the heat transfer coefficient. For more information, we advise readers to refer to more specialized references [CAR 59, BIR 60, AGA 14].

Now let us take the thermal balance equation and apply it to the cooling of a stationary semi-infinite medium. Then we obtain:

$$\rho C_p \frac{\partial T}{\partial t} = k \frac{\partial^2 T}{\partial x^2} \quad [1.54]$$

which can also be written as:

$$\frac{\partial T}{\partial t} = a \frac{\partial^2 T}{\partial x^2} \quad [1.55]$$

This equation is similar to equation [1.52]. Two important concepts can be derived from this. The first one is the penetration depth: assuming $t = 0$, we set a temperature T_s on the surface of the semi-infinite solid, with an initial temperature T_0.

The evolution of temperature over time at a distance x of the surface is given by:

$$\frac{T(x) - T_s}{T_0 - T_s} = \mathrm{erf}\left(\frac{x}{\sqrt{at}}\right) \quad [1.56]$$

where "erf" is the error function defined by:

$$\mathrm{erf}(x) = \frac{2}{\sqrt{\pi}} \int_0^x \exp(-u^2)\, du \quad [1.57]$$

Expression [1.56] links the time and depth to which the difference in temperature is felt on the surface. If we consider an order of magnitude of 50% of this difference, then we will obtain a very simple expression, allowing the order of magnitude for a cooling time to be quickly determined for a given thickness:

$$t \approx \frac{x^2}{a} \quad [1.58]$$

With the diffusivity of a molten polymer having an average value of $10^{-7}\ m^2\ s^{-1}$, we can estimate that the time required to cool a 100 μm film and a 10 mm thick plate is 0.1 s and 17 min, respectively. From this simple example, we can understand why the extruded products are generally very thin, a few millimeters only.

The second important concept is the interface temperature. If we put two semi-infinite solids of temperatures T_1 and T_2, respectively, in direct contact, the interface temperature will be determined from these temperatures and the thermal effusivity of both media:

$$T_i = \frac{b_1 T_1 + b_2 T_2}{b_1 + b_2} \qquad [1.59]$$

where b_i is the thermal effusivity defined by:

$$b_i = \sqrt{k \rho C_p} \qquad [1.60]$$

Expression [1.59] shows that the body with the highest effusivity controls the interface temperature. Thus, on contact with air, the interface temperature of a polymer will be very close to that of the polymer itself.

However, for polymer/steel or polymer/water contacts, the interface temperature will be controlled by that of the water or steel. This is why extrusion is carried out in air when it is necessary to continue deforming the polymer after exiting the die (e.g. for the stretching of films or filaments) and in water (or in a calibrator) when wanting to quickly fix the shape of an extruded product (e.g. tubes or profiles).

1.4. Bibliography

[AGA 14] AGASSANT J.-F., AVENAS P., SERGENT J.-P., *et al.*, *Mise en forme des polymères. Approche thermomécanique de la plasturgie*, Tec & Doc, Lavoisier, Paris, 2014.

[BIR 60] BIRD R.B., STEWART W.E., LIGHTFOOT E.N., *Transport Phenomena*, Wiley, New York, 1960.

[BIR 77] BIRD R.B., ARMSTRONG R.C., HASSAGER O., *Dynamics of Polymeric Liquids*, vol. 1, Fluid Mechanics, Wiley, New York, 1977.

[CAR 59] CARSLAW H.S., JAEGER J.C., *Conduction of Heat in Solids*, Oxford University Press, Oxford, 1959.

[DEA 90] DEALY J.M., WISSBRUN K.F., *Melt Rheology and Its Role in Plastics Processing*, Van Nostrand, New York, 1990.

[FER 70] FERRY J.D., *Viscoelastic Properties of Polymers*, Wiley, New York, 1970.

[SAL 88] SALENÇON J., *Mécanique des milieux continus*, Ellipses, Paris, 1988.

Chapter 2

Calculation Methods

2.1. Introduction

The aim of this chapter is to provide the basics of the methods used to calculate the polymer melt flows in extruders and extrusion dies. The main flow parameters can be calculated by solving the set of equations consisting of:

– conservation of mass;

– conservation of momentum;

– conservation of energy;

– a constitutive equation.

Although three-dimensional (3D) models increase the accuracy of calculating the flow parameters, it is often advantageous to simplify the analysis and restrict it to only 1- or 2D models [VER 86]. Therefore, at first, we present calculations of one-directional isothermal flows in simple geometries. We then discuss a 2D finite difference method, which allows flow equations in more complex geometries to be quickly solved.

For more sophisticated methods and 3D calculations often associated with finite element methods, we refer readers to more specialized sources [LUC 96, FOR 01, DON 03, CHA 04, DHA 05, ZIE 05].

2.2. 1D solutions

2.2.1. *Isothermal calculation*

For simple geometries (flow between parallel plates, flow in a circular tube, radial flow between two parallel disks and flow in an annular tube), the solution to the Poiseuille flow equations for a Newtonian fluid and a power-law non-Newtonian fluid is presented in Tables 2.1 and 2.2. In these tables, the velocity profile and the flow rate are presented as a function of the pressure drop.

For a shear flow between parallel plates with width W, separated by height H (bottom plate is stationary; top plate moving at velocity V), the velocity field and the flow rate are defined by:

$$v_z(y) = \frac{Vy}{H} \quad [2.1]$$

$$Q = \frac{VHW}{2} \quad [2.2]$$

These solutions were obtained for a flow complying with the following assumptions:

– isothermal flow: the temperature of the fluid and walls are equal and constant;

– steady-state regime: the flow is independent of time;

– laminar flow: low Reynolds number (standard assumption in the case of molten polymers);

– incompressible fluid: the density is constant;

– no-slip boundary condition: the velocity is zero at the boundary.

When the behavior of a fluid is more complex (e.g. Carreau's law), a simple analytical solution does not always exist. One way to implement the pressure drop/flow rate relation is to use the characteristic value method [MIC 84]. This technique is based on the existence of a point in the flow for which, at a fixed volumetric flow rate, the shear rate value for a power-law fluid and for a Newtonian fluid are identical (Figure 2.1).

Geometry	Solution
Parallel plates 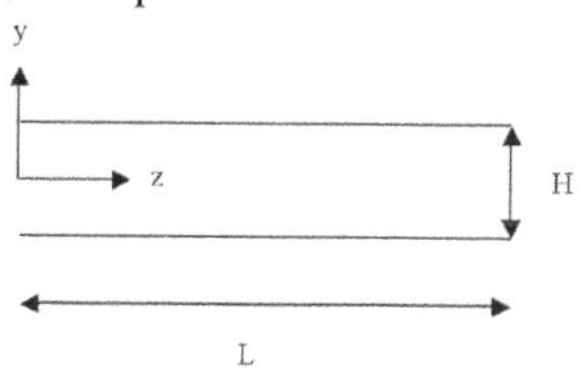 W: width	Newtonian $$v_z(y)=\frac{H^2\Delta P}{8\eta L}\left[1-\left(\frac{2y}{H}\right)^2\right]$$ $$Q=\frac{WH^3\Delta P}{12\eta L}$$
	Power law $$v_z(y)=\frac{H}{2(1+1/m)}\left[\frac{H\Delta P}{2KL}\right]^{1/m}\left[1-\left(\frac{2\|y\|}{H}\right)^{1+1/m}\right]$$ $$Q=\frac{WH^2}{2(2+1/m)}\left[\frac{H\Delta P}{2KL}\right]^{1/m}$$
Circular tube 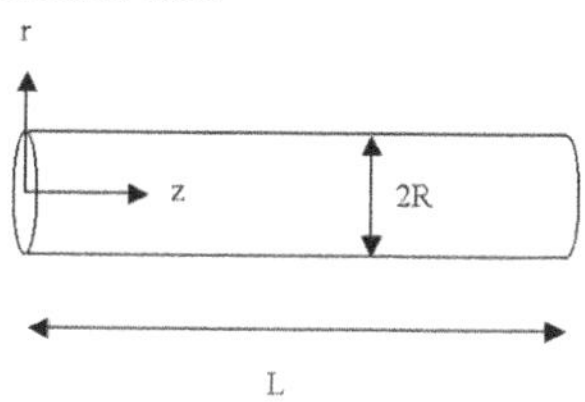	Newtonian $$v_z(r)=\frac{R^2\Delta P}{4\eta L}\left[1-\left(\frac{r}{R}\right)^2\right]$$ $$Q=\frac{\pi R^4\Delta P}{8\eta L}$$
	Power law $$v_z(r)=\frac{R}{(1+1/m)}\left[\frac{R\Delta P}{2KL}\right]^{1/m}\left[1-\left(\frac{r}{R}\right)^{1+1/m}\right]$$ $$Q=\frac{\pi R^3}{3+1/m}\left[\frac{R\Delta P}{2KL}\right]^{1/m}$$

Table 2.1. *1D solution of Poiseuille flow equations*

Geometry	Solution
Parallel disks (radial flow) 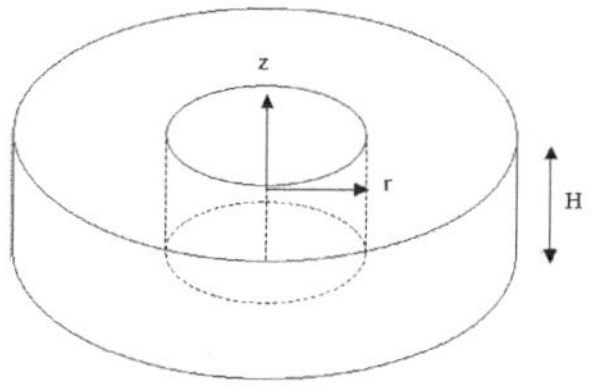 R_1: internal radius R_2: external radius	Newtonian $$v_r(r,z)=\frac{H^2\Delta P}{8\eta\, r\,\mathrm{Ln}(R_2/R_1)}\left[1-\left(\frac{2z}{H}\right)^2\right]$$ $$Q=\frac{\pi H^3}{6\eta}\frac{\Delta P}{\mathrm{Ln}(R_2/R_1)}$$

	Power law $v_r(r,z) = v_{r\max}\left[1-\left(\frac{2z}{H}\right)^{1+1/m}\right]$ $v_{r\max} = \left[\frac{(1-m)\Delta P}{R_2^{1-m}-R_1^{1-m}}\frac{H}{2K}\right]^{1/m}\left[\frac{H}{2r(1+1/m)}\right]$ $Q = \left[\frac{(1-m)\Delta P}{R_2^{1-m}-R_1^{1-m}}\frac{H}{2K}\right]^{1/m}\left[\frac{\pi H^2}{(2+1/m)}\right]$
Annular tube 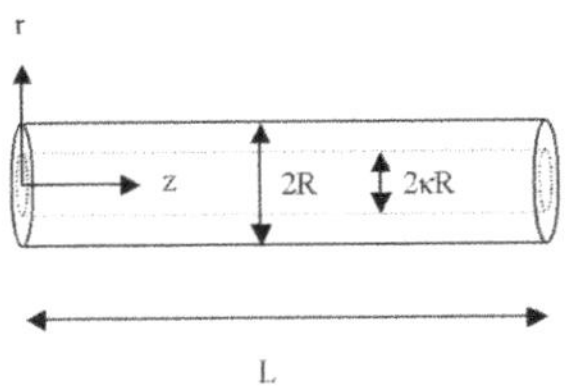 R: internal radius κR: external radius	Newtonian $v_z(r) = \frac{R^2\Delta P}{4\eta L}\left[1-\left(\frac{r}{R}\right)^2+\left(\frac{1-\kappa^2}{\mathrm{Ln}(1/\kappa)}\right)\mathrm{Ln}\frac{r}{R}\right]$ $Q = \frac{\pi R^4\Delta P}{8\eta L}\left[(1-\kappa^4)-\frac{(1-\kappa^2)^2}{\mathrm{Ln}(1/\kappa)}\right]$
	Power law No simple analytical solution

Table 2.2. *1D solution of Poiseuille flow equations*

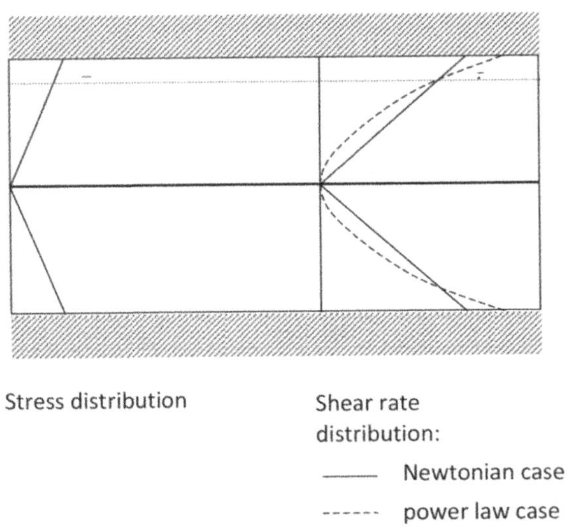

Figure 2.1. *Determination of the shear rate [MIC 84]*

Let us consider the flow of a fluid between two parallel plates. In the case of a Newtonian fluid, the shear rate is given by equation [2.3]:

$$\dot{\gamma}(y) = \left| \frac{\mathrm{d}\, v_z(y)}{\mathrm{d}\, y} \right| = \frac{\Delta P}{\eta L} y \qquad [2.3]$$

In the case of a power-law fluid, this expression becomes:

$$\dot{\gamma}(y) = \left(\frac{\Delta P}{KL} \right)^{1/m} y^{1/m} \qquad [2.4]$$

Thus, there is a characteristic value y_s, defined by:

$$\frac{\Delta P}{\eta L} y_S = \left(\frac{\Delta P}{KL} \right)^{1/m} y_S^{1/m} \qquad [2.5]$$

by using the pressure drop/flow-rate relationship from Table 2.1:

$$\frac{y_S}{H/2} = \left(\frac{3}{2 + 1/m} \right)^{\frac{m}{1-m}} \qquad [2.6]$$

This ratio is called the characteristic value and is denoted as e. Table 2.3 gives the values of e for a power-law fluid flowing in simple geometries.

Geometry	Characteristic value
Parallel plates	$e = \frac{y_s}{H/2} = \left(\frac{3}{2+1/m} \right)^{\frac{m}{1-m}}$
Circular tube	$e = \frac{r_s}{R} = \left(\frac{4}{3+1/m} \right)^{\frac{m}{1-m}}$
Radial flow between two discs	$e = \frac{y_s}{H/2} = \left(\frac{3}{2+1/m} \right)^{\frac{m}{1-m}}$

Table 2.3. *The characteristic value in different geometries for a power-law fluid*

For more complex fluids (e.g. Carreau-Yasuda fluids), the value of e can be calculated numerically. For a conventional range of m values, this value is approximately equal to 0.640 for flow between two parallel plates and 0.785 for flow in a circular tube.

Once the value of e has been determined, the shear rate can be calculated at this characteristic position. The viscosity, estimated from the constitutive equation for this particular shear rate, can then be inserted into the Newtonian solution for this problem. Therefore, this method allows the relationship between the pressure drop and the non-Newtonian flow rate to be easily estimated.

2.2.2. *Non-isothermal calculations*

The temperature evolution is an important part of the flow characterization. In fact, during polymer extrusion, thermal phenomena associated with the flow can be significant:

– the high viscosity of molten polymers leads to significant energy dissipation;

– the high thermodependence of the viscosity, on the other hand, tends to reduce this energy dissipation;

– finally, the low thermal conductivity of polymers limits heat transfer to the external environment.

Calculating the temperature within the flow is carried out by solving the energy equation (Chapter 1). Agassant *et al.* [AGA 14] designed a simple method that allows the average temperature in a transient regime to be determined for a 1D flow. For a Newtonian flow in a cylindrical pipe, and in the case where the viscosity is not thermodependent, they suggest calculating the average temperature at a longitudinal position z using the following equation:

$$\overline{T}(z) = T_0 + \frac{5}{6}\eta\frac{\overline{V}^2}{k}\left(1+\frac{6}{5Br}\right)\left[1-\exp\left(-\frac{48}{5}Ca\frac{z}{L}\right)\right] \qquad [2.7]$$

For flows between parallel plates, this becomes:

$$\overline{T}(z) = T_0 + \eta\frac{\overline{V}^2}{3k}\left(1+\frac{3}{Br}\right)\left[1-\exp\left(-12\,Ca\frac{z}{L}\right)\right] \qquad [2.8]$$

In equations [2.7] and [2.8], $\bar{T}(z)$ represents the average temperature at position z, T_0 represents the initial temperature at position $z = 0$, η represents the viscosity, $\bar{V}$ represents the average velocity in the direction of the flow, k represents the thermal conductivity, *Ca* represents the Cameron number, *Br* represents the Brinkman number and L represents the flow length.

The Cameron number primarily reflects the type of thermal regime. It is defined by:

$$Ca = \frac{k}{\rho C_p} \frac{L}{\bar{V}H^2} \qquad [2.9]$$

This number represents the ratio of the residence time to the heat diffusion time. When small (usually less than 10^{-2}), the thermal regime is adiabatic and the average temperature evolves linearly along the flow. For high *Ca* values (usually higher than 1), however, thermal equilibrium is reached: the temperature remains constant and can be partially controlled by the temperature of the walls. Between these two extremes, the thermal regime is transient.

Meanwhile, the Brinkman number characterizes the relative importance of viscous dissipation and heat conduction. It is defined by:

$$Br = \frac{\eta \bar{V}^2}{k(T_p - \bar{T})} \qquad [2.10]$$

where T_p is the wall temperature.

Of course, if the viscosity depends on temperature, equations [2.7] and [2.8] can no longer be directly applied. The slab method is, therefore, used [VER 85] where the flow is split into successive slabs of thickness Δz, and where the increase in temperature between two slabs (for a cylindrical geometry) is defined by:

$$\frac{\Delta \bar{T}}{\Delta z} = \frac{8\eta(\bar{T})\bar{V}}{\rho C_p R^2} - \frac{48}{5} Ca \frac{\bar{T}(z) - T_0}{L} \qquad [2.11]$$

The viscosity is calculated at a given slab, and the new temperature at the location $z + \Delta z$ is used to calculate the flow at the next adjacent slab.

2.3. 2D solutions

2.3.1. *Network method (or FAN method)*

2.3.1.1. *Isothermal calculations*

Tadmor *et al.* [TAD 74] developed a method that provides a solution to 2D flow in more complex geometries. This method, called the *flow analysis network* (FAN) method, applies to confined flows of variable thickness, and is based on the Hele-Shaw approximation. This means that we can neglect the transverse velocity component in comparison to the components in the flow plane [TAD 79]. For example, in Figure 2.2, v_z is negligible compared to v_x and v_y. Vergnes *et al.* [VER 84] and then Arpin *et al.* [ARP 92] applied this method to the particular case of flat coat-hanger dies (see Chapter 6). They showed that this method was valid and gave good results. Sornberger *et al.* [SOR 86] extended this method to include two-layer coextrusion in coat-hanger dies. In this case the obtained numerical results were also consistent with the experimental observations.

The basics of this method are presented below. The total width of the die is denoted by W and the length by L. The proposed method uses a conventional finite-difference mesh, subdividing the flow into N_x elements of dimension δ_x along the length, and N_y elements of dimension δ_y along the width. Each mesh is located by its coordinates (i, j) and possesses a height $H(i, j)$. The pressure nodes $P(i, j)$, situated in the center of each cell, are connected by "resistivities" $R_x(i, j)$ in the longitudinal direction and $R_y(i, j)$ in the transverse direction (Figure 2.2).

For each cell (i, j), the flow rates $Q_x(i, j)$ and $Q_y(i, j)$ are given by:

$$Q_x(i,j) = \frac{[P(i-1,j) - P(i,j)]}{R_x(i,j)} \qquad [2.12]$$

$$Q_y(i,j) = \frac{[P(i,j) - P(i,j+1)]}{R_y(i,j)} \qquad [2.13]$$

where $R_x(i, j)$ and $R_y(i, j)$ are the longitudinal and transverse resistivities, respectively.

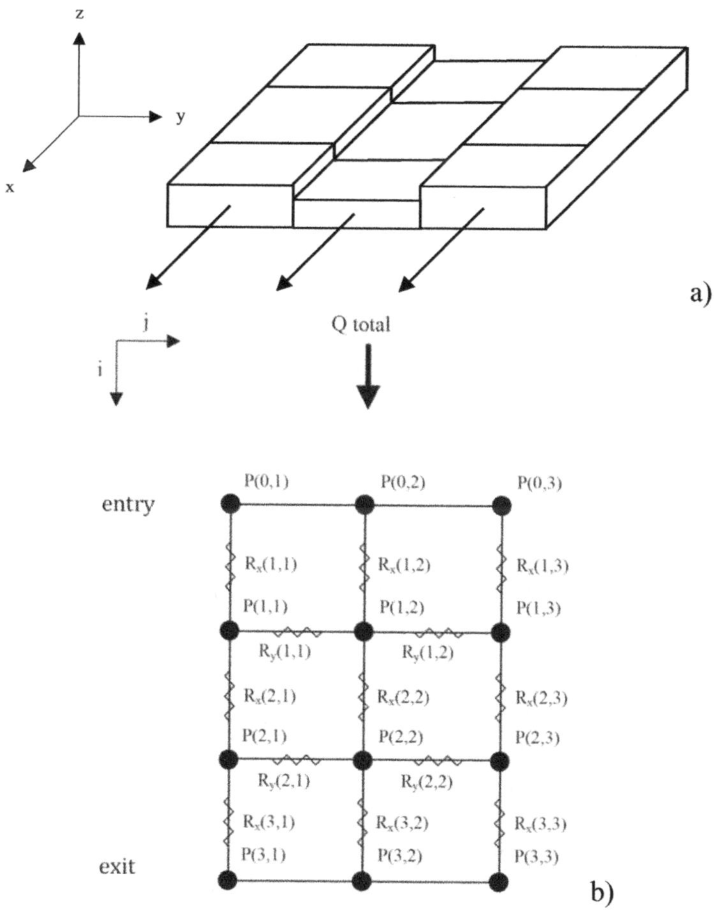

Figure 2.2. *a) Flow geometry and b) mesh for a FAN-type network model*

In the case of a Newtonian fluid, the longitudinal and transverse resitivities are simply expressed by:

$$R_x(i,j) = \frac{12\eta\,\delta_x}{\delta_y\,H(i,j)^3} \qquad [2.14]$$

$$R_y(i,j) = \frac{12\eta\,\delta_y}{\delta_x\,H(i,j)^3} \qquad [2.15]$$

In the case of a power-law fluid, these resistivities depend on the flow rate and are written as follows:

$$R_x(i,j) = \frac{2K\delta_x}{H(i,j)} \left[\frac{2(2+\frac{1}{m})}{\delta_y \, H(i,j)^2} \right]^m Q_x(i,j)^{m-1} \quad [2.16]$$

$$R_y(i,j) = \frac{2K\delta_y}{H(i,j)} \left[\frac{2(2+\frac{1}{m})}{\delta_x \, H(i,j)^2} \right]^m Q_y(i,j)^{m-1} \quad [2.17]$$

i.e.

$$R_x(i,j) = \frac{2(2+\frac{1}{m})}{\delta_y \, H(i,j)^2} \left[\frac{2K\delta_x}{H(i,j)} \right]^{\frac{1}{m}} \left[P(i-1,j) - P(i,j) \right]^{\frac{m-1}{m}} \quad [2.18]$$

$$R_y(i,j) = \frac{2(2+\frac{1}{m})}{\delta_x \, H(i,j)^2} \left[\frac{2K\delta_y}{H(i,j)} \right]^{\frac{1}{m}} \left[P(i,j) - P(i,j+1) \right]^{\frac{m-1}{m}} \quad [2.19]$$

Taking Figure 2.2 into account, the continuity equation can be transformed in a macroscopic mass balance as follows:

$$Q_x(i,j) + Q_y(i,j-1) = Q_x(i+1,j) + Q_y(i,j) \quad [2.20]$$

By replacing each of the flow rates by its equivalent in terms of pressure (equations [2.12] and [2.13]), we obtain the following equation for each cell (i, j):

$$\begin{aligned} &\frac{\left[P(i-1,j) - P(i,j) \right]}{R_x(i,j)} + \frac{\left[P(i,j-1) - P(i,j) \right]}{R_y(i,j-1)} \\ &- \frac{\left[P(i,j) - P(i+1,j) \right]}{R_x(i+1,j)} - \frac{\left[P(i,j) - P(i,j+1) \right]}{R_y(i,j)} = 0 \end{aligned} \quad [2.21]$$

Thus, for a flow field subdivided into N cells, it is possible to write a set of N equations with N unknown values of pressure, which will be linear for a Newtonian fluid. Such a system can be easily solved using a conventional Gaussian elimination method. For non-Newtonian cases, a nonlinear system is obtained, and it can be solved, for example, by an under-relaxation method [VER 85].

2.3.1.2. *Non-isothermal calculations*

The calculation of the temperature field inside simple geometries can be easily integrated into the algorithm for solving the flow calculation, in order to account for the non-isothermal conditions. This requires the thermal balance equation to be solved in terms of the local temperature $\bar{T}(i,j)$, averaged over the thickness of the flow [VER 84]:

$$\begin{aligned}&\frac{Q_x(i,j)}{\delta_x}\left[\left(\bar{T}(i,j)-\bar{T}(i-1,j)\right)+\frac{1}{\rho C_p}\left(P(i,j)-P(i-1,j)\right)\right]\\&+\frac{Qy(i,j)}{\delta_y}\left[\left(\bar{T}(i,j)-\bar{T}(i,j-1)\right)+\frac{1}{\rho C_p}\left(P(i,j)-P(i,j-1)\right)\right]\\&=2\frac{h_T}{\rho C_p}\left(T_p-\bar{T}(i,j)\right)\end{aligned} \qquad [2.22]$$

The evolution of temperature within the flow affects the rheological parameters (Chapter 1). For a power-law fluid, it is common practice to consider that the consistency K varies with temperature according to the Arrhenius equation:

$$K(\bar{T})=K_0\exp\left[\frac{E_a}{R}\left(\frac{1}{\bar{T}}-\frac{1}{T_0}\right)\right] \qquad [2.23]$$

where K_0 is the consistency at the reference temperature T_0 and E_a is the activation energy of the polymer.

The evolution of the temperature within the die can be easily estimated using the algorithm given in Figure 2.3.

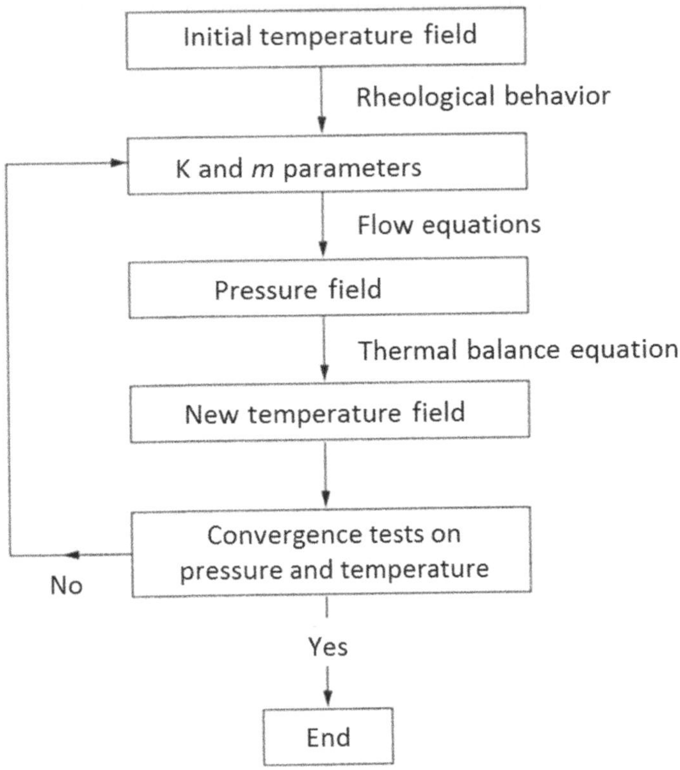

Figure 2.3. *Algorithm for solving coupled thermomechanical problems*

2.3.2. *Finite element method*

Network methods, solved by finite difference schemes, have been very popular for over 20 years as they are relatively easy to implement, without the need for large computational resources. Nowadays, they are increasingly replaced with finite element methods which, coupled with CAD software, allow a much more precise description of the geometry. However, these methods remain difficult for the non-specialist despite the rapid development of user-friendly commercial codes. Here, we provide just a brief overview of the finite element technique, and we refer the interested reader to more specialized publications [FOR 01, CHA 04, DHA 05, ZIE 05].

A flow calculation generally involves a set of partial differential equations (e.g. Stokes equations, see section 1.2.1), which result from a modeling of the physical problem.

This set of partial differential equations is formally expressed as follows:

$$A(u) = f \quad [2.24]$$

where u refers to the unknowns, A is a partial differential operator and f is a given function.

To discuss this problem, it is convenient to use a variational formulation. Many different formulations can correspond to the same problem: conform, mixed, hybrid, etc. Under certain conditions, the problem defined by equation [2.24] is written in the variational form as follows:

Find u that satisfies the boundary conditions such as:

$$a(u,v) = l(v), \forall v \in V \quad [2.25]$$

where V is the space of admissible functions, a is the bilinear form associated with operator A and l is the linear form associated with f. This variational formulation is the starting point for a solution using the finite element method.

The finite element method involves looking for an approximate solution to the problem defined by equation [2.25], by discretizing the finite-dimensional space V_h. The solution is obtained in several steps:

– mathematical analysis of the problem, written in the variational form, starting by studying the existence and uniqueness of the solution;

– implementation:

- creating the mesh M_h for the field being studied;

- definition of the finite element(s), i.e. construction of the finite-dimensional space V_h ;

- creating elementary tables corresponding to the contribution of each element M of M_h, to the matrix, to the second member of the system and to the constraints;

- creating the system to be solved (or assembly);

- taking the boundary conditions into account in the solution of the system, i.e. the calculation of the field similar to the desired solution.

In the following chapters, we will occasionally present results from finite element simulations, without going into the often complex details of the methods used.

2.4. Bibliography

[AGA 14] AGASSANT J.-F., AVENAS P., SERGENT J.-P., *et al.*, *La mise en forme des polymères. Approche thermomécanique de la plasturgie*, Tec et Doc, Lavoisier, Paris, 2014.

[ARP 92] ARPIN B., LAFLEUR P.G., VERGNES B., "Simulation of polymer flow through a coat-hanger die: a comparison between two numerical approaches", *Polymer Engineering & Science*, vol. 32, pp. 206–212, 1992.

[CHA 04] CHASKALOVIC J., *Méthode des éléments finis pour les sciences de l'ingénieur*, Tec et Doc, Lavoisier, Paris, 2004.

[DHA 05] DHATT G., THOUZOT G., LEFRANÇOIS E., *Méthode des éléments finis*, Hermes Lavoisier, Paris, 2005.

[DON 03] DONER J., HUERTA A., *Finite Element Methods for Flow Problems*, Wiley, New York, 2003.

[FOR 01] FORTIN A., *Analyse numérique pour ingénieurs*, Presses internationales Polytechnique, Montreal, 2001.

[LUC 96] LUCQUIN B., PIRONNEAU O., *Introduction au calcul scientifique,* Masson, Paris, 1996.

[MIC 84] MICHAELI W., *Extrusion Dies*, Hanser, Munich, 1984.

[SOR 86] SORNBERGER G., VERGNES B., AGASSANT J.F., "Two directional coextrusion flow of two molten polymers in flat dies", *Polymer Engineering & Science*, vol. 26, pp. 455–461, 1986.

[TAD 74] TADMOR Z., BROYER E., GUTFINGER C., "Flow analysis network: a method for solving flow problems in polymer processing", *Polymer Engineering & Science*, vol. 14, pp. 660–665, 1974.

[TAD 79] TADMOR Z., GOGOS C.G., *Principles of Polymer Processing*, Wiley, New York, 1979.

[VER 84] VERGNES B., SAILLARD P., AGASSANT J.F., "Non-isothermal flow of a molten polymer in a coat-hanger die", *Polymer Engineering & Science*, vol. 24, pp. 980–987, 1984.

[VER 85] VERGNES B., Calcul des écoulements de polymères fondus dans les filières d'extrusion, PhD Thesis, University of Nice, 1985.

[VER 86] VERGNES B., AGASSANT J.F., "Die flow computations: a method to solve industrial problems in polymer processing", *Advances in Polymer Technology.*, vol. 6, pp. 441–455, 1986.

[ZIE 05] ZIENKIEWICZ O.C., TAYLOR R.L., ZHU J.Z., *Finite Element Method, its Basis and Fundamentals*, Elsevier, Amsterdam, 2005.

Chapter 3

Single-Screw Extrusion

3.1. Introduction

The main components of a single-screw extruder are illustrated in Figure 3.1. The main part of the system involves an Archimedes' screw rotating inside a barrel.

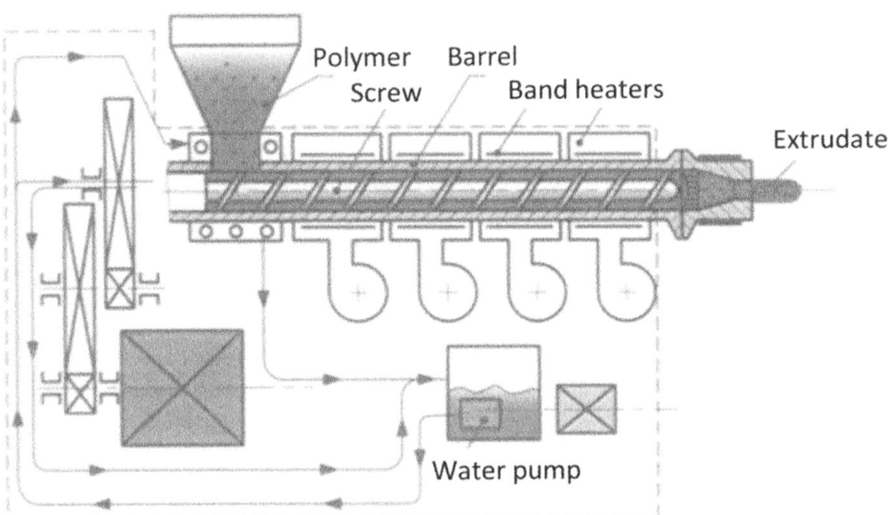

Figure 3.1. *Components of a single-screw extruder*

The screw-barrel system is at the heart of the extrusion process and determines its performance. The flow of the material results from the action of the screw flights on the polymer in contact with the inner wall of the barrel. If the polymer sticks to the screw, the extrusion does not proceed. On the other hand, the flow improves if the polymer sticks to the barrel.

The screw types currently available on the market are not very diverse. To better meet the requirements of processing companies, much progress remains to be made in the field of screw design. Nowadays, the majority of screws used are conventional screws, with the screw pitch being equal to the diameter (Figure 3.2).

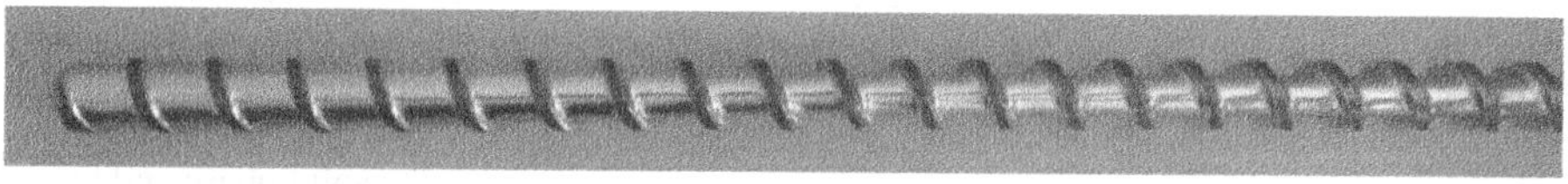

Figure 3.2. *Conventional screw*

A number of modifications, described below, have been incorporated into the conventional screw to improve its performance. Table 3.1 summarizes the main developments carried out since the 1960s, including grooved barrels, barrier screws and mixing heads.

The overall aim of this chapter is to analyze, understand and model the single-screw extrusion process using semi-analytical tools, in order to improve the screw design or, quite simply, to improve the processing conditions.

Numerous authors have taken an interest in studying the extrusion process [TAD 70, STE 95, CHU 00, RAU 01]. In general, a screw is divided into four distinct, yet interconnected, sections. These four sections, in order, are: solids conveying zone, melting delay zone, melting zone and metering zone. These different functional zones of the extruder will be analyzed individually. Integrating mathematical descriptions of each of these zones, a general model, simulating the behavior of the screw unit, is presented.

Innovation	Objective	Mechanism
Grooved barrel 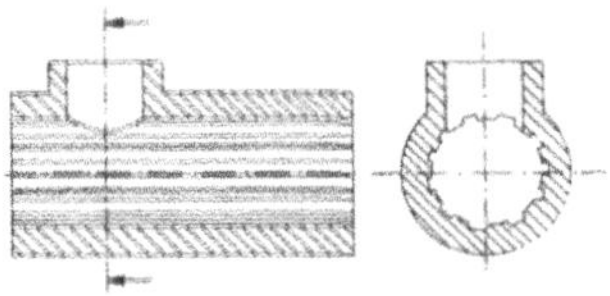	To increase the conveying capacity of the solid particles.	To increase the adherence of particles to the barrel wall.
Barrier screw 	To increase the melting capacity.	To increase surface contact between the solid phase and the barrel.
Multiflighted barrier screw 	To increase the melting capacity.	To decrease the thickness of the molten polymer film between the solid phase and the barrel.
Mixing heads 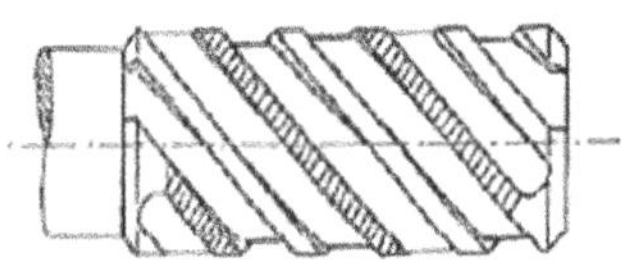	To improve mixing performance.	To divide and recombine the flow. To increase shear.

Table 3.1. *Improvements in the conventional screw-barrel system*

3.2. Geometry and approximations

Before analyzing the functional zones, it is useful to define the main geometrical characteristics of the screw-barrel system. Figure 3.3 shows the functional zones found in a screw extruder.

It is important to distinguish the functional zones, which are characteristic of the state of the material (Figure 3.3(a)), from the sections of the screw extruder, which are strictly linked to the geometry of the screw (Figure 3.3(b)). In general, the functional zones coincide well with the geometric sections, but large discrepancies can, sometimes, occur in the case of a bad design.

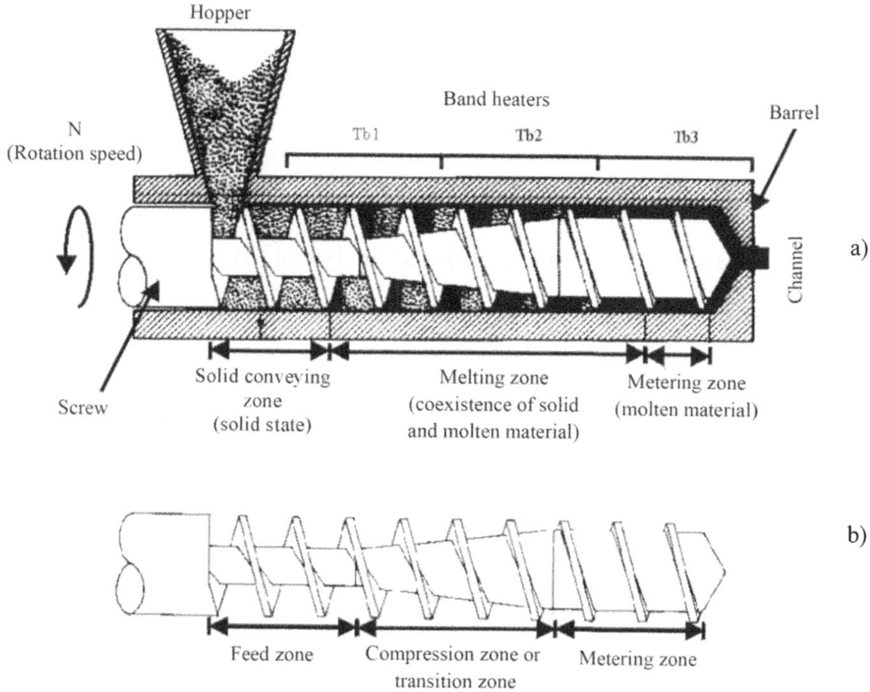

Figure 3.3. *Description of functional zones a) and geometric sections b) of a conventional screw*

The functional zones, from back to front, of the screw are:

– the solid conveying zone, where all the material is in the solid state, either in powder or pellet form;

– the melting zone, where the polymer is, progressively, converted from the solid state to the molten state;

– the metering zone, where the entire polymer is molten.

In parallel, the geometric sections for a conventional screw are:

– the feed zone, where the channel depth is constant;

– the compression zone, where the channel depth progressively gets shallower;

– the metering zone, where the channel depth is again constant, but shallower than that of the feed zone.

The channel geometry can be characterized by five independent parameters, shown in Figure 3.4:

– the screw pitch (B);

– the nominal barrel diameter (D_F);

– the channel depth (h);

– the clearance between the flights and the barrel (δ);

– the flight width, which may be constant (e) or a function of the radial position r.

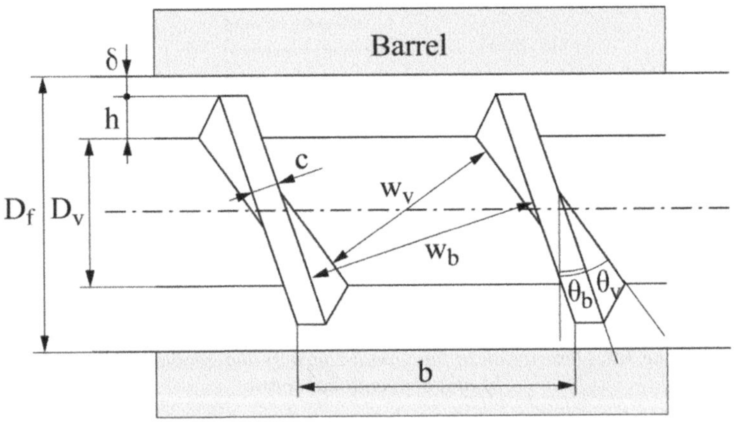

Figure 3.4. *Geometry of the screw-barrel system*

The pitch angle (θ), the channel width (W) and the screw body diameter (D_v) are other frequently used parameters, which can be calculated from the independent parameters. For the screw-body diameter (D_v), the following relation must be observed:

$$\frac{D_F - D_v}{2} = h + \delta \qquad [3.1]$$

The helical geometry, inevitably, causes the flight angle and the channel width to depend on the radius. Thus, there are different flight angles, at the screw (θ_v) and at the barrel (θ_F) and also different channel widths, at the screw (W_v) and the barrel (W_F).

To simplify the mathematical analysis of the material behavior, it is convenient to work with a reference frame attached to a stationary screw. From this perspective, the barrel rotates in the opposite direction to the screw, with the same angular velocity.

This hypothesis is best suited when the *h/W* ratio is small. However, for larger *h/W* ratios, it is essential to use a reference frame attached to a stationary barrel, which better reflects reality [CAM 01]. In fact, the velocity profiles in both systems are not identical. This assumption may cause errors when calculating the viscosity of the polymer or the flow lines.

The validity of the reference frame attached to a stationary screw, also depends on centrifugal forces. Indeed, the distribution of these forces can be very different depending on the reference frame chosen. For molten polymer flows, it is possible to show that the high viscosity of the fluid renders the centrifugal forces negligible [AGA 14]. As a result, it is appropriate to use a reference frame fixed to the screw for any operating condition. In fact, a rotation speed of tens of thousands per minute is necessary before the centrifugal forces become significant.

Even with the aforementioned approximation, the helical geometry of the channel is difficult to model. This problem can be solved by unrolling the channel through which the material is flowing, ending up with a long parallelepipedic channel. As the flight angle varies from the screw body to the barrel, the sides of the channel are unequal. The channel depth, however, is generally considered to be small in comparison with the screw diameter, and this distortion can be ignored. The screw geometry, generally used in modeling, is shown in Figure 3.5. We will use this particular geometry for the remainder of this chapter.

From Figure 3.5, it is easy to define the screw dimensions from the independent parameters. Thus, the angle θ and the channel width W are given as a function of the radius by:

$$\theta(r) = \arctan\left(\frac{B}{2\pi r}\right) \qquad [3.2]$$

$$W(r) = B\cos(\theta(r)) - e(r) \qquad [3.3]$$

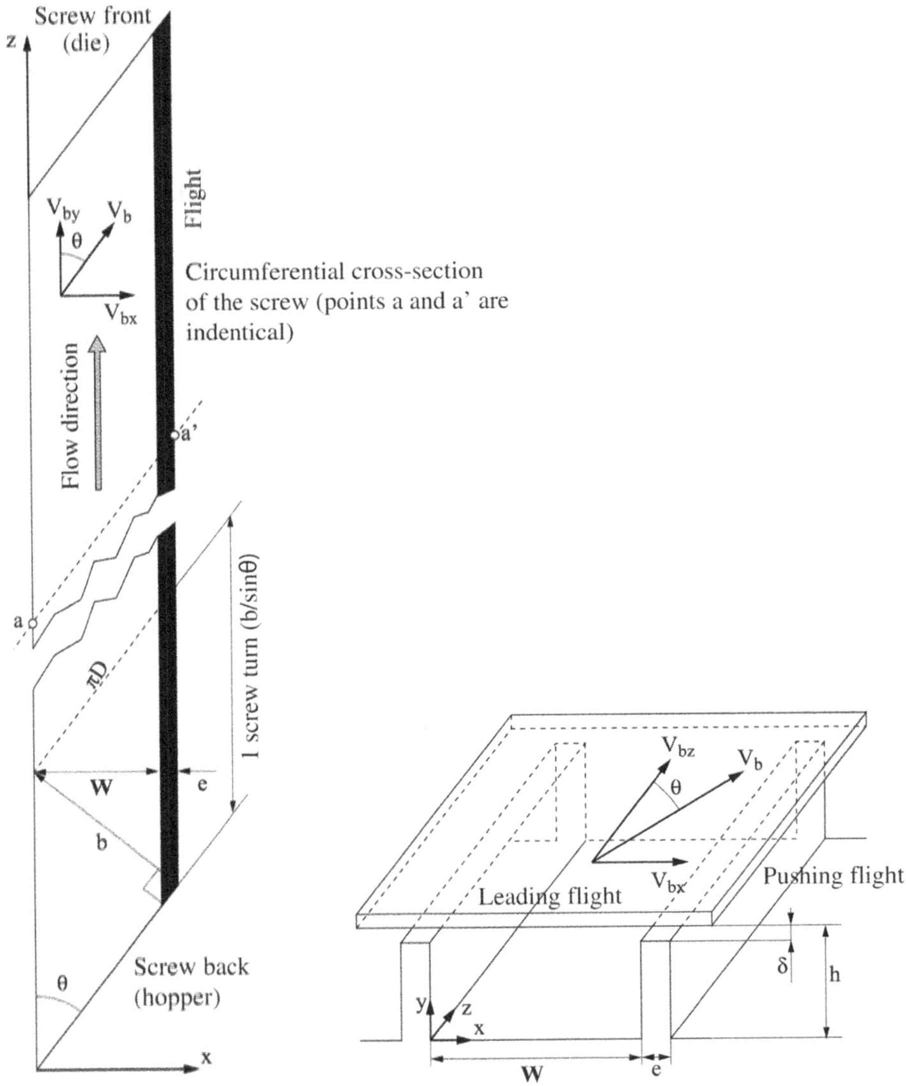

Figure 3.5. *Unrolling of the screw channel*

To facilitate the modeling, these parameters are usually considered to be constant and are calculated at the barrel level ($r = D_F/2$), which is perfectly justified if h is small compared to D_F.

The barrel velocity (V_F) can be decomposed as a function of the system axes:

$$V_{Fx} = V_F \sin\theta \qquad [3.4]$$

$$V_{Fz} = V_F \cos\theta \qquad [3.5]$$

with:

$$V_F = \pi D_F N \qquad [3.6]$$

where N is the rotation speed (in revolution/s).

On the other hand, it is common to convert the axial distance along the screw into the length of the unrolled channel (z coordinate). In this case, the following relation applies:

$$z = \frac{\text{axial distance}}{\sin\overline{\theta}} \qquad [3.7]$$

where $\overline{\theta}$ is the average angle.

The main dimensions of a conventional screw, approximately expressed in terms of the screw nominal diameter (D_F or simply D), are presented in Table 3.2.

	Dimensions
Total length	20–30 D
Length of feed zone	4–8 D
Length of metering zone	6–10 D
Screw pitch(B)	1 D
Flight width (e)	0.1 D
Depth of feed zone (h)	0.1–0.15 D
Barrel-flight clearance (δ)	0.0025 D
Compression ratio ($h_{feed}/h_{metering}$)	2–4

Table 3.2. *Dimensions of a common extrusion screw*

3.3. Solid conveying zone

The solid conveying zone (or feed zone) refers to the first few turns of the screw, during which most of the material is in the solid state. Densification takes place along this zone, and the material is progressively heated, ultimately leading to the formation of a molten polymer film at the interface between the solid phase and the inner surface of the barrel. The formation of this film marks the beginning of the melting zone.

The solid conveying zone is not easy to describe mathematically; even nowadays, a huge amount of work must still be done to analyze and model this zone. There are several reasons for this. First, the sensors traditionally used in extrusion, such as thermocouples and pressure transducers, cannot be used when dealing with solids. In fact, a non-compacted solid does not generate hydrostatic pressure and the temperature is incredibly difficult to measure due to the presence of air inbetween the pellets or powder particles. Secondly, unlike the case for the zone where melting begins, it is not possible to visually observe the movement of the particles. This makes it difficult to verify any model attempting to describe this zone.

Another important issue to consider is the difficulty in obtaining accurate data for the polymer–metal friction coefficients for the range of temperatures present in the solid conveying zone. These tribological data are in fact necessary in order to accurately describe particle behavior. The relevance of the obtained values is, however, difficult to assess, since these measurements can vary depending on the type of measuring device and protocol used.

Proper modeling of the feed zone involves being able to predict:

– the extrusion volumetric flow rate;

– the pressure along the screw channel;

– the temperature profile in the material (necessary to calculate the friction coefficients);

– the point at which the feed zone ends and the melting process begins.

These parameters are particularly influenced by the following factors:

– the nature of the material (friction coefficients, density curves and lateral compaction coefficients);

– the processing conditions (barrel temperature, rotation speed and back pressure);

– the geometry of the barrel and of the screw.

3.3.1. *Phenomenological description*

A phenomenological description of the feed zone can be found in manuals by Chung [CHU 00] and Rauwendaal [RAU 01].

A natural way to deal with the feed zone involves dividing it into three different sections (Figure 3.6):

– the compaction zone (sections 1–5);

– the plug flow zone with frictional driving forces (sections 6–8);

– the plug flow zone with viscous driving forces (sections 9–10).

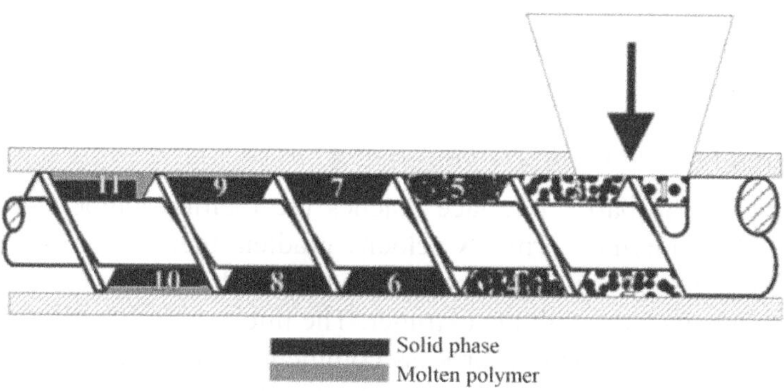

Figure 3.6. *Subzones in the feed zone*

3.3.1.1. *Compaction zone*

Immediately at the entry of the screw, under the hopper, a mixture of air and polymer pellets, of low apparent density, enters the extruder and becomes gradually denser as it flows. The degree of densification varies according to the granulometry of the used material (pellets, flakes, powder, etc.) and the cohesion between the particles. Taking polyvinyl chloride (PVC) as an example, the value of the density can, often, double between the hopper and a few lengths further down the extrusion path.

Often responsible for low volumetric flow rates, the compaction zone is of a major industrial significance. However, its analysis is highly complex due to the three dimensional (3D) relative movements between particles, and the need to have detailed information about the material (tribological data, rigidity, elasticity of the particles, etc.). The experimental intricacies present in the compaction zone, combined with the mathematical challenges associated with finding a solution, make the modeling of this zone very difficult.

3.3.1.2. *Plug flow zone with frictional driving forces*

Once the material is compressed, a plug flow develops in the screw extruder. The friction on the barrel is the driving force behind the transport of the raw material. In this type of regime, pressure and temperature at the interface between the polymer and the barrel rapidly increase. When the interface temperature reaches the melting temperature, the plug flow zone with viscous driving forces begins. Since analyzing a plug flow is much simpler than analyzing the compaction zone, the plug flow zone with friction forces is, by far, the most documented and most frequently modeled subzone of the feed zone.

3.3.1.3. *Plug flow zone with viscous driving forces*

When the solid-barrel interface reaches the melting temperature, a thin layer of molten polymer forms. A velocity gradient then develops across this film, creating a shear stress on the solid bed, which, in turn, causes the raw material to advance through the extruder. The thickness of the film gradually increases up until the melting process commences. This zone is also called the delay zone or melting delay zone, according to the terminology used by Tadmor and Klein [TAD 70].

3.3.2. *Modeling*

To better understand the mechanisms involved in the solid conveying zone, we begin by describing the behavior of the solid in plug flows with both friction and viscous forces. Afterward, we will describe the compaction zone.

3.3.2.1. *Plug flow with friction forces*

3.3.2.1.1. Darnell and Mol model [DAR 56]

The first complete method for modeling the solid conveying zone was described by Darnell and Mol [DAR 56], who studied the plug flow with friction forces.

The general assumptions were as follows:

– the solid polymer behaves like a continuous medium;

– the solid is in contact with the barrel, the screw and the flights;

– the depth of the channel is constant;

– the clearance above the flights is negligible;

– the regime is steady-state;

– the pressure only increases in the direction of the flow;

– the force of gravity is negligible;

– the density of the material is constant.

This model assumes that the barrel is rotating at velocity V_F around a stationary screw. The polymer is considered as a continuous solid moving in the direction of the channel at velocity V_{sz}. The relative angle of the barrel velocity, viewed by an observer situated on the solid bed, is denoted Φ. A diagram of velocity vectors for this problem is shown in Figure 3.7.

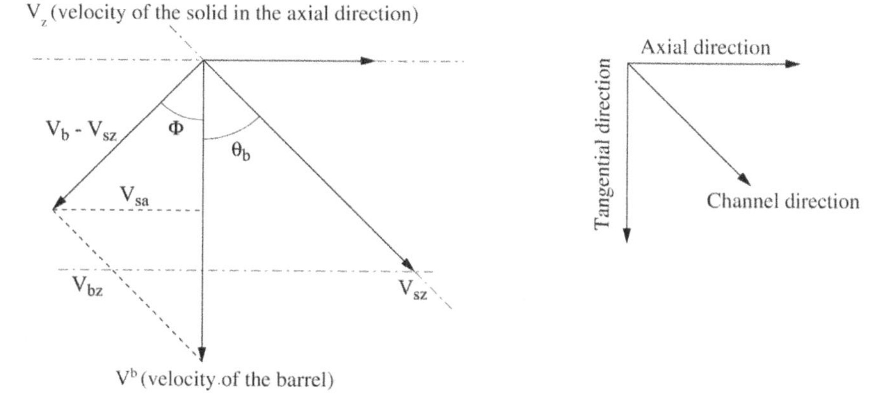

Figure 3.7. *Velocity diagram*

The first stage of this model involves calculating the volumetric flow rate of the solid polymer. This flow rate, denoted Q_s, is given by:

$$Q_s = \left[\frac{\pi}{4}\left(D_F^2 - D_v^2\right) - \frac{e h}{\sin \bar{\theta}}\right] V_s \rho_s \qquad [3.8]$$

where V_s is the velocity of the solid bed in the axial direction. It can be expressed in terms of the angle Φ, as follows:

$$\tan \Phi = \frac{V_s}{V_F - V_s / \tan \theta_F} \qquad [3.9]$$

i.e.:

$$V_s = V_F \frac{\tan \Phi \tan \theta_F}{\tan \Phi + \tan \theta_b} \qquad [3.10]$$

The barrel velocity V_F is defined by equation [3.6].

We can also write:

$$D_F^2 - D_v^2 \approx 4h(D_F - h) \qquad [3.11]$$

and the average width of the channel can be written as:

$$\bar{W} = \pi D_F \sin \bar{\theta} - e \qquad [3.12]$$

By combining equations [3.8], [3.10], [3.6], [3.11] and [3.12], the flow rate of the solid is expressed as follows:

$$Q_s = \pi^2 N\, h\, D_F (D_F - h) \frac{\tan \Phi \tan \theta_F}{\tan \Phi + \tan \theta_F} \left[\frac{\bar{W}}{\bar{W} + e}\right] \rho_s \qquad [3.13]$$

We find that the flow rate of the solid depends on the geometry of the feed zone and the friction coefficients (*via* the angle Φ, which is explained later) and is proportional to the rotation speed of the screw.

The second step involves calculating the angle Φ. To do this, let us consider a section of the solid bed perpendicular to the flow direction through the channel.

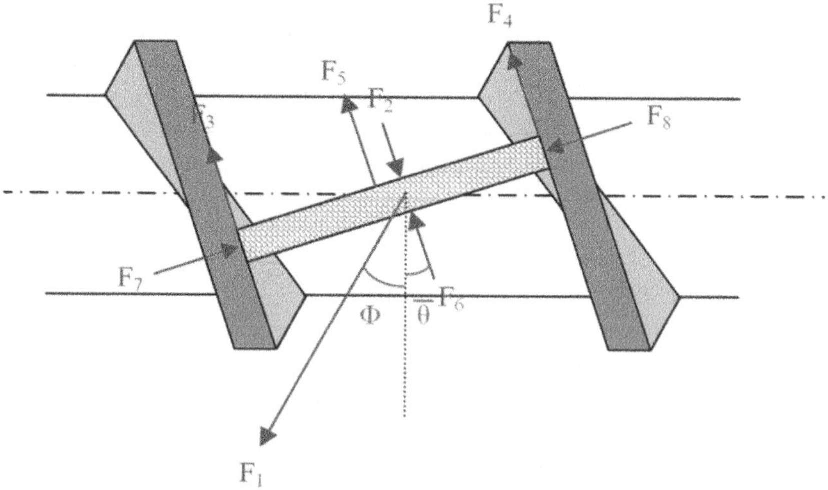

Figure 3.8. *Forces acting upon a section of the solid bed*

As shown in Figure 3.8, this section is subjected to eight forces:

– F_1 is the driving force due to friction on the barrel;

– F_5 is the force due to friction on the screw body;

– $(F_6 - F_2)$ is the force resulting from the pressure gradient in direction z;

– F_7 and F_8 are normal forces exerted by the flights;

– F_3 and F_4 are the forces due to friction on the screw flights.

These forces can be, simply, expressed in terms of the geometric parameters of the screw, the friction coefficients between the polymer and the screw (f_v), as well as between the polymer and the barrel (f_F), the pressure P and the position z_b along the flow channel (Table 3.3). Here, it is assumed that the friction stresses τ follow Coulomb's law, i.e. $\tau = fP$.

	Expression	Longitudinal component	Transverse component
F_1	$F_1 = f_F\, PW_b\, dz_b$	$F_{1l} = f_F\, PW_b\, dz_b \sin\Phi$	$F_{1\theta} = f_F\, PW_b\, dz_b \cos\Phi$
F_5	$F_5 = f_v\, PW_v\, dz_v$	$F_{5l} = f_v\, PW_v\, dz_v \sin\theta_v$	$F_{5\theta} = f_v\, PW_v\, dz_v \cos\theta_v$
$(F_6 - F_2)$	$F_6 - F_2 = h\overline{W}\, dP$	$(F_6 - F_2)_l = h\overline{W}\, dP \sin\overline{\theta}$	$(F_6 - F_2)_\theta = h\overline{W}\, dP \cos\overline{\theta}$
F_7	$F_7 = Ph\, d\overline{z} + F^*$	$F_{7l} = (Ph\, d\overline{z} + F^*)\sin\overline{\theta}$	$F_{7\theta} = (Ph\, d\overline{z} + F^*)\cos\overline{\theta}$
F_8	$F_8 = Ph\, d\overline{z}$	$F_{8l} = Ph\, d\overline{z} \sin\overline{\theta}$	$F_{8\theta} = Ph\, d\overline{z} \cos\overline{\theta}$
F_3	$F_3 = F_7\, f_v$	$F_{3l} = F_7\, f_v \sin\overline{\theta}$	$F_{3\theta} = F_7\, f_v \cos\overline{\theta}$
F_4	$F_4 = F_8\, f_v$	$F_{4l} = F_8\, f_v \sin\overline{\theta}$	$F_{4\theta} = F_8\, f_v \cos\overline{\theta}$

Table 3.3. *Expression of forces acting upon the solid bed*

With the balance of forces and moments, it is possible to obtain the following implicit equation for Φ:

$$\begin{aligned}\cos\Phi &= \frac{\overline{D}}{D_F}\left(\frac{\sin\overline{\theta} + f_v \cos\overline{\theta}}{\cos\overline{\theta} - f_v \sin\overline{\theta}}\right)\sin\Phi \\ &+2\frac{h}{W_F}\frac{f_v}{f_F}\sin\theta_F\left(K + \frac{\overline{D}}{D_F}\operatorname{cotan}\overline{\theta}\right) \\ &+\frac{W_v}{W_F}\frac{f_v}{f_F}\sin\theta_F\left(K + \frac{D_v}{D_F}\operatorname{cotan}\theta_v\right) \\ &+\frac{\overline{W}}{W_F}\frac{H}{z_b}\frac{1}{f_F}\sin\overline{\theta}\left(K + \frac{\overline{D}}{D_F}\operatorname{cotan}\overline{\theta}\right)\ln\frac{P}{P_0}\end{aligned} \qquad [3.14]$$

where P_0 and P denote the pressure at the start of the solid conveying zone ($z = 0$), and at a distance z_b from this starting point.

Equation [3.14] can be written in a simplified form as follows:

$$\cos\Phi = K\sin\Phi + M \qquad [3.15]$$

with:

$$K = \frac{\bar{D}}{D_F}\left(\frac{\sin\bar{\theta} + f_v \cos\bar{\theta}}{\cos\bar{\theta} - f_v \sin\bar{\theta}}\right) \tag{3.16}$$

$$\begin{aligned} M = {} & 2\frac{h}{W_F}\frac{f_v}{f_F}\sin\theta_F\left(K + \frac{\bar{D}}{D_F}\operatorname{cotan}\bar{\theta}\right) \\ & + \frac{W_v}{W_F}\frac{f_v}{f_F}\sin\theta_F\left(K + \frac{D_v}{D_F}\operatorname{cotan}\theta_v\right) \\ & + \frac{\bar{W}}{W_F}\frac{H}{z_b}\frac{1}{f_F}\sin\bar{\theta}\left(K + \frac{\bar{D}}{D_F}\operatorname{cotan}\bar{\theta}\right)\ln\frac{P}{P_0} \end{aligned} \tag{3.17}$$

The solution is written as:

$$\sin\Phi = \frac{\sqrt{1 + K^2 - M^2} - KM}{1 + K^2} \tag{3.18}$$

Equation [3.19] allows the pressure P at a flow distance z_b to be calculated:

$$\frac{P}{P_0} = \exp\left[\left(\frac{B_1 - A_1 K}{B_2 + A_2 K}\right) z_b\right] \tag{3.19}$$

with:

$$A_1 = f_F W_F \sin\Phi + 2h f_v \sin\theta_F + W_v f_v \sin\theta_F \tag{3.20}$$

$$A_2 = h\bar{W}\sin\bar{\theta} \tag{3.21}$$

$$\begin{aligned} B_1 = {} & f_F W_F \cos\Phi - 2h f_v \sin\theta_F \operatorname{cotan}\left(\bar{\theta}\frac{\bar{D}}{D_F}\right) \\ & - W_v f_v \sin\theta_F \operatorname{cotan}\left(\theta_v \frac{D_v}{D_F}\right) \end{aligned} \tag{3.22}$$

$$B_2 = h\bar{W}\cos\left(\bar{\theta}\frac{\bar{D}}{D_F}\right) \tag{3.23}$$

The expression [3.19] shows an exponential increase in pressure with distance along the channel. This increase in pressure is strongly linked to the ratio between the friction coefficients on the barrel and the screw body (Figure 3.9).

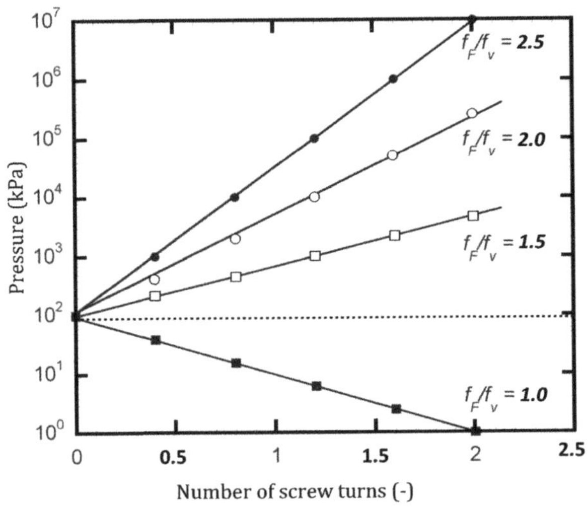

Figure 3.9. *Evolution of the pressure along the screw channel in the solid conveying zone for different f_F/f_v values*

To calculate this pressure, the initial value must be positive at the beginning of the solid conveying zone. As we will see in the following sections, this conclusion is subject to discussion. The effect of the channel depth and the flight angle on the flow rate of the solid can be calculated from this analysis, for fixed friction coefficients (Figures 3.10 and 3.11).

Figure 3.10 shows that the flow rate is sensitive to increases in pressure. If the pressure is low, the flow rate is proportional to the channel depth. By contrast, in the case where the pressure increases significantly, there is an optimal depth that allows the highest flow rate.

Finally, Figure 3.11 shows one of the most significant findings to come from the Darnell and Mol model [DAR 56]. If the friction coefficients on the barrel and the screw are identical, a maximum flow rate can be reached with a flight angle of approximately 20°. The same occurs when the screw pitch is equal to the diameter (here, an angle of 17°); this is a very widespread configuration in single-screw extrusion. However, when the coefficients are different, higher on the barrel, the potential flow rate is greater, but the optimal flight angle must be increased. This explains the increasingly frequent use of grooved cylinders (see section 3.7.1).

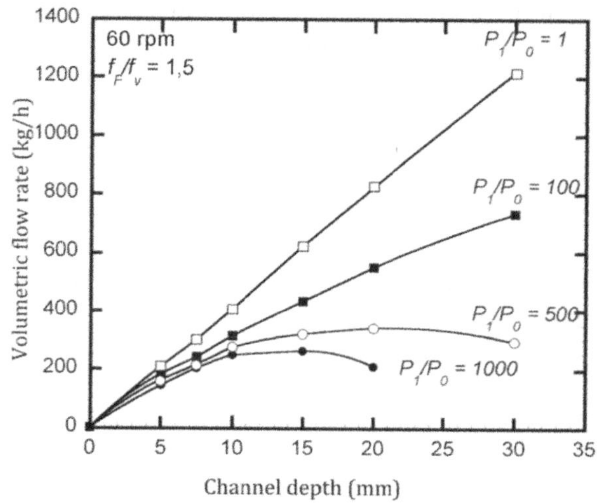

Figure 3.10. *Evolution of flow rate with screw channel depth, for different pressure values at the end of the solids conveying zone*

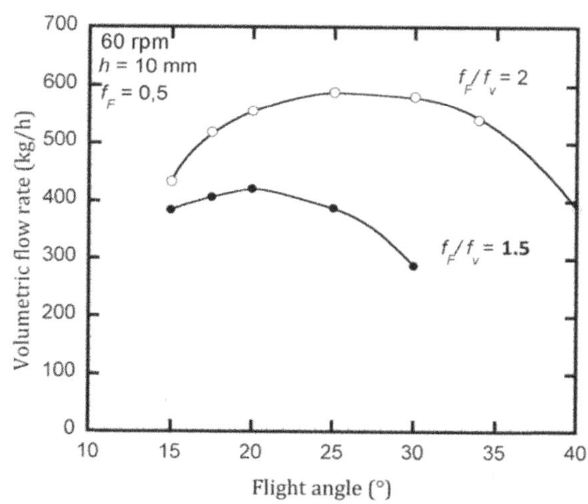

Figure 3.11. *Evolution of flow rate with flight angle for different f_F/f_v values*

3.3.2.1.2. The Broyer and Tadmor isothermal model [BRO 72]

The Darnell and Mol model [DAR 56] is useful for a channel with a constant depth; however, in the solid conveying zone, the channel depth can decrease. It is for this reason that Broyer and Tadmor [BRO 72] developed a model very similar to that of Darnell and Mol [DAR 56]. Their model, however, incorporates the case where the channel depth varies linearly along the solid conveying zone:

$$h(z) = h_0 - z\tan\alpha \qquad [3.24]$$

where h_0 is the initial channel depth, α is the angle between the screw body and the barrel. In this case, it is appropriate to include the normal force in the balance of forces and moments exerted by the screw body upon the solid polymer; also, the components following the axial and tangential directions are no longer negligible (Figure 3.12).

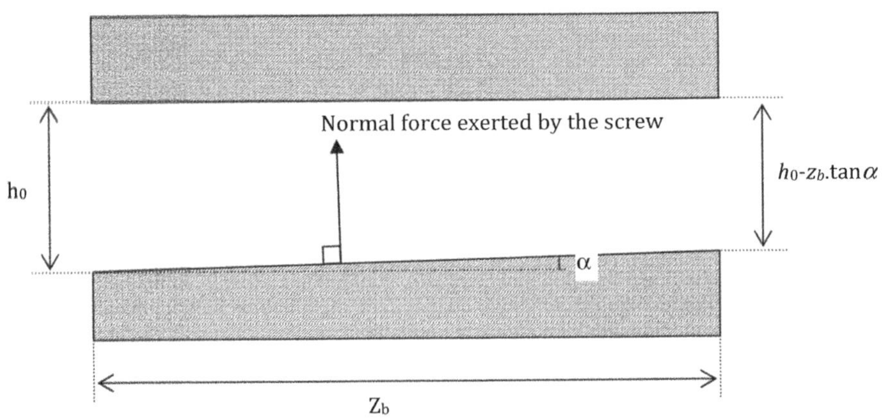

Figure 3.12. *Normal force exerted by the screw*

This force is denoted by F_9 and is expressed as follows:

$$F_9 = PW_v \sin\alpha\, dz_v \qquad [3.25]$$

The axial and tangential components are:

$$F_{9l} = PW_v \sin\alpha\, dz_v \sin\theta_v \qquad [3.26]$$

$$F_{9\theta} = PW_v \sin\alpha\, dz_v \cos\theta_v \qquad [3.27]$$

Moreover, Broyer and Tadmor [BRO 72] included the effect of pressure anisotropy, building on the work by Schneider [SCH 69]. The pressures P_f, P_v and P_F at the level of the flights, the screw body and the barrel can be obtained from the pressure P, which is calculated in the channel, using the following relations:

$$P_f = K_f P \qquad [3.28]$$

$$P_v = K_v P \qquad [3.29]$$

$$P_F = K_F P \qquad [3.30]$$

where K_f, K_v and K_F are the lateral pressure coefficients, valid at the level of the flights, screw and barrel, respectively.

Taking the variable depth of the channel and the pressure anisotropy into account, the expression [3.31] can be written:

$$\frac{P}{P_0} = \exp\left[\int_{z=0}^{z=z_b} \frac{B_1 - A_1 K}{A_2 K + B_2} dz\right] \qquad [3.31]$$

with:

$$A_1 = f_F W_F K_F \sin\Phi - W \tan\alpha \sin\theta + 2\bar{h} f_v K_f \sin\theta_F + W_v f_v K_v \sin\theta_F \left(\cos\alpha + \frac{\sin\alpha}{f_v}\right) \qquad [3.32]$$

$$A_2 = h\bar{W}\sin\bar{\theta} \qquad [3.33]$$

$$B_1 = f_F W_F K_F \cos\Phi + \bar{W}\tan\alpha\cos\left(\bar{\theta}\frac{\bar{D}}{D_F}\right) - 2\bar{h} f_v K_f \sin\left(\theta_F \frac{\bar{D}}{D_F}\right)\operatorname{cotan}\bar{\theta}$$
$$-W_v f_v K_v \left(\cos\alpha + \frac{\sin\alpha}{f_v}\right)\sin\theta_F \operatorname{cotan}\left(\theta_v \frac{D_v}{D_F}\right) \qquad [3.34]$$

$$B_2 = h\bar{W}\cos\left(\bar{\theta}\frac{\bar{D}}{D_F}\right) \qquad [3.35]$$

For a given flow rate, integrating equation [3.31] requires the use of numerical methods because of the variations in the local geometry and the physical properties used.

Note that, with a polymer of constant density, flowing in a channel of uniform depth, and under isotropic pressure conditions ($K_f = K_v = K_F = 1$), equation [3.31] reduces to the Darnell and Mol model [DAR 56].

3.3.2.1.3. The Broyer and Tadmor non-isothermal model [BRO 72]

The analysis presented below was developed by Broyer and Tadmor [BRO 72] in order to take the thermal phenomena in the solid conveying process into account.

In the feed zone, as the solid polymer comes into contact with different metallic surfaces, friction is produced, which significantly heats up the solid polymer. The physical properties of the material vary as a function of temperature. This phenomenon can significantly affect how the pressure evolves.

Therefore, the solid conveying zone cannot be considered as being isothermal. It is for this reason that Broyer and Tadmor [BRO 72] decided to couple fluid mechanics and heat transfer in the solid conveying zone.

The heat generated per unit area at the barrel level is:

$$g = P_F\, f_F\, \pi N D_F \frac{\sin\theta_F}{\sin(\theta_F + \Phi)} \qquad [3.36]$$

Thermal analysis is carried out on a section of the solid polymer perpendicular to the flights, as shown in Figure 3.13.

It is possible to show that the heat flux at the level of the flights and the screw body are negligible. Therefore, the heat transfer problem presented in this section becomes a 1D conduction problem. The variations in the temperature T_s of the solid polymer over time are described by the energy equation:

$$\frac{\partial T_s}{\partial t} = \frac{k_s}{\rho_s C_{ps}} \frac{\partial^2 T_s}{\partial y^2} \qquad [3.37]$$

where k_s and C_{ps} are the thermal conductivity and the specific heat capacity of the solid polymer, respectively.

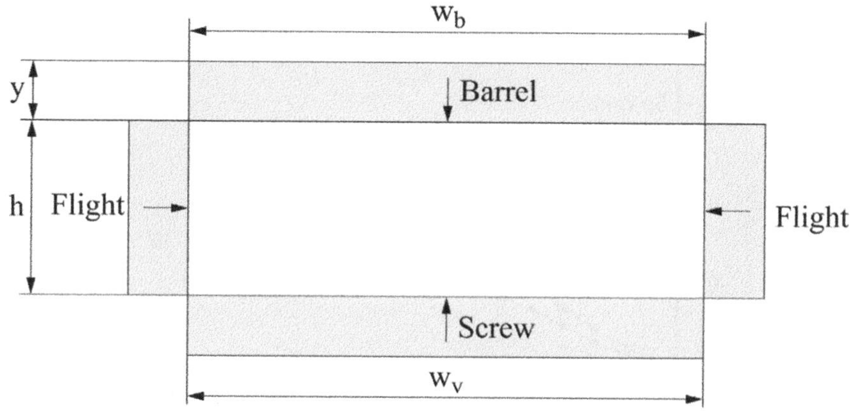

Figure 3.13. *Section used for Broyer and Tadmor thermal analysis [BRO 72]*

Equation [3.37] is solved using the following boundary conditions:

– the temperature is constant at the interface between the solid polymer bed and the barrel;

– the heat generated by friction at the level of the barrel is dissipated in two fluxes:

$$g = -k_s \left(\frac{\partial T_s}{\partial y} \right)_{y=0} + k_F \left(\frac{\partial T_F}{\partial y} \right)_{y=0} \qquad [3.38]$$

The solution to this problem can be obtained numerically. If we know the velocity of the solid bed, the evolution of temperature at a given distance along the channel can be calculated.

The complete calculation of the solid conveying parameters (temperature and pressure profiles) involves an iterative process.

The Broyer and Tadmor non-isothermal method [BRO 72] can account for variations in the physical parameters of the material with temperature and pressure, and produces a rather realistic pressure profile (Figure 3.14). Moreover, it enables the calculation of the distance at which the surface temperature of the solid reaches the melting temperature, i.e. this method allows us to determine the length of the solid conveying zone.

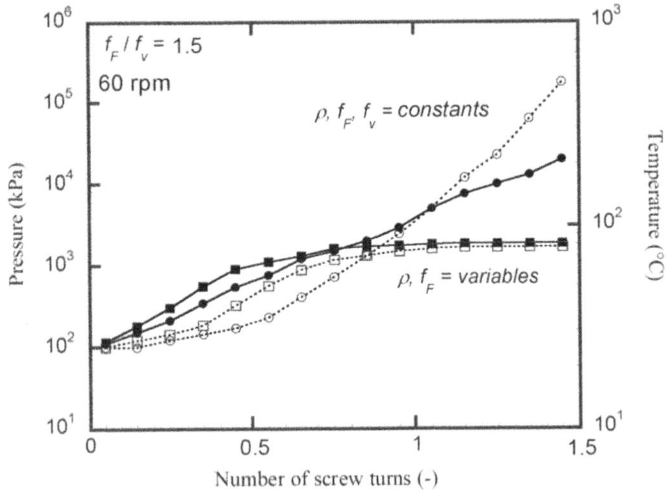

Figure 3.14. *Calculated pressure (full lines) and temperature profiles (dotted lines)*

3.3.2.2. *Plug flow with viscous forces*

From the previous examples, Kacir and Tadmor [KAC 72] developed a model applicable to the flow observed in the melting delay zone. In this model, the friction produced by the barrel in contact with the solid bed is replaced by the shear stress τ exerted by the liquid film upon the solid bed. For a film with a viscosity η and thickness δ, this stress is calculated as follows:

$$\tau = \eta \bar{\dot{\gamma}} \qquad [3.39]$$

with: $$\bar{\dot{\gamma}} = \frac{V_F \sin\theta_F}{\delta \sin(\theta_F + \Phi)} \qquad [3.40]$$

Taking the pressure anisotropy into account, the Kacir and Tadmor method [KAC 72] leads to the following pressure profile:

$$P_1 = \left(P_0 + \frac{A_3}{B_1 - A_1 K} \right) \exp\left[\int_{z=z_0}^{z=z_1} \frac{B_1 - A_1 K}{A_2 K + B_2} dz \right] - \left(\frac{A_3}{B_1 - A_1 K} \right) \qquad [3.41]$$

with:

$$A_1 = -W \tan\alpha \sin\theta + 2\bar{h}\, f_v K_f \sin\theta_F + W_v f_v K_v \sin\theta_F \left(\cos\alpha + \frac{\sin\alpha}{f_v} \right) \qquad [3.42]$$

$$A_2 = h\bar{W} \sin\bar{\theta} \qquad [3.43]$$

$$A_3 = \tau W_F \left(\cos\Phi - K \sin\Phi \right) \qquad [3.44]$$

$$B_1 = \bar{W} \tan\alpha \cos\bar{\theta} \frac{\bar{D}}{D_b} - 2\bar{h}\, f_v K_f \sin\theta_F \operatorname{cotan}\bar{\theta} \frac{\bar{D}}{D_F} - W_v f_v K_v \sin\theta_F \operatorname{cotan}\theta_v \frac{D_v}{D_F} \left(\cos\alpha + \frac{\sin\alpha}{f_v} \right) \qquad [3.45]$$

$$B_2 = h\bar{W} \cos\bar{\theta} \frac{\bar{D}}{D_F} \qquad [3.46]$$

In equation [3.41], P_0 and P_1 are the pressures for $z = z_0$ (start of the melting delay zone) and $z = z_1$, respectively.

The thickness of the film plays a major role in determining the shear stress. However, determining this parameter is complex, given the nonlinearity of the temperature and pressure profiles. The calculation used to determine the film thickness will be described in detail during the analysis of the melting zone.

Also, note that a method similar to that of Kacir and Tadmor [KAC 72] was proposed by Chung [CHU 75], for the case of power-law fluids.

3.3.2.3. *Flow in the compaction zone*

Unlike the plug flow zones, there is no widely accepted model used on a routine basis to describe the compaction zone. This zone is often analyzed in a similar way to a piston flow, by using the appropriate apparent densities. However, this method is still a long way from physical reality.

Attala and Podio-Guidugli [ATT 80] modeled particle behavior in the compaction zone as a planar flow of an incompressible viscous fluid in a steady-state regime with undeformable spheres.

Zhang and Sernas [ZHA 02] introduced a velocity component in a direction perpendicular to the channel, in order to account for the two-directional helical flow. The solid particles (in powder form) are treated as an incompressible fluid whose viscosity is calculated from the parameters of a Bingham solid. The model also includes an analysis of the heat transfer in the flow. This approach fits well with the rheological modeling of a molten polymer in the metering zone. However, it proves to be particularly applicable to powder flows, whose behavior is closest to that of a continuous medium.

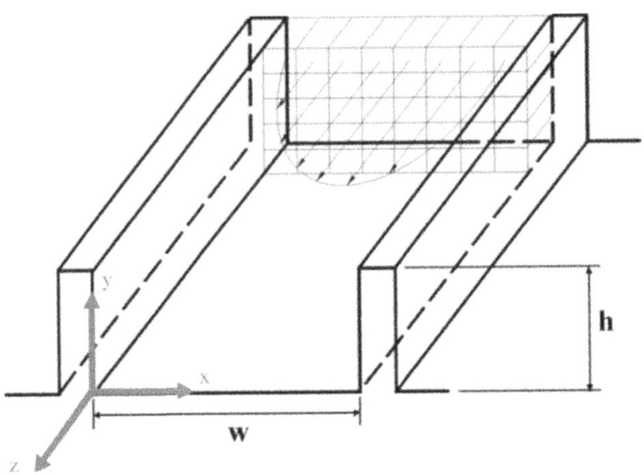

Figure 3.15. *The solid conveying model proposed by Fang et al. [FAN 91]*

A rather interesting approach is that of Fang *et al.* [FAN 91]; from experiments carried out on an extruder equipped with a transparent barrel, they provided evidence for the existence of relative movements between particles and built a model capable of explaining this phenomenon. This model studies the one-directional flow of polymer pellets. Particle size can be modeled from the mesh dimension used. The velocity of the solid varies in the x and the y directions, as shown in Figure 3.15. The model is based on a 3D analysis of the stresses. For element (i, j), the stress state is shown in Figure 3.16.

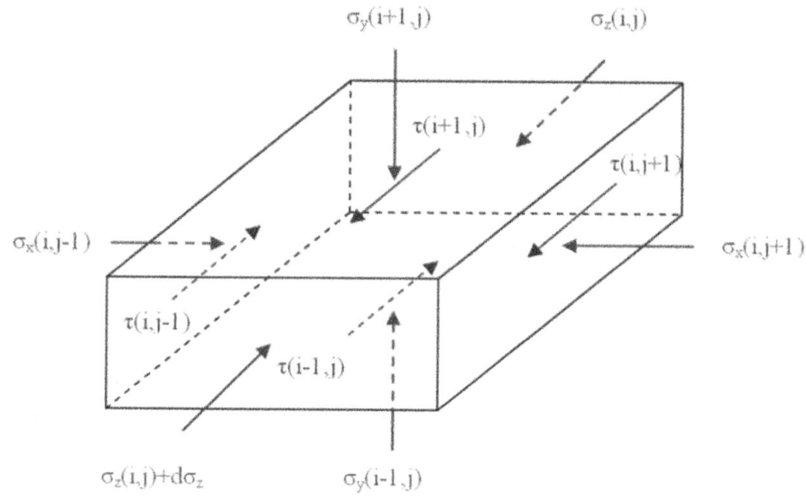

Figure 3.16. The *stress state of an element in the Fang et al. [FAN 91] model*

The relationship between the normal stress σ and the shear stress τ is given by Coulomb's law of friction:

$$\tau = f\sigma \qquad [3.47]$$

For each element position (i, j), the friction coefficient f depends on friction on the barrel (f_F), friction on the screw (f_v) or on internal friction (f_i), due to relative movement between the solid particles.

The stress and velocity profiles are determined from the theory of linear elasticity and the virtual power principle [TED 71], using a finite element method.

Huang [HUA 02] developed a similar model to that of Fang *et al.* [FAN 91], based on a 2D analysis of stresses in the solid bed. The model takes the compaction behavior of the polymer and the dynamic friction into account.

3.3.3. *Synthesis*

The majority of models devoted to solid conveying are based on a plug flow. The initial model, proposed by Darnell and Mol [DAR 56], has undergone successive improvements. In particular, Broyer and Tadmor [TAD 72] accounted for the non-uniformity in the pressure distribution and incorporated an analysis of heat transfer.

Previously, we highlighted that, for this type of model, the calculation of the pressure profile along the channel requires an initial pressure P_0 that is positive. For a long time, this value has been calculated using the hydrostatic pressure produced by the material in the hopper.

However, Lovegrove and Williams [LOV 74] showed experimentally that a significant change in the level of the material in the hopper did not affect the buildup of pressure. Taking the gravitational forces into consideration, they proposed a model capable of predicting the pressure profile in the screw for a polymer of a negligible height in the hopper. The simulation carried out predicted a behavior similar to that observed experimentally, for a low filled hopper.

Recently, models proposing different approaches to that of Darnell and Mol [DAR 56] have been developed. In particular, the work by Fang *et al.* [FAN 91], led to lower flow rates and pressures than those predicted by rigid block models, and seem to be closer to reality. For instance, Tables 3.4 and 3.5 compare the experimental and numerical results obtained by Darnell and Mol [DAR 56], Tedder [TED 71] and Fang *et al.* [FAN 91] models.

N (rpm)	Volumetric flow rate Q_s (kg/h)			
	Experimental	Darnell and Mol model	Tedder model	Fang *et al.* model
31	9.45	20.79	19.68	9.41
41	12.48	27.45	26.01	12.39
51	15.90	34.17	32.26	15.68
61	19.56	40.89	38.71	19.24
71	22.03	47.57	45.05	21.52
82	26.86	54.94	52.03	26.45
102	33.03	68.38	64.74	32.69
122	39.36	81.73	77.42	38.99

Table 3.4. *Comparison between the calculated and measured volumetric flow rates*

N (rpm)	Pressure (MPa)			
	Experimental	Darnell and Mol model	Tedder model	Fang *et al.* model
31	13.8	2.13×10^5	5.44×10^{39}	15.63
41	14.1	1.92×10^5	1.07×10^{37}	13.28
51	14.1	1.83×10^5	1.94×10^{36}	13.07
71	14.1	2.02×10^5	7.88×10^{37}	15.49
82	13.2	2.30×10^5	1.32×10^{40}	19.92
102	13.8	1.77×10^5	5.92×10^{35}	12.55
122	13.8	1.58×10^5	1.34×10^{34}	10.67

Table 3.5. *Comparison between the calculated and measured pressures ($\rho_s = 595$ kg/m^3)*

The comparison was done using an LDPE resin, and an extruder with a diameter of $D_F = 45$ mm, an L/D_F ratio of 5.5 and a channel depth of $h = 7$ mm. The friction coefficients between the polymer and the screw, and between the polymer and the barrel, were $f_v = 0.25$ and $f_F = 0.35$, respectively. The results provided by the Fang *et al.* [FAN 91] model for flow rate appear to be much more realistic than the others (Figure 3.17).

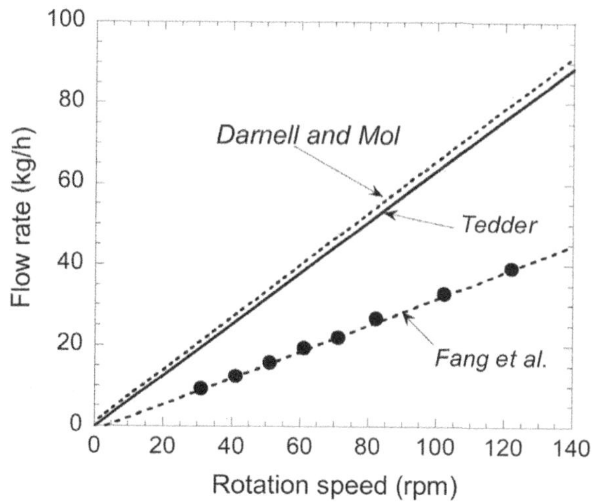

Figure 3.17. *Comparison of different model predictions with experimental results (•)*

Moreover, with regard to pressure, Table 3.5 shows that the only values with a realistic order of magnitude come from the Fang *et al.* [FAN 91] model. The unrealistic values given by the other models are very likely due to the flow length (5.5 D_F).

A different approach to analyze the solid bed was proposed by Potente and Pohl [POT 02], who used the discrete element method to represent the behavior of granular materials. This method was used by Moysey and Thompson [MOY 04] and Michelangelli *et al.* [MIC 11] to study, in 3D, the flow of solids in the transport zone of an extruder. This approach allows us to determine the relative particle movement and the shape of the solid bed. However, this kind of model involves the use of extensive computing resources and is particularly affected by the quality of tribological data.

3.3.4. *Determination of physical parameters*

Describing the flow inside a single-screw extruder using models requires the physical parameters to be previously determined. The friction, compaction and densification properties are particularly critical when modeling the behavior of a material in the solid conveying and melting zones. That is why, in the next section, we will clarify the measuring procedures used to determine these parameters.

3.3.4.1. *Friction coefficients*

The friction coefficient is generally given by the ratio between the friction force F_s on a body and the normal force F_a, exerted at the surface A_0 of this body at equilibrium (Figure 3.18):

$$f = \frac{F_s}{F_a} \quad [3.48]$$

The previous relation can also be expressed in the following form:

$$f = \frac{F_s / A_0}{F_a / A_0} \quad [3.49]$$

i.e.: $f = \dfrac{\tau_s}{P_a}$ [3.50]

where τ_s is the shear stress at the interface and P_a is the apparent pressure on the sample.

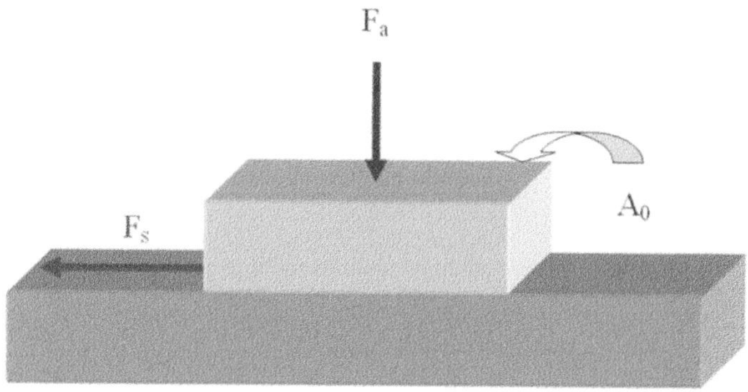

Figure 3.18. *Calculation of the friction coefficient*

When the contact surfaces are stationary, we mention the static friction coefficient (f_s) to which the friction force F_s corresponds when the body begins to move. When the body is mobile, the friction force F_s decreases until the velocity is uniform: the friction coefficient, in this case, is referred to as dynamic (f_d). When modeling single-screw extrusion in a steady-state regime, only the dynamic friction coefficient is important. This coefficient is affected, in a rather complex manner, by the chemical structure of the polymer, the pressure, the temperature, the nature of the interface and the velocity.

There are several types of devices that can be used to measure the friction coefficient. Chung *et al.* [CHU 77] developed a device to simulate the setup of a rotating cylinder and a stationary screw. This device, called the extrusion simulator, is shown in Figure 3.19. The solid sample (in the form of a rigid block, powder or pellets) is inserted into a chamber and pressed against a rotating steel cylinder. A maximum normal force of 5,600 kg can be applied. The friction coefficient is determined by measuring the friction force corresponding to the applied load at the chosen rotation speed and temperature conditions.

The advantage of this device is that it can accurately represent the behavior, associated with the friction of the material, for plug flows, in the feed zone. Moreover, in the case where temperature at the interface is greater than the melting temperature, a film of molten polymer forms on the steel cylinder. The measured force is, therefore, a shear force and, as a result, it becomes possible to characterize the viscous plug flows observed in the melting delay zone.

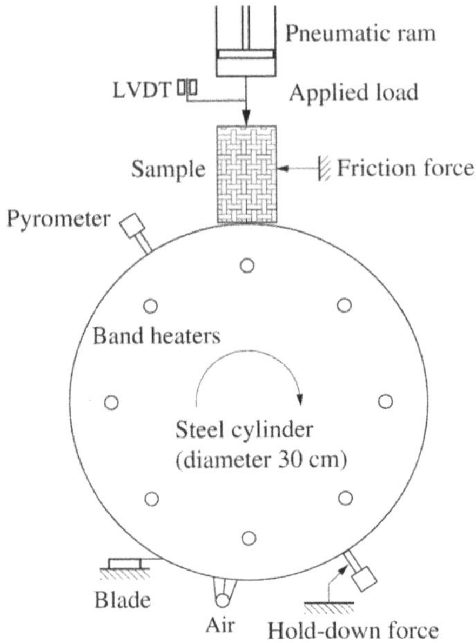

Figure 3.19. *Schematic representation of the tribometer by Chung et al. [CHU 77]*

A much more refined tribometer was developed by Hwang and McKelvey [HWA 89] to study friction in the case where the solid polymer is not completely compacted. This device, shown in Figure 3.20, allows the friction coefficient to be determined by measuring the torque, corresponding to the applied load, exerted on the stationary part of the assembly, under given rotation speed and temperature conditions. The operating range allows conditions typically found in the compaction zone of the extruder to be reproduced. Note that, with such a device, the more the material is compacted, the harder the radial pressure is to measure and interpret. For this type of regime, it is therefore appropriate to refer to the Chung *et al.* tribometer [CHU 77].

A third kind of tribometer was developed by Liu and O'Brien [LIU 91] to measure the interparticle friction coefficient (coefficient of internal friction) within the solid bed. This tribometer is shown in Figure 3.21. The internal friction coefficient is determined by measuring the force needed for the lateral movement of the top ring, depending on the applied load, and the temperature of the system.

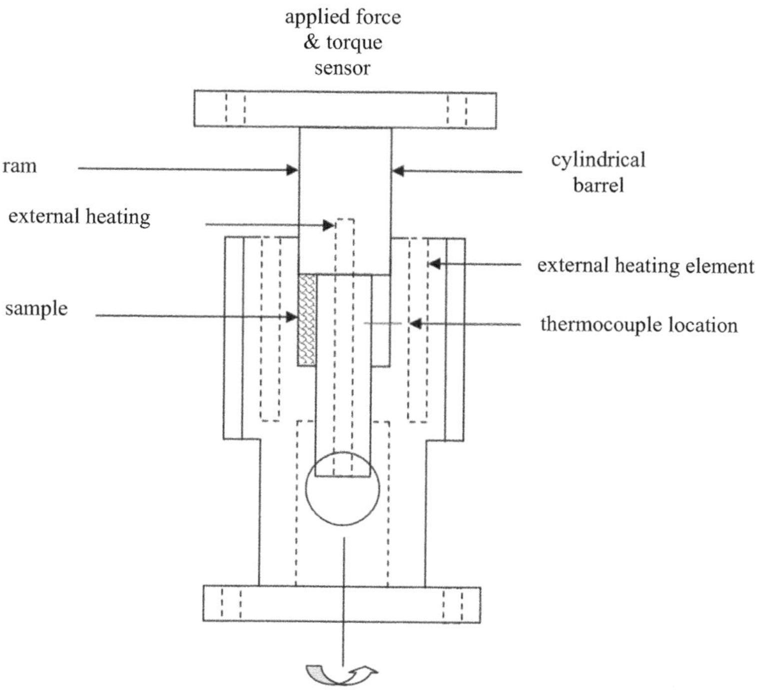

Figure 3.20. *Schematic representation of the tribometer by Hwang and McKelvey [HWA 89]*

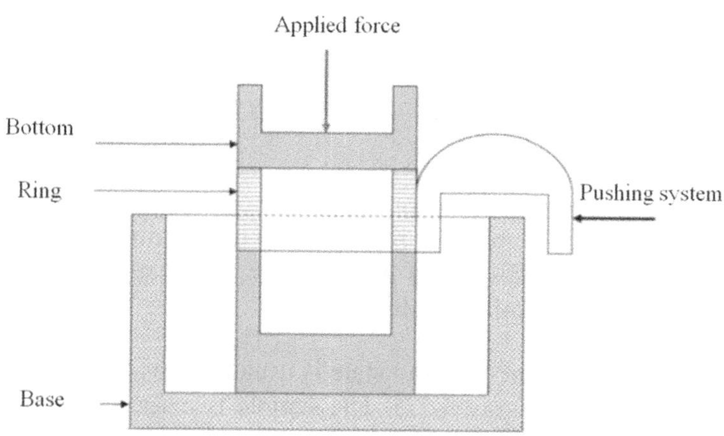

Figure 3.21. *Schematic representation of the tribometer by Liu and O'Brien [LIU 91]*

However, note that irrespective of the measuring device used, the main experimental difficulty resides in determining the temperature at the interface between the sample and the friction surface. In practice, it is impossible to directly measure this temperature, due to the heat generated by friction or viscous dissipation. One way to obtain a reasonable estimate of the actual temperature involves combining experimental temperature readings with a numerical analysis of the temperature field, as proposed by Spalding and Hyun [SPA 95] and by Gamache *et al.* [GAM 99].

Table 3.6 compares the friction coefficient measured by certain authors for four polymer families.

Polymers	[SPA 95]	[SCH 69]	[GAM 99]
HDPE	0.15–0.25	0.10–0.20	0.15–0.19
PP	----	0.25–0.40	0.13–0.17
PVC	----	0.40	0.18–0.21
LDPE	0.20–0.50	0.40	0.37–0.41

Table 3.6. *Friction coefficients of four commodity polymers*

3.3.4.2. *Density*

The density of continuous solid media (or solid density) essentially depends on the nature of the monomer and the molecular structure of the polymer (branched or linear). However, before entering the extruder, the material is in a granular form and consists of a mixture of air and matter. This gives rise to the concept of apparent density (bulk density).

The compaction of the material in the screw channel depends on temperature and pressure: obviously, an increase in pressure further compacts the material and an increase in temperature also enhances compaction. When the pressure is high enough, the maximum compaction density tends toward the solid density.

The apparent density in the solid state is usually measured using a closed vessel under pressure (Figure 3.22). The sample is inserted into a chamber situated between the base and the piston. The entire box is placed in a convection oven, which allows the temperature to be controlled. The piston is moved down toward the base using a hydraulic machine. Gradually compressing the sample in the solid state allows force-displacement curves to be obtained.

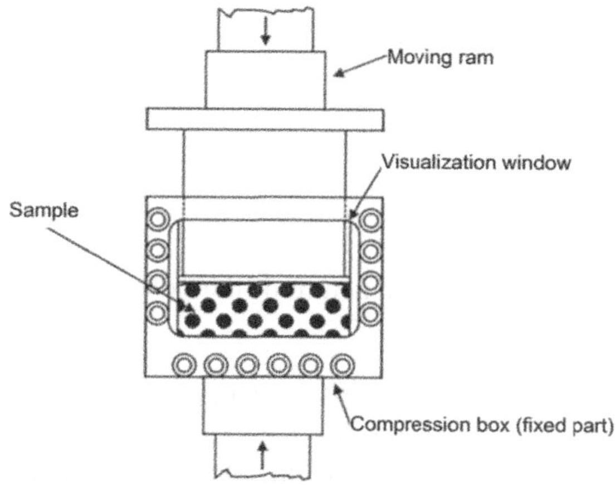

Figure 3.22. *Schematic diagram of a compression box*

The density can be deduced, for a given temperature and pressure, just by knowing the cross-section of the cell base as well as the sample mass. A typical compaction curve obtained for rigid PVC is shown in Figure 3.23.

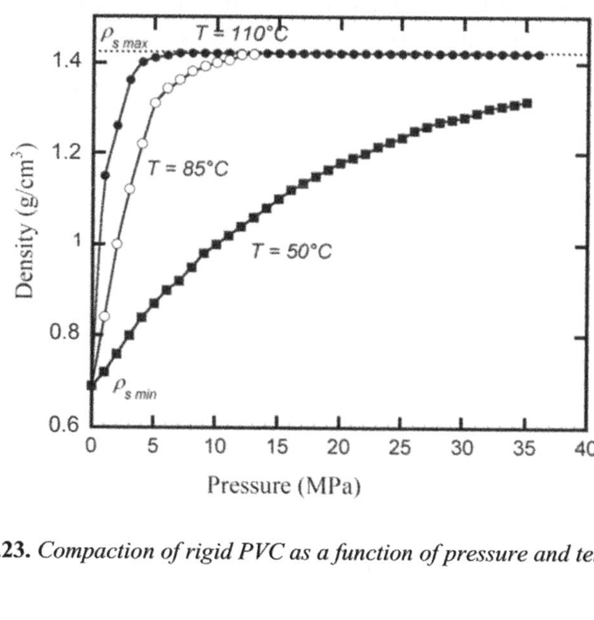

Figure 3.23. *Compaction of rigid PVC as a function of pressure and temperature*

Hyun and Spalding [HYU 90] described densification using the following formula:

$$\frac{\rho_s - \rho_{s\,max}}{\rho_{s\,min} - \rho_{s\,max}} = \exp(-F\,P) \qquad [3.51]$$

with:

$$F = c_1 + c_2 T + c_3 T^2 \qquad [3.52]$$

where c_1, c_2 and c_3 depend on the nature of the polymer as well as the shape and arrangement of the particles.

Potente and Schöppner [POT 95] calculated the effective density of spherical particles of diameter d for a face-centered cubic (FCC) arrangement in a finite volume of width W and height h, resembling the channel in the extruder screw. The smaller the volume in relation to the diameter of the constituent spheres, the more the boundary effects are present and they tend to decrease the maximum compaction for this kind of structure. Figure 3.24 shows this result. This study shows that the volume available for the material in the screw channel can significantly affect the apparent density. However, this reasoning is only valid for spherical particles not undergoing a significant strain (in the case of low pressure combined with a temperature lower than the glass transition temperature).

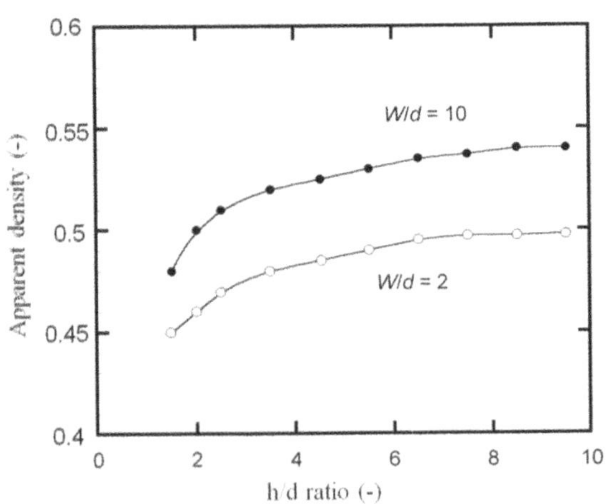

Figure 3.24. *Apparent density (FCC arrangement of spherical particles) as a function of channel geometry (h, height; W, width; d, particle diameter)*

Using densification curves to calculate the density in the extruder is valid, only, when the material is subjected to compaction and density increases. Any calculation of density in the case of decompaction of a material should be interpreted cautiously. In fact, during a transition from a region of high pressure to a region of low pressure, the particles subjected to high pressure can deform and even interpenetrate, thus, significantly affecting the density obtained as the pressure decreases.

3.3.4.3. *Lateral pressure coefficients*

The lateral pressure coefficient K_i is defined in Figure 3.25, as the σ_2/σ_1 ratio. This coefficient allows the anisotropy of a solid medium of particles to be calculated in a rather simple way. The closer the K_i value is to 1, the more isotropic the behavior becomes, and the more it resembles that of a liquid. The anisotropy of a particulate medium is related to the internal resistance to strain, and depends on the granulometry, the type of packing, and the internal friction coefficient.

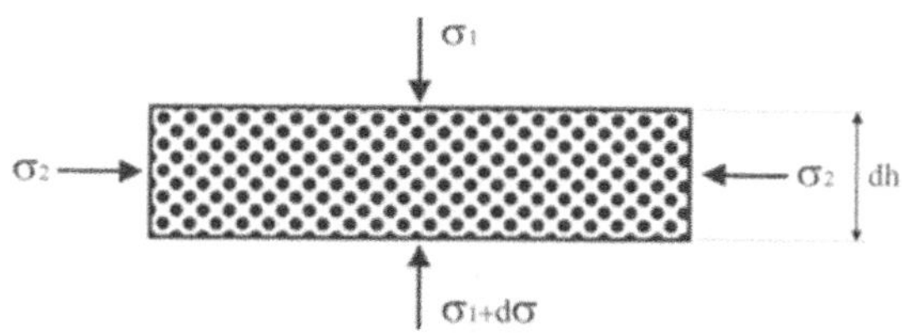

Figure 3.25. *Lateral pressure coefficient*

Spalding *et al.* [SPA 97] developed a simple apparatus shown in Figure 3.26, which allows the lateral pressure coefficients to be measured for polymer particles, as a function of pressure and temperature.

Let P_θ be the radial pressure measured by the pressure sensor and P_z be the axial pressure applied at the level of the sensor. The lateral pressure coefficient is given by:

$$K = \frac{P_\theta}{P_z} \qquad [3.53]$$

The pressure P_z is obtained from averaging the pressures P_1 (the pressure measured by the top load cell) and P_2 (the pressure measured by the bottom load cell). A major disadvantage of this device is how the radial pressure is measured. In fact, this pressure is measured by means of a diaphragm pressure sensor, whose surface can be small compared to that of solid pellets when the material is not in powder form. This makes the results difficult to interpret, since the lateral pressure is only deduced from the force transfer of a very small number of pellets, or even a single one. An average reading over the entire lateral surface would be much more realistic. This device, however, has the advantage of being very versatile, since it is also capable of measuring the density and the static friction coefficient.

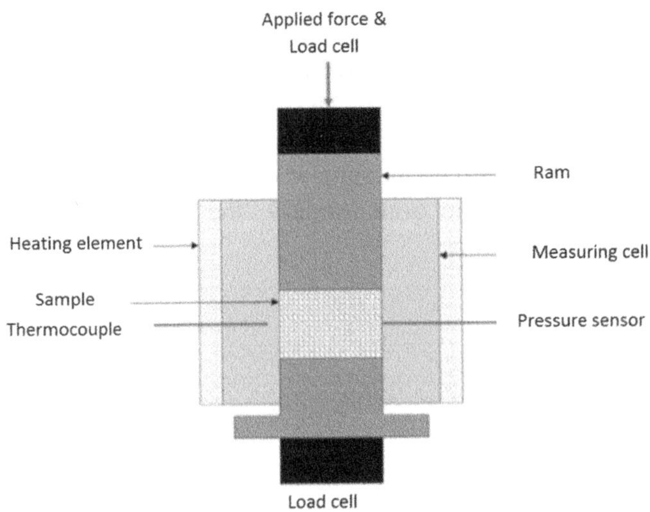

Figure 3.26. *Device for measuring the lateral pressure coefficient [SPA 97]*

3.4. Melting zone

3.4.1. *Phenomenological description*

Polymers, generally, have low thermal conductivity. Therefore, during the extrusion process, the solid bed melts very slowly. Two mechanisms of heat transfer are involved at this stage: a dissipative mechanism and a conduction mechanism.

3.4.1.1. *Melting by dissipation*

The melting mechanism in a single-screw extruder was first observed by Maddock [MAD 59] in the late 1950s. When the temperature of the screw or the barrel surpasses the melting temperature of the polymer, the surface of the solid bed melts immediately upon contact with the metal surfaces. This is usually accompanied by a significant decrease in pressure (Figure 3.27). The solid bed starts to melt and a thin film is formed at the interface with the metal surfaces. At the level of the barrel surface, this liquid film is subjected to significant shear stress. Due to the high viscosity of polymers, this shear generates significant heat by viscous dissipation. This heat is conducted from the liquid film to the solid bed, causing melting at the film-solid interface. Thus, the liquid film gradually thickens, until a melt pool is formed at the rear of the channel (Figure 3.28).

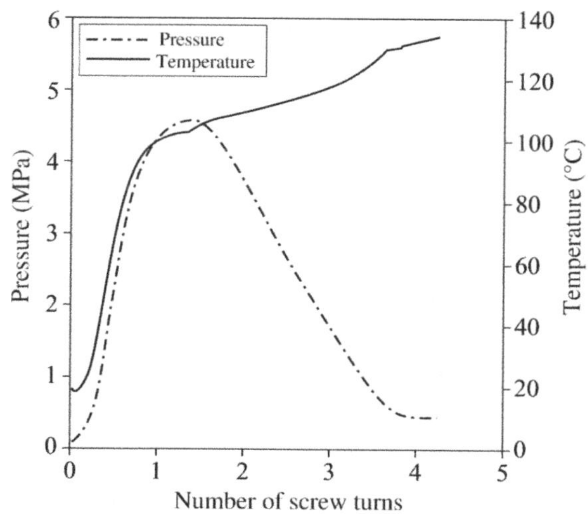

Figure 3.27. *Evolution of pressure and temperature at the start of the melting zone*

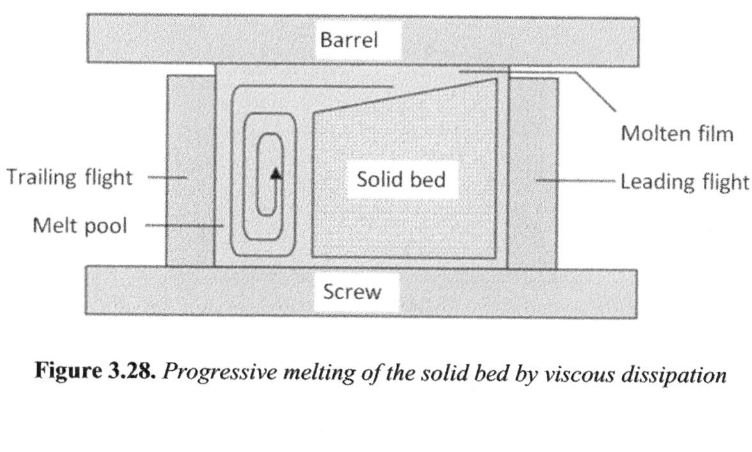

Figure 3.28. *Progressive melting of the solid bed by viscous dissipation*

The creation of the melt pool is due to the pressurizing mechanism of the molten polymer as it passes over the trailing flight. We can show [AGA 14] that the creation of this melt pool is even faster when the screw-barrel clearance δ is low. In practice, the pool formation, generally, starts as soon as the thickness of the molten polymer film reaches 5 µm [TAD 70]. The transition between the formation of the molten polymer film at the interface and that of the molten polymer pool corresponds to the delay zone.

Viscous dissipation is the most important mechanism involved during the melting of the polymer. This mechanism is primarily controlled by the viscosity of the molten polymer, being only slightly influenced by the size of the particles making up the solid bed.

3.4.1.2. *Melting by conduction*

As the melting proceeds, the cross-sectional area of the solid bed decreases as the melt pool enlarges. Below a certain critical size, the solid bed can no longer maintain its integrity and the different forces to which it is subjected cause it to rupture. It is, generally, accepted that this occurs at the distance where the value of L/D is, approximately, equal to 20 [CHU 00]. The minimum particle size as a result of this rupture is that of the material introduced at the feed zone entrance.

The subsequent melting occurs as heat is conducted from the surrounding molten polymer to the solid particles (Figure 3.29). Due to the low thermal conductivity of polymers, this mechanism is slow and depends on the size and shape of the particles: the time needed to melt a solid particle immersed in a molten bed increases exponentially with the particle size.

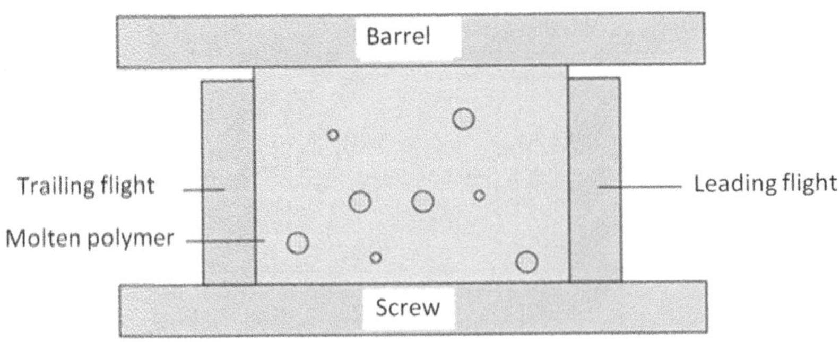

Figure 3.29. *Melting of solid particles by heat conduction*

3.4.2. *Modeling*

3.4.2.1. *Summary of the problem*

Tadmor and Klein [TAD 70] were the first to propose a melting model based on the experimental observations of Maddock [MAD 59]. This approach makes it possible to understand the basic mechanisms involved in the melting zone, and forms the basis of more sophisticated models.

Let δ be the thickness of the liquid film formed between the solid bed and the barrel, let T_F be the barrel temperature and let T_s be the temperature of the solid bed far from the interface. Assume that the solid bed has two velocity components: V_{sz} in the direction of the channel and V_{sy} in the direction of the solid–liquid interface. The barrel velocity V_F also has two components, V_{Fx} and V_{Fz}. At the interface between the solid bed and the liquid film, the temperature is equal to the melting temperature T_f. The unknowns, outlined in Figure 3.30, are δ and V_s.

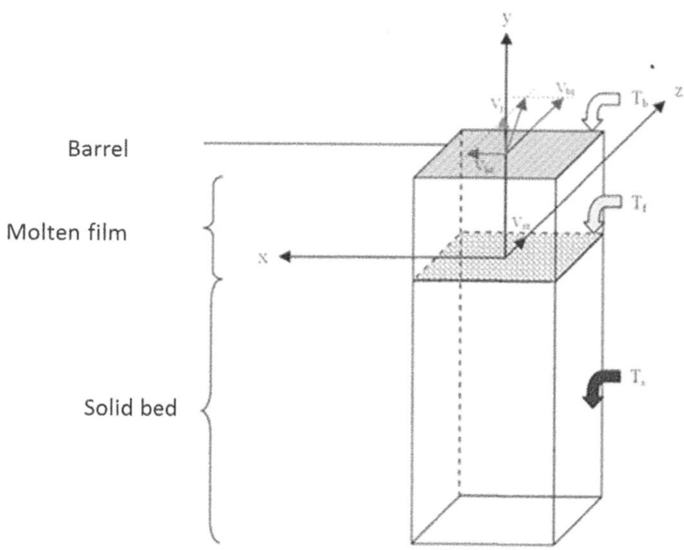

Figure 3.30. *A representation of the melting problem*

3.4.2.2. *Newtonian fluid*

Let us consider a Newtonian fluid with a constant viscosity η. At a given time t, the position of the solid–liquid interface is indicated by y and the movement of the liquid film can be considered as a flow between two parallel plates: the bottom plate is the solid–liquid interface and the top plate is the barrel surface. With regard to the solid–liquid interface, the bottom plate is stationary, whereas the top plate is moving at a speed V_j such as:

$$V_j = V_F - V_{sz} \qquad [3.54]$$

The liquid film flows in the j direction, defined by the angle α_j (angle between V_j and the channel direction z), such that:

$$\tan\alpha_j = \frac{V_{Fx}}{V_{Fz} - V_{sz}} \qquad [3.55]$$

i.e.:

$$\tan\alpha_j = \frac{V_F \sin\theta_F}{V_F \cos\theta_F - V_{sz}} \qquad [3.56]$$

The velocity profile in the liquid film is linear. From the stated velocity boundary conditions, we obtain:

$$v_j(y) = \frac{V_j}{\delta} y \qquad [3.57]$$

In the liquid film, assuming the thermal regime is in equilibrium, the energy equation takes the form:

$$k_l \frac{d^2T}{dy^2} + \eta\left(\frac{dv_j}{dy}\right)^2 = 0 \qquad [3.58]$$

By solving the set of equations [3.57] and [3.58], it is possible to obtain a temperature profile in the liquid film:

$$\frac{T - T_f}{T_F - T_f} = \frac{\eta V_j^2}{2k_l\left(T_F - T_f\right)} \frac{y}{\delta}\left(1 - \frac{y}{\delta}\right) + \frac{y}{\delta} \qquad [3.59]$$

The heat flux from the liquid at the interface (per unit area) is:

$$-q_y\Big|_{y=0} = k_l\left(\frac{dT}{dy}\right)_{y=0} \qquad [3.60]$$

i.e. according to equation [3.59]:

$$-q_y\Big|_{y=0} = \frac{k_l}{\delta}\left(T_F - T_f\right) + \frac{\eta V_j^{\,2}}{2\delta} \qquad [3.61]$$

In the solid bed, the energy equation becomes:

$$\rho_s C_{ps} V_{sy} \frac{dT}{dy} = k_s \frac{d^2T}{dy^2} \qquad [3.62]$$

Integrating this equation leads to the following temperature profile:

$$\frac{T - T_s}{T_f - T_s} = \exp\left(\frac{V_{sy}}{k_s/\rho_s C_{ps}} y\right) \qquad [3.63]$$

The heat flux from the solid at the interface (per unit area) is:

$$-q_y\Big|_{y=0} = k_s\left(\frac{dT}{dy}\right)_{y=0} \qquad [3.64]$$

i.e. according to equation [3.63]:

$$-q_y\Big|_{y=0} = \rho_s C_{ps} V_{sy}\left(T_f - T_s\right) \qquad [3.65]$$

At the interface, the energy balance gives the following relationship between δ and V_{sy}:

$$\left[\frac{k_l}{\delta}\left(T_F - T_f\right) + \frac{\eta V_j^{\,2}}{2\delta}\right] - \left[\rho_s C_{ps} V_{sy}\left(T_f - T_s\right)\right] = \rho_s V_{sy}\, \Delta H_f \qquad [3.66]$$

where ΔH_f is the enthalpy of melting. The right side represents the melting rate at the interface per unit area.

Let X be the width of the solid bed in direction x. A second relationship between the unknowns can be obtained from a mass balance on the liquid film in the channel direction:

$$\left(\frac{V_{Fz} + V_{sz}}{2}\right)\rho_f \frac{d\left(X\delta\right)}{dz} + V_{sy}\, \rho_s\, X - \frac{V_{Fx}}{2}\rho_f\, \delta = 0 \qquad [3.67]$$

In this equation, the first term represents the mass flow in the z direction, the second term represents the mass flow rate crossing the solid–liquid interface and the third term represents the mass flow rate moving toward the melt pool. We can show that the first term is negligible compared to the other two. Thus, equation [3.67] becomes:

$$V_{sy}\,\rho_s\,X - \frac{V_{Fx}}{2}\rho_f\,\delta = 0 \qquad [3.68]$$

Solving equation [3.68] and replacing V_{sy} in equation [3.66], it is possible to determine the thickness of the film:

$$\delta = \sqrt{\frac{\left[2k_l\left(T_F - T_f\right) + \eta V_j^2\right]X}{V_{Fx}\,\rho_f\left[C_{ps}\left(T_f - T_s\right) + \Delta H_f\right]}} \qquad [3.69]$$

The melting rate per unit length of the channel is:

$$\varpi = V_{sy}\,\rho_s\,X \qquad [3.70]$$

i.e.:

$$\varpi = \phi\sqrt{X} \qquad [3.71]$$

with:

$$\phi = \sqrt{\frac{V_{Fx}\,\rho_f\left[k_l\left(T_F - T_f\right) + \frac{\eta}{2}V_j^2\right]}{2\left[C_{ps}\left(T_f - T_s\right) + \Delta H_f\right]}} \qquad [3.72]$$

The parameter ϕ characterizes the overall melting efficiency under the given thermo-mechanical conditions. The numerator includes the terms corresponding to the heat sources, i.e. heat conduction and viscous dissipation. The denominator includes the terms corresponding to heat consumption, that is to say, the specific heat of melting.

One solution to increase the melting rate, for a polymer of a given viscosity, is to raise its temperature. Increasing the barrel temperature within the thermal stability limit of the polymer, greatly favors melting. Increasing the rotation speed of the screw, however, should be considered carefully.

Indeed, an increase in V_F also leads to an increase in the feed rate (equation [3.13]). As a result, it is not certain whether the melting is improved in this case.

Another solution includes lowering the denominator. For example, by preheating the pellets, it is possible to reduce the gap between the melting temperature T_f and the temperature of the solid bed T_s, thus, increasing the value of the parameter ϕ.

Now, we are left with determining the evolution of the solid bed profile inside the channel, that is to say, the change in the width X as a function of z. This relation can be obtained by performing a mass balance on the solid bed in the z direction. The depth of the solid bed is equal to the difference between the depth of the channel and the thickness of the melted film. As the thickness of the film varies slowly with z, the depth h of the solid can be considered as the difference between the channel depth and a certain constant value, which can be reasonably considered as the flight depth. We obtain:

$$-\frac{d(hX)}{dz}=\frac{\varpi}{\rho_s V_{sz}} \qquad [3.73]$$

Assuming that the depth of the channel is constant and that when $z = 0$ the solid bed occupies the whole width W of the channel, equation [3.73] can be integrated to obtain:

$$\frac{X}{W}=\left(1-\frac{\phi}{2\rho_s V_{sz} h\sqrt{W}}z\right)^2 \qquad [3.74]$$

According to the above equation, the width of the solid polymer bed decreases in a parabolic fashion along the channel. Complete melting is achieved when $X = 0$, i.e. for a length z_f such as:

$$z_f=\frac{2\rho_s V_{sz} h\sqrt{W}}{\phi} \qquad [3.75]$$

The velocity component V_{sz} of the solid polymer is assumed to be constant throughout the extruder and can be deduced from the solid mass flow rate Q_s:

$$Q_s=\rho_s V_{sz} hW \qquad [3.76]$$

Equations [3.74] and [3.75] can be expressed in the following form:

$$\frac{X}{W} = \left(1 - \frac{\phi\sqrt{W}}{2Q_s} z\right)^2 \qquad [3.77]$$

$$z_f = \frac{2Q_s}{\phi\sqrt{W}} \qquad [3.78]$$

Equation [3.78] shows that, if the rotation speed is increased to speed up melting (via the parameter ϕ), the flow rate Q_s also increases (equation [3.13]). The value of the Q_s/ϕ ratio ultimately defines whether this operation is efficient or not.

Now, assuming that the channel depth varies linearly (Figure 3.12), the depth h of the solid is expressed as follows:

$$h = h_0 - z\tan\alpha \qquad [3.79]$$

Consider the dimensionless number ψ:

$$\Psi = \left(\frac{\phi}{\rho_s V_{sz}\sqrt{W}}\right)^2 \qquad [3.80]$$

For boundary conditions identical to those used in the case of a channel with a constant height, the solution to equation [3.73] is:

$$\frac{X}{W} = \left[\frac{\Psi}{\tan\alpha} - \left(\frac{\Psi}{\tan\alpha} - 1\right)\sqrt{\frac{h_0}{h_0 - z\tan\alpha}}\right]^2 \qquad [3.81]$$

The total length of the melting zone is expressed by:

$$z_f = \frac{h_0}{\Psi}\left(2 - \frac{\tan\alpha}{\Psi}\right) \qquad [3.82]$$

which can also be written as:

$$z_f = z_{f(\alpha=0)}\left(1 - \frac{\mathrm{tg}\,\alpha}{4h_0} z_{f(\alpha=0)}\right) \qquad [3.83]$$

This shows that compression reduces the length required for melting, especially when α is large. We cannot, however, go beyond the limit set by:

$$\operatorname{tg}\alpha = \frac{4h_0}{z_{f(\alpha=0)}} \qquad [3.84]$$

since under stronger compression beyond this limit, the mechanism proposed, up to this point, can no longer be applied.

3.4.2.3. *Non-Newtonian fluid*

The previous model applies to a Newtonian fluid; however, the majority of polymers have a non-Newtonian behavior. As viscosity is one of the main parameters affecting the dissipative melting mechanism, the Newtonian hypothesis is the main source of deviation of model predictions from experimental results.

For this reason, Tadmor and Klein [TAD 70] built a specific model for power-law fluids. The adopted approach is identical to that for a Newtonian fluid. This model, however, has no simple analytical solution. The temperature and pressure profiles must, therefore, be solved, independently, and the solution to the system is obtained using an iterative process.

The full development of this model can be found in books by Tadmor and Klein [TAD 70] and Chung [CHU 00].

3.4.2.4. *Other models*

Tadmor and Klein's model [TAD 70] is the best known for melting in single-screw extrusion. It is based on the experimental observations of Maddock [MAD 59]. It served as the basis for more sophisticated models, such as that developed by Lindt [ELB 84, LIN 85]. This model accounts for the polymer melt pool, a film of molten polymer between the barrel and the solid, and molten polymer films between the solid bed and the screw. Figure 3.31 shows the solid bed profiles obtained using the Lindt model under various conditions. The rigid solid bed with recirculation seems to be the most realistic.

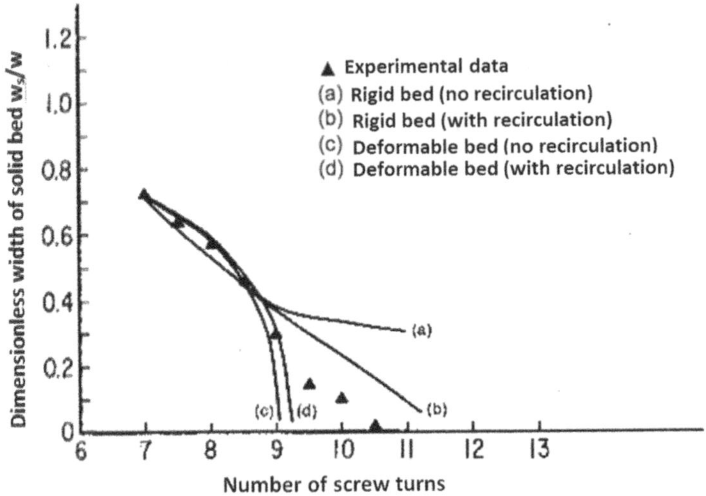

Figure 3.31. *Evolution of the width of the solid bed along the screw. Comparison of experimental results (▲) and results from the Lindt model, under various conditions [LIN 85]*

There are other models that do not consider the melt pool on the trailing flight. The most recent is a 3D finite element model, by Altinkaynak *et al.* [ALT 11], that considers a biphasic flow in the solid conveying zone of an extruder. The viscosity is represented by a Cross-WLF model, with a solid phase with a very large viscosity. A series of screw extraction experiments enables us to visualize the solid bed profile and to compare it with model predictions. Although the model is more descriptive in nature, a melting behavior, similar to that obtained by the Maddock model, can be observed (Figure 3.32). In addition, the melting behavior calculated by this 3D model gives results comparable to both the experimental results and those from the traditional Tadmor model (Figure 3.33). We can, therefore, consider that, despite its simplicity, the Tadmor analytical model yields satisfactory qualitative results taking into account the required physical parameters and properties.

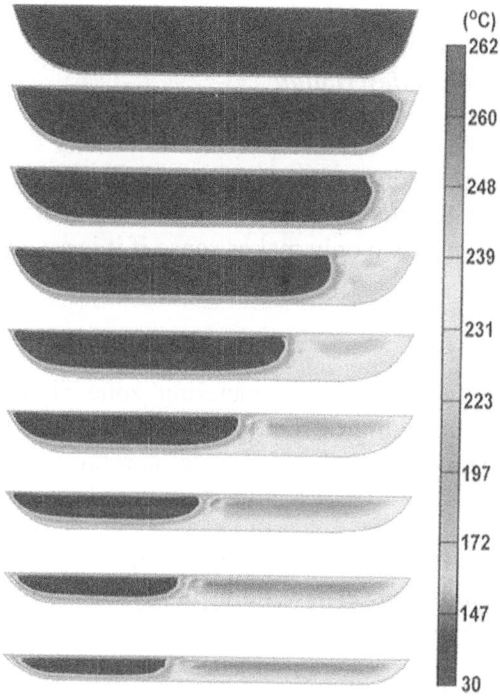

Figure 3.32. *The temperature field in different sections of the extruder, obtained by 3D finite element calculation of melting [ALT 11] (for a color version of this figure, see www.iste.co.uk/lafleur/polymer.zip)*

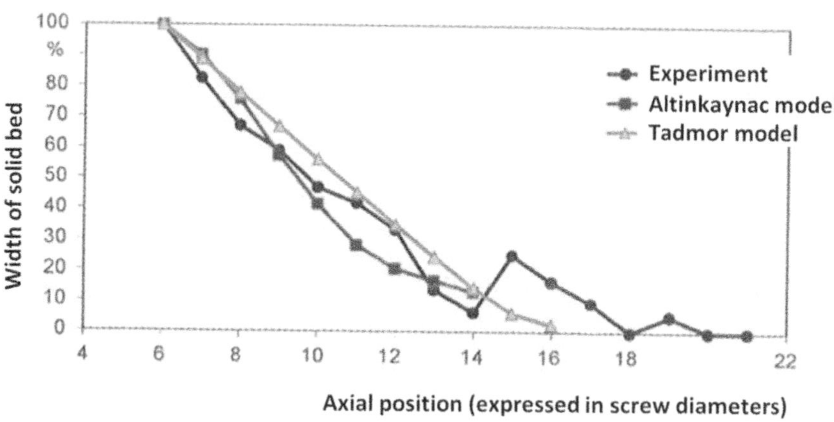

Figure 3.33. *A comparison of experimental results with predictions of the models of Tadmor and Altinkaynac et al. [ALT 11]*

3.5. Metering zone

3.5.1. *Phenomenological description*

In the metering zone, the entire polymer is usually in the molten state. This zone provides the polymer with the pressure required to ensure flow through the die at a desired flow rate. As emphasized by Agassant *et al.* [AGA 14], the metering zone only makes sense if it is associated with a die.

3.5.2. *Modeling*

Analysis of the flow inside the metering zone is much simpler than modeling the solid conveying or melting zones. The modeling of this zone can be simplified by assuming that the barrel is rotating around a stationary screw.

In this section, we discuss an isothermal 1D flow in the metering zone for a Newtonian fluid and a power-law fluid. In the case of a Newtonian fluid, the more realistic 2D flow will also be discussed.

3.5.2.1. *Isothermal flow of a Newtonian fluid*

Let us consider the isothermal flow of a Newtonian fluid within a channel of constant geometry. The regime is assumed to be steady state and established along the z axis, and the fluid to be incompressible. The clearance between the barrel and the flights, as well as the curvature of the channel, are negligible.

According to these assumptions, the flow equation in the z direction of the channel is written as:

$$\rho_f\left[v_x\frac{\partial v_z}{\partial x}+v_y\frac{\partial v_z}{\partial y}\right]=-\frac{\partial P}{\partial z}+\eta\left[\frac{\partial^2 v_z}{\partial x^2}+\frac{\partial^2 v_z}{\partial y^2}\right] \quad [3.85]$$

The left-hand side represents the inertial forces. For a fluid of high viscosity, which is the case for the majority of molten polymers, this term is negligible compared to the viscous forces. The previous equation, therefore, becomes:

$$\frac{\partial P}{\partial z}=\eta\left[\frac{\partial^2 v_z}{\partial x^2}+\frac{\partial^2 v_z}{\partial y^2}\right] \quad [3.86]$$

In the same way, along the x and y directions we, respectively, obtain:

$$\frac{\partial P}{\partial x}=\eta\left[\frac{\partial^2 v_x}{\partial x^2}+\frac{\partial^2 v_x}{\partial y^2}\right] \quad [3.87]$$

$$\frac{\partial P}{\partial y}=\eta\left[\frac{\partial^2 v_y}{\partial x^2}+\frac{\partial^2 v_y}{\partial y^2}\right] \quad [3.88]$$

The flow in the metering zone is a 3D flow. In the case of a Newtonian fluid, however, it is possible to split it into a longitudinal flow, described by equation [3.86], and a transverse flow, described by equations [3.87] and [3.88].

3.5.2.1.1. Longitudinal flow

The simplest way of considering the flow problem within the metering zone is to consider a channel with a parallel plate geometry. In this case, the velocity of the fluid along the z axis is a function of y only.

Using this assumption, the flow equation in the z direction is written as:

$$\frac{dP}{dz}=\eta\frac{d^2 v_z}{dy^2} \quad [3.89]$$

Assuming a no-slip condition at the barrel and screw walls, and integrating this equation leads to the following velocity profile:

$$v_z(y)=V_{Fz}\frac{y}{h}-\frac{y(h-y)}{2\eta}\frac{dP}{dz} \quad [3.90]$$

The mass flow rate is obtained by integrating the velocity profile:

$$Q=\frac{1}{2}V_{Fz}Wh-\frac{Wh^3}{12\eta}\frac{dP}{dz} \quad [3.91]$$

i.e.:

$$Q=Q_c-Q_p \quad [3.92]$$

with:

$$Q_c = \frac{1}{2} V_{Fz} W h \qquad [3.93]$$

$$Q_p = \frac{W h^3}{12\eta} \frac{dP}{dz} \qquad [3.94]$$

The first term on the right-hand side represents the shear flow (Q_c), induced by the movement of the barrel relative to the screw. The second term represents the pressure flow (Q_p). The shear flow is proportional to the width and depth of the channel, as well as to the rotation speed of the screw. The pressure flow is identical to that calculated for a 1D flow between two parallel plates. It is proportional to the channel width and to the pressure gradient, and inversely proportional to viscosity. It also depends on the depth of the channel.

Using equations [3.93] and [3.94], the longitudinal velocity component can also be written:

$$v_z(y) = V_{Fz} \frac{y}{h} \left[1 - \frac{3y(h-y)}{h} \frac{Q_p}{Q_c} \right] \qquad [3.95]$$

This expression shows that the velocity is highly influenced by the pressure flow, along the metering zone.

Equation [3.91] shows that the maximum flow rate is obtained if there is no die and that the maximum pressure corresponds to a die of infinite resistance. The operating curve – also called the characteristic curve (the flow rate Q versus the pressure drop ΔP) depends on the screw rotation speed (Figure 3.34).

Characteristic curves for dies can also be calculated. They are of the form $Q = k\Delta P$, where k only depends on the viscosity of the material, and on the geometry of the die. As shown in Figure 3.34, the intersection of the screw and the die characteristics curves define the operating condition of the extruder, i.e. the actual flow rate, and the pressure required. The use of characteristic curves is important, whether it is about understanding how a machine operates, or how to optimize performance.

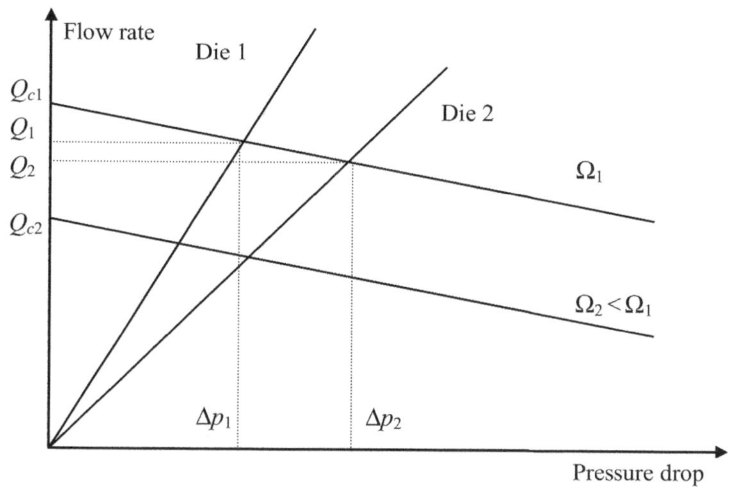

Figure 3.34. *Characteristic curves of the metering zone*

3.5.2.1.2. Transverse flow

In this case, we consider the flow in the transverse plane. The velocity field components along the x and y axes can be obtained from equations [3.87] and [3.88].

For a flow channel of a low aspect ratio:

$$\frac{\partial^2 v_x}{\partial x^2} << \frac{\partial^2 v_x}{\partial y^2} \tag{3.96}$$

Equation [3.87] is reduced to:

$$\frac{\partial P}{\partial x} = \eta \frac{\partial^2 v_x}{\partial y^2} \tag{3.97}$$

Assuming a no-slip condition at the walls, this equation can be integrated into:

$$v_x(y) = V_{Fx}\frac{y}{h} - \frac{y(h-y)}{2\eta}\frac{\partial P}{\partial x} \tag{3.98}$$

As the net flow in the transverse direction x is zero, the component v_x must satisfy the following relation:

$$\int_{y=0}^{y=H} v_x(y)dy = 0 \quad [3.99]$$

i.e.:

$$\frac{\partial P}{\partial x} = \frac{6\eta V_{Fx}}{h^2} \quad [3.100]$$

By substituting equation [3.100] in equation [3.98], we obtain:

$$v_x(y) = V_{Fx}\frac{y}{h}\left[3\frac{y}{h} - 2\right] \quad [3.101]$$

3.5.2.1.3. Coupled flow

To have a more precise idea about the flow in the metering zone, the two previous flows must be coupled. This results in helical trajectories, such as those shown in Figure 3.35. We find that, according to the starting position of the particle in the transverse plane, these trajectories will vary in length; their residence times can, thus, be significantly different.

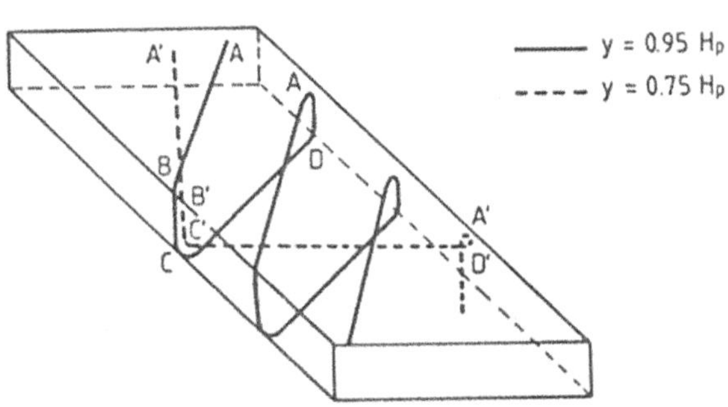

Figure 3.35. *Helical trajectories in the metering zone for two different starting positions [AGA 14]*

In the case where the existence of lateral walls is taken into account, the flow rate can be written as:

$$Q = F_d Q_c - F_p Q_p \quad [3.102]$$

where Q_c and Q_p are given by equations [3.93] and [3.94]. F_d and F_p are shape factors associated with shear flow and pressure flow, respectively. These factors have been calculated by Squires [SQU 58]:

$$F_d = \frac{16}{\pi^3} \frac{1}{(h/W)} \sum_{i=1,3}^{\infty} \frac{1}{i^3} \tanh\left(\frac{i\pi}{2}(h/W) \right) \qquad [3.103]$$

$$F_p = 1 - \frac{192(h/W)}{\pi^5} \sum_{i=1,3}^{\infty} \frac{1}{i^5} \tanh\left(\frac{i\pi}{2(h/W)} \right) \qquad [3.104]$$

F_d and F_p are presented graphically in Figure 3.36.

For the majority of screws used in industry, the aspect ratio *h/W* is set between 0.05 and 0.1. As a result, the values of F_d and F_p are, as a general rule, always higher than 0.94 and 0.93, respectively.

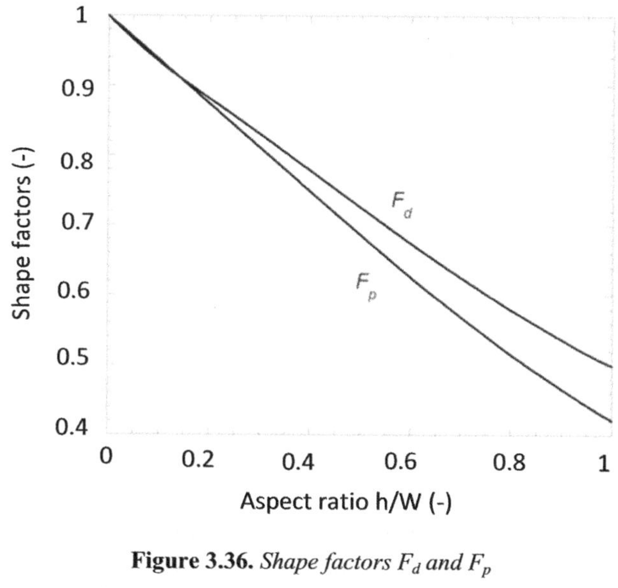

Figure 3.36. *Shape factors F_d and F_p*

3.5.2.2. *Isothermal flow of a power-law fluid*

Strictly speaking, it is no longer possible in this case to separate the longitudinal and transverse flows. However, it can be useful to calculate the longitudinal flow disregarding the transverse components, by using the assumptions previously made for a Newtonian and isothermal fluid. The component of the flow equation in the z direction, in the case of a power-law fluid, reduces to the following expression:

$$\frac{dP}{dz} = \frac{d\tau_{yz}}{dy} \qquad [3.105]$$

For a power-law fluid, the shear stress τ_{yz} acting in the z direction on the plane perpendicular to the direction y is given by the constitutive equation:

$$\tau_{yz} = K\left|\frac{dv_z}{dy}\right|^{m-1}\frac{dv_z}{dy} \qquad [3.106]$$

The boundary conditions are identical to those used for a Newtonian and isothermal fluid (no-slip boundary conditions).

The solution to the flow equation can be simplified by using the following dimensionless parameters:

$$\xi = \frac{y}{H} \qquad [3.107]$$

$$\varphi = \frac{v_z}{\Gamma V_{Fz}} \qquad [3.108]$$

$$\Gamma = \frac{h}{V_{Fz}}\left(\frac{h}{K}\frac{dP}{dz}\right)^{\frac{1}{m}} \qquad [3.109]$$

where ξ is the dimensionless position in the y direction, φ is the dimensionless velocity in the flow channel direction and Γ is the normalized pressure. By introducing these parameters, equation [3.105] becomes:

$$\frac{d}{d\xi}\left(\left|\frac{d\varphi}{d\xi}\right|^{m-1}\frac{d\varphi}{d\xi}\right) = 1 \qquad [3.110]$$

Depending on the value of Γ, two cases must be distinguished (Figure 3.37).

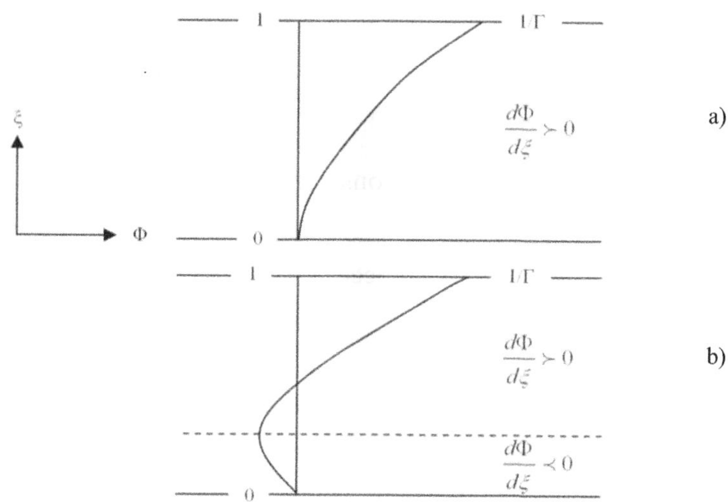

Figure 3.37. *Velocity profiles in the metering zone: a)* $\Gamma \leq \frac{1+m}{m}$ *; b)* $\Gamma > \frac{1+m}{m}$

If $\Gamma \leq \frac{1+m}{m}$, equation [3.110] becomes:

$$\frac{d}{d\xi}\left(\frac{d\varphi}{d\xi}\right)^m = 1 \qquad [3.111]$$

The solution to this equation is:

$$\varphi(\xi) = \frac{1}{s}\left[(\xi + C_1)^s - C_2\right] \qquad [3.112]$$

with:

$$s = \frac{1+m}{m} \qquad [3.113]$$

The constants of integration C_1 and C_2 are obtained from the boundary conditions. In particular, C_1 is obtained using an iterative calculation. The solution to the flow equation is therefore:

$$Q = V_{Fz} W H \frac{\Gamma}{s}\left[\frac{(1+C_1)^{1+s} - C_1^{1+s}}{1+s} - C_1^s\right] \qquad [3.114]$$

If $\Gamma > \frac{1+m}{m}$, two cases must be considered:

For $0 \leq \xi \leq \xi^*$, equation [3.110] becomes:

$$-\frac{d}{d\xi}\left(-\frac{d\varphi}{d\xi}\right)^m = 1 \qquad [3.115]$$

The solution to this equation is:

$$\varphi(\xi) = \frac{1}{s}\left[\left(\xi^* - \xi\right)^s - \left(\xi^*\right)^s\right] \qquad [3.116]$$

For $\xi^* \leq \xi \leq 1$, equation [3.110] becomes:

$$\frac{d}{d\xi}\left(\frac{d\varphi}{d\xi}\right)^m = 1 \qquad [3.117]$$

The solution to this equation is:

$$\varphi(\xi) = \frac{1}{s}\left[\left(\xi - \xi^*\right)^s - \left(1 - \xi^*\right)^s\right] + \frac{1}{\Gamma} \qquad [3.118]$$

For $\xi = \xi^*$, the expressions [3.116] and [3.118] have the same value. As a result, ξ^* must satisfy the following equation:

$$\frac{s}{\Gamma} = \left(1 - \xi^*\right)^s - \left(\xi^*\right)^s \qquad [3.119]$$

The value of ξ^* is obtained from equation [3.119], by iteration.

It can be shown that the mass flow rate, obtained by integration of the velocity profile, is given by:

$$Q = V_{Fz} W H \frac{\Gamma}{s} \left[\frac{\left(\xi^*\right)^{1+s} - \left(1-\xi^*\right)^{1+s}}{1+s} - \left(\xi^*\right)^s \right] \qquad [3.120]$$

In this case, we see that the overall flow rate is no longer simply the sum of the shear and pressure flows, although it depends both on the pressure gradient (via Γ) and on the barrel velocity V_{Fz}.

3.5.2.3. *Numerical 3D models*

Several numerical models including a non-isothermal 3D analysis by finite volumes or finite elements have been developed over the past few years [LAW 94, KWA 02]. The majority are based on the unrolled geometry and on a reference frame fixed on a stationary screw, with a rotating barrel. Table 3.7 shows a comparison of the results from an unrolled geometry model with these obtained in a helical geometry, along with some experimental results [GHO 04]. From the table, we can see that the helical geometry better represents the experimental results.

Rotation speed (rpm)	Experimental flow rate (g/s)	Helical geometry flow rate (g/s)	Unrolled geometry flow rate (g/s)
20	0.168	0.166	0.159
40	0.333	0.330	0.317
60	0.487	0.484	0.464

Table 3.7. *A comparison between experimental data and results from 3D models [GHO 04]*

It is now possible to solve the actual problem of a 3D non-isothermal and non-Newtonian flow in a complex helical geometry, with a rotating screw, by using the immersed boundary method. This was made possible by using parallelization calculation techniques [VAL 09, ILI 10]. These techniques, however, are limited to the flow section where the entire polymer is molten, they are not easy to use, and, usually, require significant calculation time and computer resources.

3.6. Overall model

The function of a single-screw extruder is to melt the solid polymer and pump the molten polymer across a die, which shapes it to produce a profile, a sheet or a film.

We can define an overall model by connecting the models, presented here so far, for each of the functional zones; this will allow us to describe the complete single-screw extrusion process, from the moment the pellets enter the hopper until the polymer exits the die.

A number of commercial software packages are available to analyze single-screw extrusion. They differ in their ease of use and the assumptions made to represent each zone in the extruder. Without giving a full description, the main softwares used to date are as follows:

– EXTRUD, SPR, United States;

– SSD, Stevens Institute, United States;

– REX, University of Paderborn, Germany;

– VEL Extruder Module, Compuplast, Czech Republic.

To illustrate the behavior of a screw extruder under various processing conditions, we are going to use a model, developed by Amellal *et al.* [AME 91], which allows each zone of the extruder to be simulated, both for barrier and conventional screws.

The first two results, presented in Figure 3.38, show that the pressure gradient in the metering zone can be positive or negative depending on the flow rate imposed through the die by changing the rotation speed.

To verify that the model can simulate extreme extrusion conditions, a die with a very low resistance to flow was used. Figure 3.39 shows a low pressure at the exit and a strongly negative pressure gradient in the metering zone.

Figure 3.40 shows the results for extrusion of linear low-density polyethylene (LLDPE), with a Barr barrier screw (see section 3.7.2), with a rotation speed of 15 rpm.

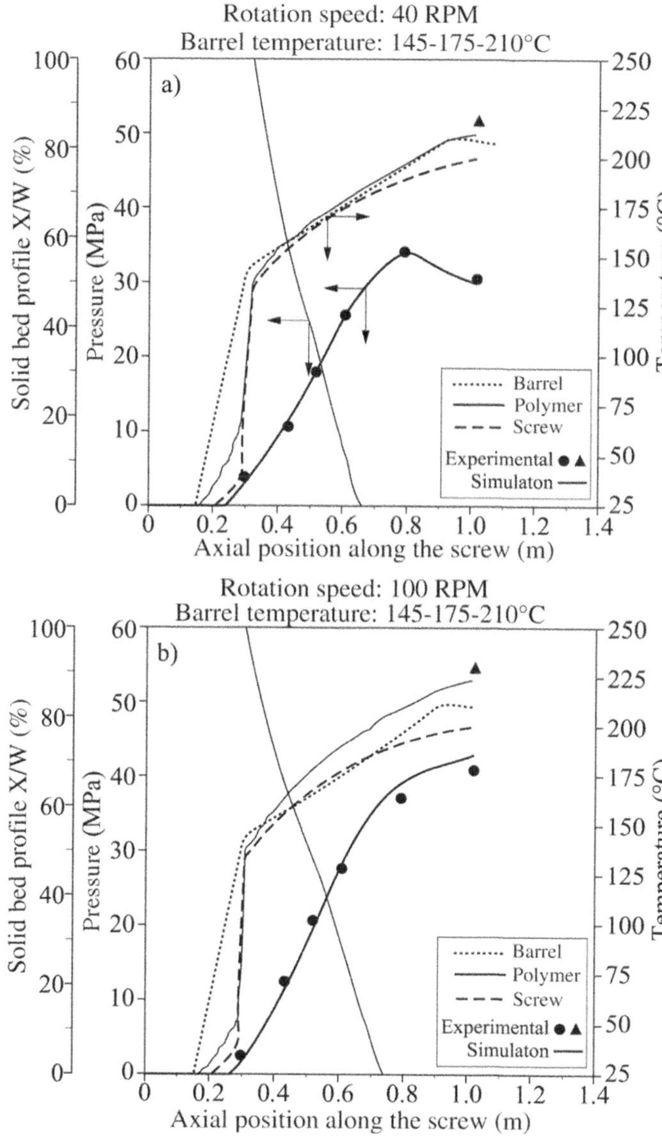

Figure 3.38. *Evolution of temperature, width of solid bed and pressure along the screw for two rotation speeds: a) 40 rpm, b) 100 rpm. The symbols (●, ▲) are experimental points [AME 91]*

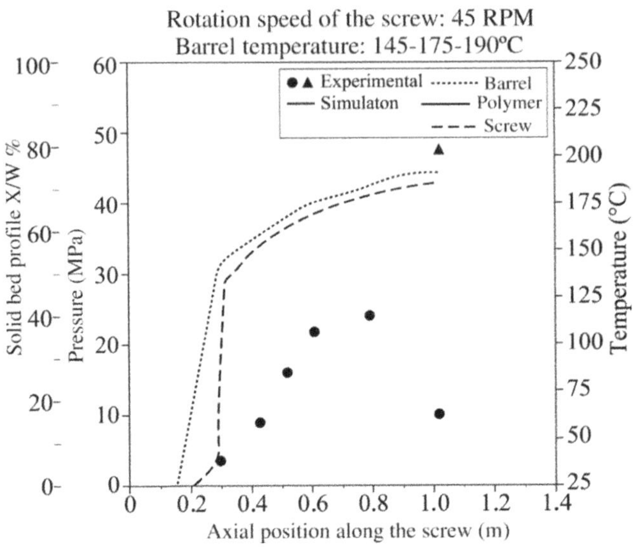

Figure 3.39. *Evolution of temperature, width of the solid bed and pressure along the screw for a low restrictive die. The symbols (●, ▲) are experimental points [AME 91]*

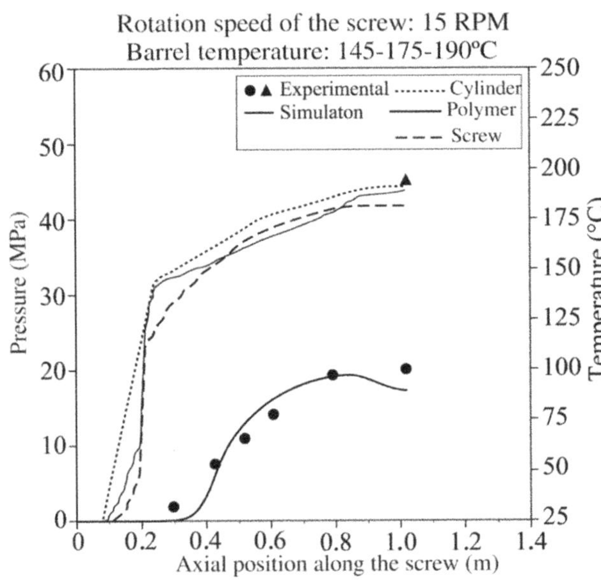

Figure 3.40. *Evolution of temperature and pressure along the screw for a barrier screw. The symbols (●, ▲) are experimental points [AME 91]*

Figure 3.41 shows the same screw but with a different polymer, whose properties are poorly suitable to the barrier screw geometry; since the melting of the polymer is slower, the channel is not filled as quickly.

As mentioned earlier, accounting for the effect of temperature and pressure on the physical properties of the material can influence the quality of a model's predictions. Figure 3.42 shows that, for the same barrier screw, the obtained pressure profile is different when the effect of temperature and pressure on the molten polymer is taken into account.

The models presented above are very useful in understanding how the process works; in practice, these models also have other uses.

For example, a challenge in extrusion is the scale-up from laboratory size (process development) to industrial size (industrial production): if the geometry and processing conditions are defined for a 15 mm diameter screw, producing 5 kg/h, how can we estimate the proper flow rate to obtain a product of the same quality, if a 30 mm screw is to be used?

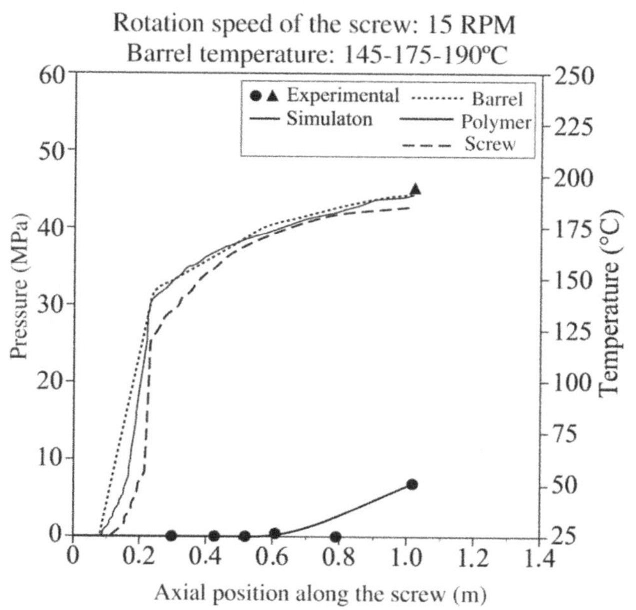

Figure 3.41. *Evolution of temperature and pressure along the screw for a barrier screw with a misfit polymer. The symbols (●, ▲) are experimental points [AME 91]*

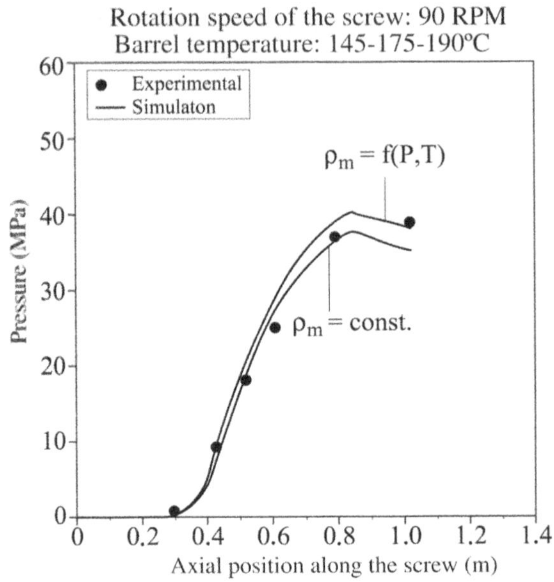

Figure 3.42. *Evolution of pressure along the screw for a barrier screw with density constant or varying with temperature and pressure, where (•) are experimental points [AME 91]*

The scale-up criteria will vary according to the parameter, i.e. the specific energy does not obey the same rules as the shear rate. Furthermore, only a simulation device enables the rapid testing of different solutions, keeping only the most relevant ones.

Similarly, the optimization of an existing process can be done by trial and error, which is always long and costly. Alternatively, this can be achieved by coupling a simulation tool with some optimization algorithms, thereby enabling us to obtain, not only the optimal processing conditions for a given screw geometry, but also the optimal screw design itself, according to the criteria set by the user [GAS 01, COV 09].

3.7. Technological improvements

Two main limitations of the single-screw extruder, compared to a twin-screw extruder, are a relatively low flow rate and a limited mixing capacity. To improve these two limitations, different modifications and/or improvements of conventional screws have been suggested.

3.7.1. *Grooved barrels*

In the solid conveying section, we discussed the effects of the friction coefficient on the screw and the barrel. The movement of the solid polymer along the screw depends on these friction coefficients. The flow rate increases with the f_F/f_v ratio. The majority of extruders have a section containing a smooth barrel underneath the hopper. This section is as deep as possible so that the pellets can be compacted and the screw can be fed properly. In grooved barrels, several grooves are carved into the inside of the barrel, beneath the hopper and along part of the solid conveying zone (Figure 3.43).

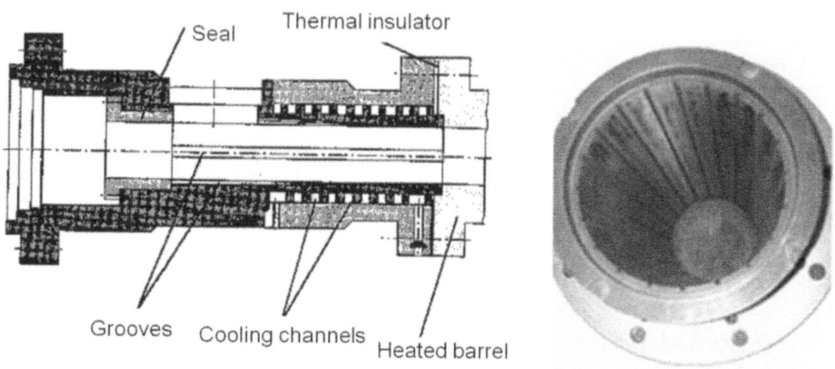

a) b)

Figure 3.43. *A cross-section a) and a photo b) of a longitudinally grooved barrel*

These grooves can be either longitudinal or spiralled, with a pitch opposite to that of the screw flights. The operating principle is that certain pellets are captured by the grooves, and rather than circulating along the flight angle, they circulate within the grooves, pushed by the flights. This considerably increases friction and the transport of solids [VER 02]. A cooling system is implemented in this section to evacuate the heat produced and to prevent melting. The grooves have a depth that decreases along the solid conveying zone and can reach a length of 6*D*. The literature suggests [GOF 90] using a number of grooves equal to *D*/10 (*D* expressed in mm), with a depth varying from 0.5 to 1 cm and an angle of 3° (the angle determining the decreasing depth of the grooves).

In a conventional extruder, the pressure increases steadily, reaching its maximum at the level of the die. The pressure drop in the die is, thus, generated by the metering zone of the extruder. For a given screw, with a fixed rotation speed, the flow rate decreases as the restriction of the die increases (characteristic curve, Figure 3.44(a)). For an extruder with a grooved barrel, due to the high friction forces, the maximum pressure is reached in the solid conveying zone, thereby allowing a constant flow rate even in the case of an increase in die pressure. Figure 3.44(b) shows this behavior [DAV 98].

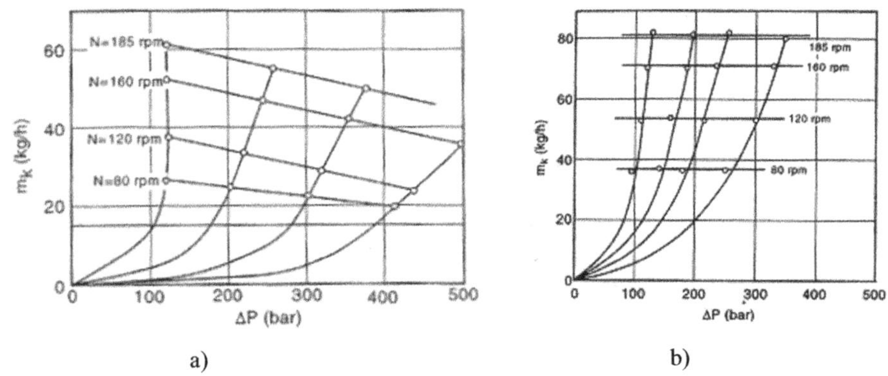

Figure 3.44. *Characteristic curves of an extruder with a) a smooth barrel and b) a grooved barrel [DAV 98]*

Another advantage of grooved barrels is the significant lowering of temperature at the extruder exit, due to a much lower rotation speed for a given flow rate. One of the disadvantages, however, is the premature wear of the barrel, which is caused by the increased friction and pressure. This problem can be minimized by implementing a decompression zone at the end of the grooved section.

For very viscous polymers with very low friction coefficients, the grooved barrels increase the productivity and stability of the process. However, to ensure that melting takes place under optimal conditions, it is recommended to use barrier screws, as we will see in the next section. In addition, as screws with grooved barrels require a lower compression rate than conventional screws, the mixing capacity is reduced and it is often necessary to use mixing heads at the extruder exit.

3.7.2. *Barrier screws*

The operating principle of barrier screws involves introducing a second flight (barrier flight) at the start of the melting zone, which separates the solid and molten polymer. The first patent for such a screw was filed by Maillefer in 1959 [MAI 59] (Figure 3.45). The gap (clearance) between the barrier flight and the barrel must be larger than the gap between the main flight and the barrel (five to six times). The gap must be large enough to allow the molten polymer, but not the solid, to flow through, thereby separating the two. It is possible to independently vary the geometric parameters of the section containing the molten polymer and that containing the solid. This allows the optimization of the melting process by varying the depth and width of the channel.

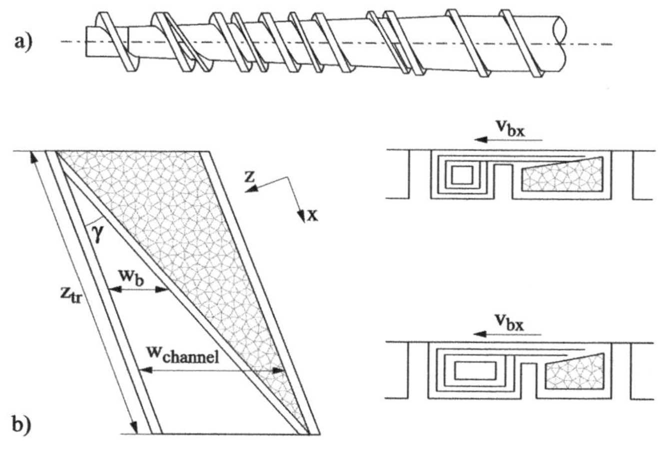

Figure 3.45. *a) Diagram of a Maillefer screw; b) top view (left), and cross-section (right) of the channel in the zone with two flights*

There are many types of barrier screws, based on the number of secondary flights, flight angle and depth of the channels [RAU 01]. Two classic examples are the Maillefer screw, already mentioned, and the Barr screw. For the Maillefer screw, the secondary flight angle is greater than the primary angle, thereby rapidly reducing the width of the solid bed. However, the depth of these two channels is identical.

In the case of the Barr screw (Figure 3.46), the initial part is identical to that of the Maillefer screw. However, when the section containing the molten polymer is large enough, the barrier flight becomes parallel to the main flight; the depth of the channel containing the solid polymer decreases, whereas that containing the molten polymer increases. In the case of the Barr screw, there is a reduction of approximately 72% in the length needed for melting compared to a conventional screw.

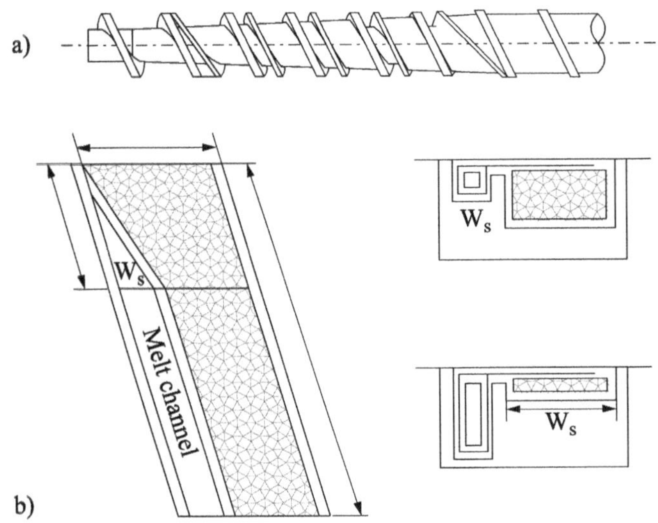

Figure 3.46. *a) Diagram of a Barr screw; b) top view (left), and cross-section (right) of the channel in the zone with two flights*

3.7.3. *Mixing heads*

Due to its geometry and mode of operation, the single-screw extruder does not always ensure good homogeneity of the polymer at the end of the metering zone, and even more so if fillers or additives are present. Moreover, using a grooved barrel and a barrier screw increases the flow rate and, thus, accentuates the mixing quality problem.

One technique used to increase the mixing action in a single-screw extruder is to put a mixing head at the end of the screw. Depending on the intended application, distributive or dispersive mixing may be desired.

Both mixing types are obtained by creating high stresses and strains in the molten polymer. Dispersive mixing favors the rupture of components, whereas distributive mixing ensures uniform distribution of constituents. Figure 3.47 illustrates both mixing types [MAN 09a].

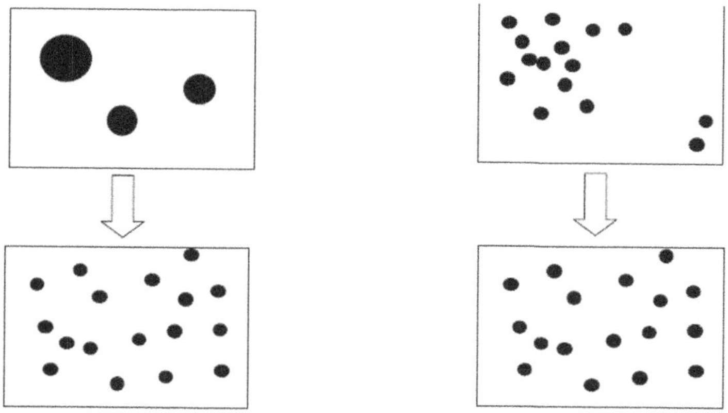

a) b)

Figure 3.47. *Examples of a) dispersive mixing and b) distributive mixing*

Figure 3.48 illustrates four types of mixing heads that favor distributive mixing [RAU 09]. In this case, we aim to separate and recombine the flow, which is the most efficient way to homogenize the filler distribution. The mixing heads are composed of pins or channels designed for this purpose. These tools generally cause limited pressure drops and temperature increases.

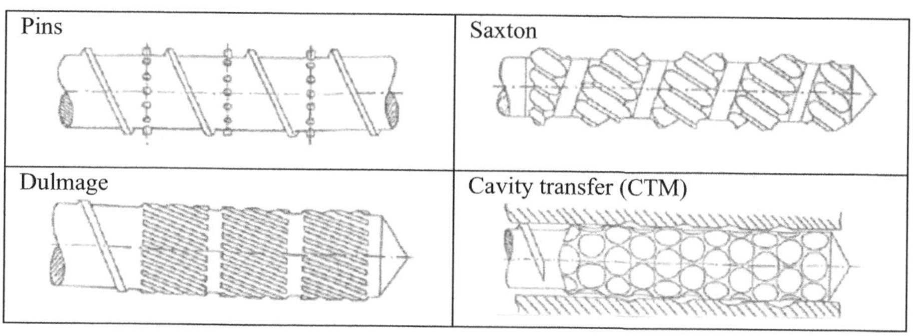

Figure 3.48. *Mixing heads used for distributive mixing [RAU 09]*

Figure 3.49 shows four types of mixing heads that ensure the dispersion of the components. For this aspect of the problem, the component must be subjected to a large enough stress so that it ruptures [MAN 09b]. Thus, the material is forced to pass through a narrow gap. The Maddock mixing elements are certainly the most common for dispersive mixing, and unlike distributive elements, they cause additional pressure drops, and significant increases in temperature, at the end of the screw.

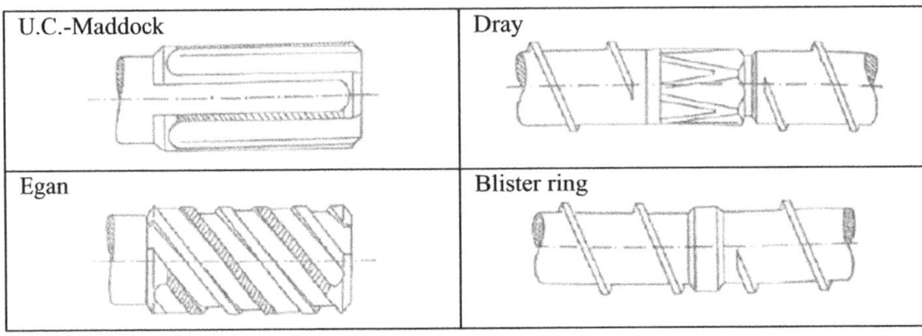

Figure 3.49. *Mixing heads used for dispersive mixing [RAU 09]*

3.8. Conclusion

Single-screw extrusion is still the primary method for processing and shaping polymers. Because of the studies carried out for over 60 years, it is now well understood and the main mechanisms involved have been identified and modeled. Technological improvements including grooved barrels, barrier screws or mixing heads have significantly improved performance of the screw-barrel system. Comprehensive models for calculating and predicting the parameters involved in the process are commercially available. They allow the user, not only to better understand the process, but also to solve the production problems and optimize the screw geometry and the processing conditions.

3.9. Bibliography

[AGA 14] Agassant J.-F., Avenas P., Sergent J.-P., *et al.*, *La mise en forme des polymères. Approche thermomécanique de la plasturgie*, Tec et Doc, Lavoisier, Paris, 2014.

[ALT 11] Altinkaynac A., Gupta M., Spalding M.A., *et al.*, "Melting in a single screw extruder: experiments and 3D finite element simulations", *International Polymer Processing*, vol. 26, pp. 182–196, 2011.

[AME 91] AMELLAL K., ARPIN B., LAFLEUR P.G., "A computer-model for the simulation of conventional and barrier screws", *Polymer-Plastics Technology and Engineering*, vol. 30, pp. 655–664, 1991.

[ATT 80] ATTALLA G., PODIO-GUIDUGLI P., "On modeling the solids conveying zone of a plasticating extruder", *Polymer Engineering & Science*, vol. 20, pp. 709–715, 1980.

[BRO 72] BROYER E., TADMOR Z., "Solids conveying in screw extruders. Part I: a modified isothermal model", *Polymer Engineering & Science*, vol. 12, pp. 12–24, 1972.

[CAM 01] CAMPBELL G.A., WANG C., CHENG H., *et al.*, "Investigation of flow rate and viscous dissipation in a single screw pump-extruder", *International Polymer Processing*, vol. 17, pp. 323–333, 2001.

[CHU 75] CHUNG C.I., "Maximum pressure developed by solid conveying force in screw extruders", *Polymer Engineering & Science*, vol. 15, pp. 29–34, 1975.

[CHU 77] CHUNG C.I., HENNESSEY W.J., TUSIM M.H., "Frictional behavior of solid polymers on a metal surface at processing conditions", *Polymer Engineering & Science*, vol. 17, pp. 9–20, 1977.

[CHU 00] CHUNG C.I., *Extrusion of Polymers. Theory and Practice*, Hanser, Munich, 2000.

[COV 09] COVAS J.A., GASPAR-CUNHA A., "Extrusion scale-up: an optimization-based methodology", *International Polymer Processing*, vol. 24, pp. 67–82, 2009.

[DAR 56] DARNELL W.H., MOL E.A.J, "Solids conveying in screw extruders", *Society of Plastics Engineers Journal*, vol. 12, pp. 20–29, 1956.

[DAV 98] DAVIS B.A., GRAMANN P.J., NORIEGA M.E., *et al.*, "Grooved feed single screw extruders improving productivity and reducing viscous heating effects", *Polymer Engineering & Science*, vol. 38, pp. 1199–1204, 1998.

[ELB 84] ELBIRLI B., LINDT J.T., GOTTGETREU S.R., *et al.*, "Mathematical modeling of melting of polymers in a single-screw extruder", *Polymer Engineering & Science*, vol. 24, pp. 988–999, 1984.

[FAN 91] FANG S., CHEN L., ZHU F., "Studies on the theory of single screw plasticating extrusion. Part II: non-plug flow solid conveying", *Polymer Engineering & Science*, vol. 31, pp. 1117–1122, 1991.

[GAM 99] GAMACHE E., LAFLEUR P.G., PEITI C., *et al.*, "Coefficients of dynamic friction measured at extrusion processing conditions", *Polymer Engineering & Science*, vol. 39, pp. 1604–1613, 1999.

[GAS 01] GASPAR-CUNHA A., COVAS J.A., "The design of extrusion screws: an optimization approach", *International Polymer Processing*, vol. 16, pp. 229–240, 2001.

[GHO 04] GHOREISHY M.H.R., RAZAVINOURI M., "Modeling the flow in the metering zone of single screw extruders using a helical geometry model", *International Polymer Processing*, vol. 19, pp. 95–100, 2004.

[GOF 90] GOFFREDA F., "The application of groove feed extruders to produce mono-layer and coextruded medical films", *SPE ANTEC Technical Papers*, pp. 1121–1125, 1990.

[HUA 02] HUANG F., *Study of the Solids Conveying in Single Screw Extrusion*, Mémoire de maîtrise, Ecole polytechnique de Montréal, Canada, 2002.

[HWA 89] HWANG C.G., MCKELVEY J.M., "Solid bed compaction and frictional drag during melting in a simulated plasticating extruder", *Advances in Polymer Technology*, vol. 9, pp. 227–251, 1989.

[HYU 90] HYUN K.S, SPALDING M.A, "Bulk density of solid polymer resins as a function of temperature and pressure", *Polymer Engineering & Science*, vol. 30, pp. 571–576, 1990.

[ILI 10] ILINCA F., HÉTU J.F., "Three-dimensional finite element solution of the flow in single and twin-screw extruders", *International Polymer Processing Journal of the Polymer Processing*, vol. 25, pp. 275–286, 2010.

[KAC 72] KACIR L., TADMOR Z., "Solids conveying in screw extruders. Part III: The delay zone", *Polymer Engineering & Science*, vol. 12, pp. 387–395, 1972.

[KWA 02] KWAG D.S., LYU M.Y., KIM W.S., "A 3D numerical study of fluid flow and heat transfer in a single screw extruder", *International Polymer Processing*, vol. 17, pp. 95–101, 2002.

[LIN 85] LINDT J.T., "Mathematical modeling of melting of polymers in a single-screw extruder: a critical review", *Polymer Engineering & Science*, vol. 25, pp. 585–588, 1985.

[LIU 91] LIU C.Y., O'BRIEN K.T., "The interparticulate friction behavior of acetal copolymer pellets with bilobal and trilobal cross-sections", *Polymer Engineering & Science*, vol. 31, pp. 1637–1646, 1991.

[LOV 74] LOVEGROVE J.G.A., WILLIAMS J.G., "Pressure generation mechanisms in the feed section of screw extruders", *Polymer Engineering & Science*, vol. 14, pp. 589–594, 1974.

[MAD 59] MADDOCK P.G., "A visual analysis of flow and mixing in extruder screws", *Society of Plastics Engineers Journal*, vol. 15, pp. 383–389, 1959.

[MAI 59] MAILLEFER C., "Boudineuse à vis", Swiss Patent no. 363149, 1959.

[MAN 09a] MANAS-ZLOCZOWER I., "Basic concepts", in MANAS-ZLOCZOWER I. (ed.), *Mixing and Compounding of Polymers*, Hanser, Munich, 2009.

[MAN 09b] MANAS-ZLOCZOWER I., FEKE D.L., "Dispersive mixing of solid additives", in MANAS-ZLOCZOWER I. (ed.), *Mixing and Compounding of Polymers*, Hanser, Munich, 2009.

[MIC 11] MICHELANGELLI O.P., YAMANOI M., GASPAR-CUNHA A., *et al.*, "Modelling pellet flow in single screw extrusion with DEM", *Proceedings of the Institution of Mechanical Engineers, Part E: Journal of Process Mechanical Engineering,* vol. 225, pp. 255–268, 2011.

[MOY 04] MOYSEY P.A., THOMPSON M.R., "Investigation of solids transport in a single-screw using a 3-D discret particle simulation", *Polymer Engineering & Science,* vol. 42, pp. 2203–2215, 2004.

[POT 02] POTENTE H., POHL T.C., "Polymer pellet flow out of the hopper into the first section of a single screw", *International Polymer Processing*, vol. 17, pp. 11–21, 2002.

[POT 95] POTENTE H., SCHOPPNER V., "Bulk-density of plastic pellets in a screw channel", *International Polymer Processing*, vol. 10, pp. 10–14, 1995.

[RAU 01] RAUWENDAAL C., *Polymer Extrusion*, 4th ed., Hanser, Munich, 2001.

[RAU 09] RAUWENDAAL C., "Mixing in single screw extruders", in MANAS-ZLOCZOWER I. (ed.), *Mixing and Compounding of Polymers*, Hanser, Munich, 2009.

[SCH 69] SCHNEIDER K., "Frictional properties of granulated plastics", *Kunststoffe*, vol. 59, pp. 97–102, 1969.

[SPA 95] SPALDING M.A., HYUN K.S., "Coefficients of dynamic friction as a function of temperature, pressure and velocity for several polyethylene resins", *Polymer Engineering & Science*, vol. 35, pp. 557–563, 1995.

[SPA 97] SPALDING M.A., HYUN K.S., HUGHES K.R., "Stress distribution in solid polymer compacts", *The Journal of Reinforced Plastics and Composites*, vol. 16, pp. 1231–1241, 1997.

[SQU 58] SQUIRES P.H., "Screw extruder pumping efficiency", *SPE J.*, vol. 14, pp. 24–30, 1958.

[STE 95] STEVENS M.J., COVAS J.A., *Extrusion, Principles and Operation,* Chapman & Hall, London, 1995.

[TAD 70] TADMOR Z., KLEIN I., *Engineering Principles of Plasticating Extrusion,* Van Nostrand Rheinhold, New York, 1970.

[TED 71] TEDDER W., "A new theory of solid conveying in single-csrew extruders", *SPE J.*, vol. 27, pp. 68–73, 1971.

[VAL 09] VALETTE R., COUPEZ T., DAVID C., *et al.*, "A direct 3D numerical simulation code for extrusion and mixing processes", *International Polymer Processing*, vol. 24, pp. 141–147, 2009.

[VER 02] VERGNES B., PUISSANT S., "Extrusion monovis. Partie 1", *Techniques de l'Ingénieur*, vol. AM 3 650, pp. 1–16, 2002.

[ZHA 02] ZHANG S., SERNAS V., "Incompressible model of solids conveying in a single-screw extruder", *The Journal of Reinforced Plastics and Composites*, vol. 21, pp. 1399–1409, 2002.

Chapter 4

Co-rotating Twin-Screw Extrusion

4.1. Twin-screw extrusion

The twin-screw extrusion process was developed in the beginning of the 20th Century alongside the single-screw extrusion process. It has, however, experienced rapid growth over the past 30 years, especially with regard to co-rotating extrusion, with specific applications in plastic compounding, and in the manufacture of complex materials [WHI 10]. Before detailing its different forms, let us begin by providing some basic information about twin-screw extrusion: a twin-screw extruder includes, by definition, two parallel screws, rotating inside a barrel with a figure-eight cross-section. Unlike the single-screw extruder, the twin-screw extruder comes in many different types. Therefore, it is essential to classify them.

4.1.1. *The different types of extruders*

The two main criteria used for classifying twin-screw extruders are illustrated in Figure 4.1:

– The screw rotation direction: the extruders are referred to as *co-rotating* if both screws rotate in the same direction, and *counter-rotating* if they rotate in opposite directions.

– Interpenetration: the screws are said to be *interpenetrated* if the flight of one screw penetrates into the channel of the other. This interpenetration can be partial or complete. *Tangential and non-interpenetrated configurations are also possible.*

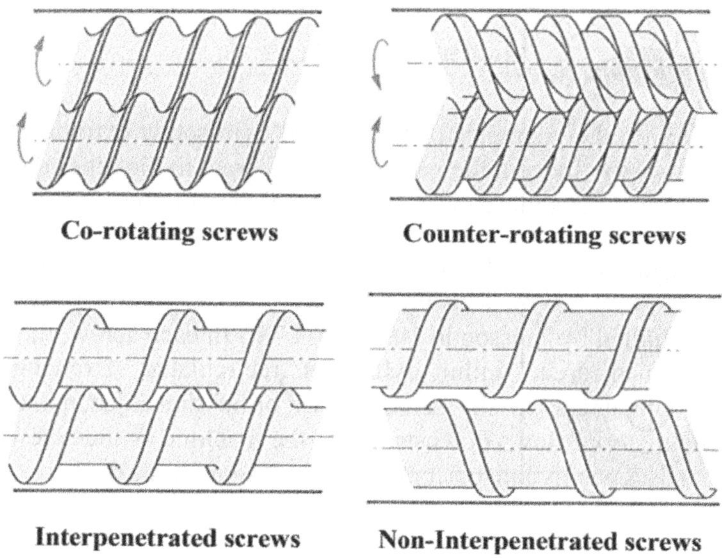

Figure 4.1. *Classification of twin-screw systems [VER 01a]*

Other differentiation criteria play a less important role in defining the flow conditions. For example, the screws can be tapered or parallel and the design can be conjugated (self-cleaning), or not. A design is said to be conjugated when the surface of one screw fits perfectly on the surface of the other. In this case, the flow of the material between both screws, from one side of a plane passing through the axes to the other, is very limited. Unlike the case in a single-screw extruder, the screws in co-rotating twin-screw extruders often have several flights. Screws with two flights are most common.

Numerous combinations of these different criteria are possible, and the chosen configuration usually depends on the desired application. Interpenetrated screws are used to convey and pressurize, whereas tangential or non-interpenetrated screws are used to mix fluids of low viscosity. In our area of interest, i.e. the extrusion of polymers, the main applications include the extrusion of polyvinyl chloride (PVC), (manufacture of profiles and pipes), for which we mainly use counter-rotating twin-screw extruders, and the manufacture of complex materials (filled products, polymer blends, reactive extrusion, etc.), for which the most efficient system is interpenetrated co-rotating twin-screw extrusion, which will be the focus of this chapter.

4.1.2. *Different flow types*

Depending on the geometry of the studied twin-screw extruder, the flow characteristics vary. Indeed, the path from one screw to the other through the interpenetration zone is more or less open depending on the configuration.

In the case of very closed geometries (for example, a completely interpenetrated counter-rotating profile), we find "C-chamber"-type flows, which were studied by Janssen in 1978 [JAN 78]. In each screw, the material fills a C-shaped area, within which it recirculates. Exchanges with neighboring chambers, in the same screw, or in the adjacent screw, are limited. The material flows downstream in the machine as the screw rotates, acting, here, as a positive displacement pump (Figure 4.2(a)).

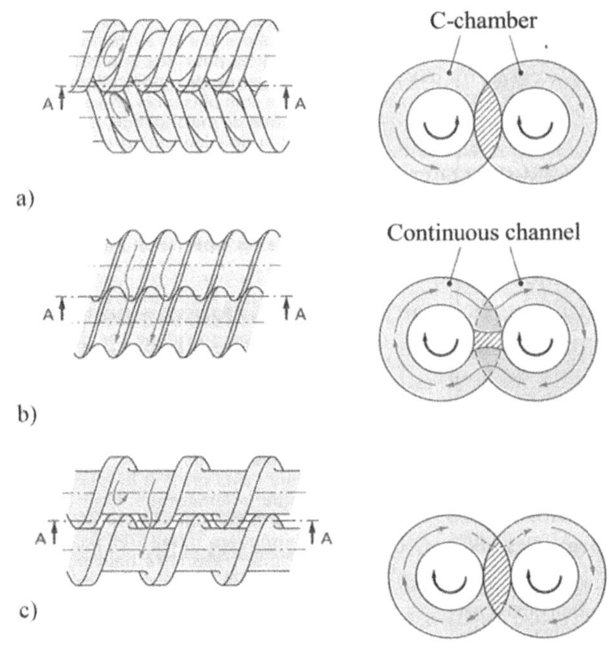

Figure 4.2. *Different flow types: a) C-chamber, b) continuous flow and c) intermediate case [VER 01a]*

For geometries with a large opening in the interpenetration zone (for example, an interpenetrated co-rotating conjugate profile), we encounter "continuous channel" flows, where the material easily passes from one screw to the other in a figure-eight trajectory, along the machine (Figure 4.2b). This type of flow is similar to that discussed in Chapter 3, in the metering zone of a single-screw extruder.

Of course, apart from these two extreme cases, any intermediate situation is possible: for example, the interpenetration zone may allow the material to pass from one screw to the other, but by significantly restricting the flow (Figure 4.2(c)). In practice, each flow type, depending on the assumed simplifications, requires a different theoretical approach. This explains why the different twin-screw systems must correctly be classified, and shows how complex a conventional approach, such as the one developed for single-screw extrusion, can become.

In this chapter, we will study the co-rotating extruders in detail, which are currently the most commonly used extruders. We advise the readers to refer to more detailed references on counter-rotating extruders [JAN 78, WHI 01, RAU 01, WHI 10]. Table 4.1 compares the general characteristics of both types [RAU 01].

	Type of extruder		
	Interpenetrated co-rotating	Interpenetrated counter-rotating	Tangential counter-rotating
Flow rate	++	+	++
Conveying	0	++	–
Rotation speed	++	0	++
Pumping capacity	0	++	–
Self-cleaning capacity	++	+	–
Width of the residence time distribution	0	++	0
Distributive mixing	++	+	++
Dispersive mixing	++	0	–

Notes: ++ very good, + good, 0 average, – poor.

Table 4.1. *Main functional characteristics of twin-screw extruders [RAU 01]*

4.2. General overview of co-rotating twin-screw extruders

In this section, we only discuss fully interpenetrated systems. We refer readers to the literature for tangential and partially interpenetrated systems [RAU 01, WHI 10], which are rarely used in the processing of molten polymers.

4.2.1. *Characteristic features of co-rotating twin-screw extruders*

Between the entry and the exit of the machine, aside from some differences, the functional zones are the same as those in single-screw extruders: that is to say, the conveying of solid pellets, the melting zone and the molten flow zone. Figure 4.3 illustrates this situation, obtained once the machine has stopped, the material has cooled and the barrel has been removed.

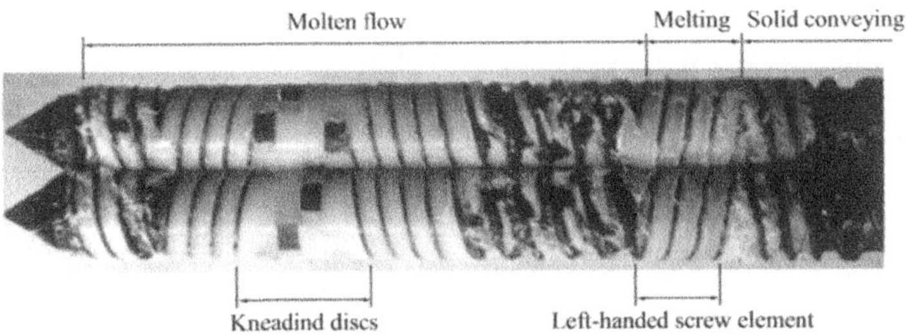

Figure 4.3. *State of the material inside a twin-screw extruder*

The first difference between twin-screw extruders and single-screw extruders is their more complex geometry. In fact, most twin-screw extruders are modular, that is to say, they are composed of small parts of a given length, which can be arranged on a fluted shaft to obtain a given profile. These elements are primarily of two types:

– screw elements, with right- or left-handed flights;

– kneading elements (see Figure 4.4).

The kneading elements are non-helical elements of a certain thickness and have a cross-section identical to that of the screw elements. They are mounted onto the shafts, one staggered by a certain angle in relation to the other, which produces complex flows during the rotation and allows efficient mixing. This staggering is said to be direct (or right-handed) if the tips create a pseudo-channel and are analogous to that of a screw element with a direct flight. Otherwise, we refer to an inverse (or left-handed) staggering. The mixing elements and the left-handed screw elements are restrictive elements, which tend to hinder the "natural" flow of the polymer towards the die. Of course, flow characteristics differ depending on the type of geometric elements, and each requires a specific theoretical approach.

The second point of interest is the melting zone. Unlike with single-screw extrusion (Chapter 3, section 3.4), melting seems to occur quickly. We will see in section 4.4 that the melting mechanisms in twin-screw extrusion are rather particular.

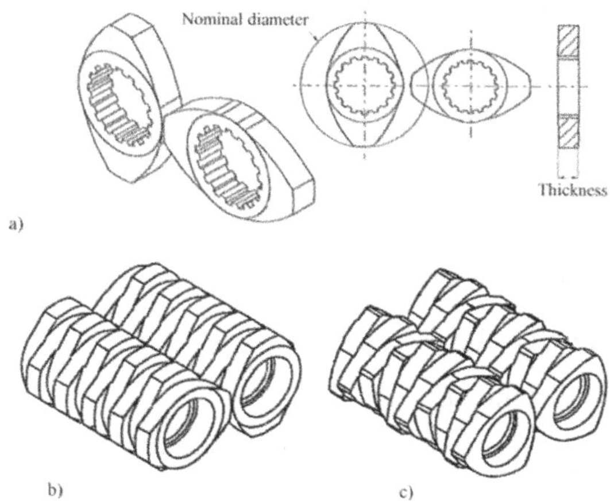

Figure 4.4. *a) Two-tips kneading discs, b) block composed of two-tips kneading discs at 90° and c) block composed of three-tips kneading discs at 30° [VER 01a]*

Figure 4.3 shows that the extruder is not completely filled. In fact, the conveying of the solid material occurs due to the positive displacement of the interpenetrated screws, rather than due to the effect of friction forces, as is the case in single-screw extrusion. The feed zone, and more generally the conveying zones, along the screws (except for restrictive elements), can only be slightly filled, which decouples the two processing parameters, namely the feed rate Q_m and the rotation speed N. As a result, the twin-screw extrusion process is incredibly versatile in terms of processing conditions. The local filling ratio is another parameter that must be taken into account during the modeling. In practice, the Q_m / N ratio is often used to characterize the overall filling ratio of twin-screw extruders.

4.2.2. *Geometry of the screws and barrel*

The geometry of conjugated interpenetrated co-rotating twin-screw extruders was studied in detail by Booy [BOO 78]. It should be noted that the geometric analysis presented here is valid for both screw elements, as well as, for kneading elements, which, in fact, are screw elements without helicity.

The geometric analysis by Booy is made by assuming that there is no leakage gap between the two screws, or between the screws and the barrel. The screw cross-section, in a plane perpendicular to the axes, is obtained assuming that the screws are in contact while one slides on top of the other. This suggests the following relation between the centerline distance E, the channel depth H, the internal diameter D_i and the external diameter D_e of the screw:

$$E = D_i + H = D_e - H \qquad [4.1]$$

The cross-section of the screw can be described using three parameters:

– the screw diameter D_e;

– the centerline distance E;

– the number of flights n.

In fact, the shape of the cross-section in a plane perpendicular to the screw axis is directly dependent on the number of flights (Figure 4.4).

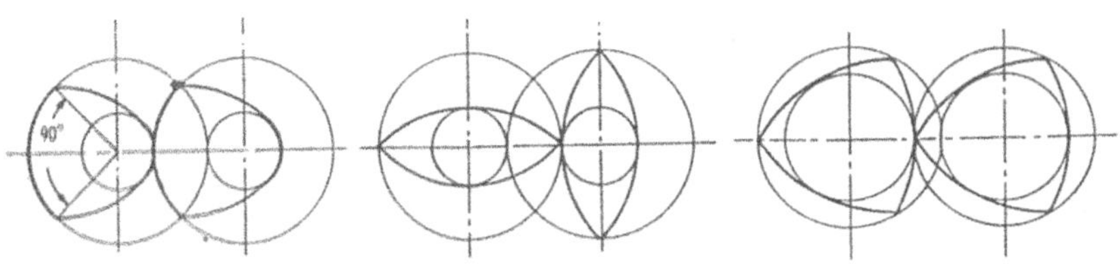

Figure 4.5. *Cross-sections of a screw with one, two and three flights [BOO 78]*

Booy [BOO 78] showed that this cross-section is defined by arcs of a circle, of the respective radii $D_e/2$, E, and $D_i/2$. These arcs are limited by two angles ψ and α:

– ψ defines the interpenetration zone (Figure 4.6). It is given by:

$$\cos\psi = \frac{E}{D_e} \tag{4.2}$$

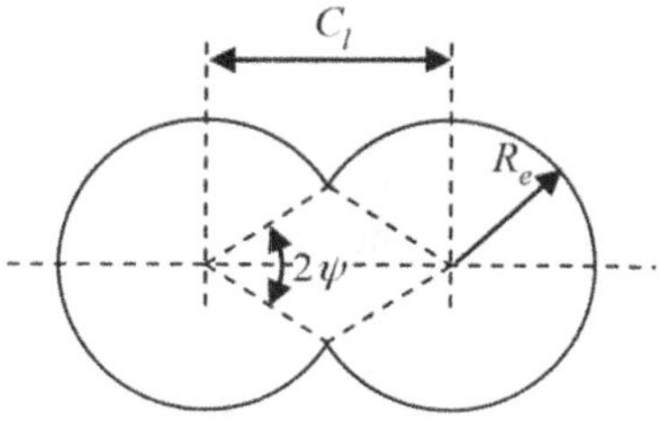

Figure 4.6. *Interpenetration zone (section perpendicular to the screw axis) and definition of angle ψ*

– α defines the flight thickness in the section perpendicular to the screw axis (Figure 4.7). It is given by:

$$n(2\alpha + 4\psi) = 2\pi \quad \text{i.e.} \quad \alpha = \frac{\pi}{n} - 2\psi \tag{4.3}$$

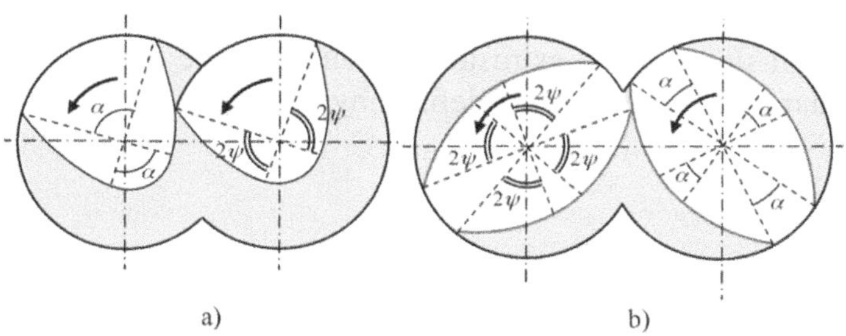

Figure 4.7. *Geometric parameters in the plane perpendicular to the screw axis. For a screw with a) one and b) two flights [Boo 78]*

We can see that when the centerline distance and diameter are identical, a two-flighted screw has a thinner flight, and, therefore, the cross-section of the channel is less obstructed in the interpenetration zone, which allows the material to flow more freely between the screws.

For α to be positive, it must comply with certain conditions regarding the centerline distance, the screw diameter and the number of flights:

– for a single flight screw, the E/D_e ratio can be of any value;

– it must be greater than $\sqrt{2}/2$ for screws with two flights;

– it must be greater than $\sqrt{3}/2$ for screws with three flights.

From this analysis, we can calculate the main geometric parameters. For example, the open cross-section of the barrel is:

$$S_F = (\pi - \psi)\frac{D_e^2}{2} + ED_e\frac{\sin\psi}{2} \tag{4.4}$$

The cross-section of the screw is defined by:

$$S_V = nE\left(\psi E - \frac{D_e}{2}\sin\psi\right) + \frac{n\alpha}{2}\left[\frac{D_e^2}{4} + \left(E - \frac{D_e}{2}\right)^2\right] \tag{4.5}$$

We can show that, since E and D_e are fixed, this cross-sectional area decreases as the number of flights n increases. The cross-section filled by the material and defined by $S_L = S_F - 2S_V$ will, therefore, be much larger when there are a large number of flights.

Just like in single-screw extrusion, the angle of the flight θ with a plane perpendicular to the screw axes is defined by:

$$\tan\theta(r) = \frac{B}{2\pi r} \tag{4.6}$$

where r is the radial position.

For screws with fairly shallow channels, $\tan\theta = B/\pi D_1$ is a good approximation (note that, in general, the channels are often deeper than in single-screw extrusion).

We can then calculate the flight thickness e in a plane perpendicular to the screw channel:

$$e = \alpha\frac{D_e}{2}\sin\theta \tag{4.7}$$

The channel depth along a section perpendicular to the axis is constant and at its maximum between points B and C (Figure 4.8), where it equals: $H = D_e - E$. It is negligible, by definition, between D and E. Between B and A, for angle γ varying between 0 and 2ψ, it is equal to:

$$H(\gamma) = \frac{D_e}{2}(1+\cos\gamma) - \sqrt{E^2 - \frac{D_e^2}{4}\sin^2\gamma} \quad [4.8]$$

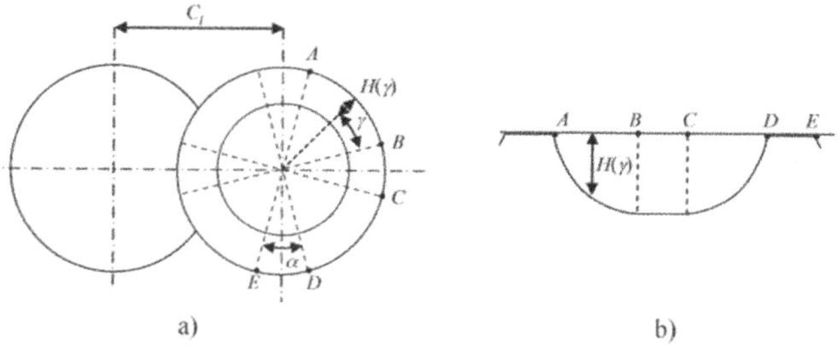

Figure 4.8. *Evolution of channel depth: a) cross-section and b) axial section [BOO 78]*

This depth does not depend on the number of flights. However, the number of flights will determine the width and the number of "independent" channels in an interpenetrated configuration. In fact, the material flows along the screws through different parallel channels, whose number depends on the number of flights. Figure 4.9 shows the cross-section of a screw with two flights. It shows that the surface of the screws defines three channels: one on screw 1 (channel 1), one on screw 2 (channel 2) and the last one straddling both screws and the interpenetration zone (channel 3). There are, therefore, three imbricated independent channels, along which the flow conditions will be considered to be identical. More specifically, we can show that the theoretical number of independent channels is:

$$m = 2n - 1 + \frac{\alpha n}{\pi} \quad [4.9]$$

When α is small, we can consider $2n - 1$ independent channels, i.e. 1 for a single-flighted screw, 3 for a two-flighted screw and 5 for a three-flighted screw, which is the maximum encountered in practice. For modeling, we may, thus, limit ourselves to the study of a single channel with a flow rate of $Q_{ch} = Q / m$.

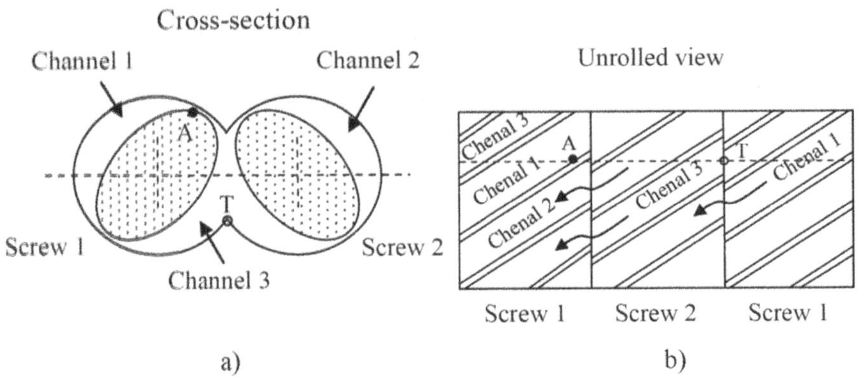

Figure 4.9. *Geometry of two-flighted screws: a) cross-section and b) top view after channel unrolling*

Of course, the channel width of the screw depends on the number of flights. At the barrel level, it is defined, just like in single-screw extrusion, by:

$$W = \frac{B\cos\theta}{n} - e \quad [4.10]$$

4.2.3. *Conventional approximations*

One of the challenges when modeling flows in a twin-screw extruder is that the geometry cyclically varies with time and that there is no simple reference frame for considering flow as an independent of time. By fixing a reference frame onto one of the two screws, just like in single-screw extrusion, the barrel will turn around the stationary screw and cause the second screw to rotate around the first, whose channel will be regularly swept by the flight of the neighboring screw [TAY 92]. Nevertheless, to obtain a stationary geometry, most authors consider both screws as fixed, with a rotating barrel. The major drawback of this hypothesis is that it disregards the relative movement of the two screws and, therefore, does not account for the flow in the interpenetration zone. However, this is the only straightforward approach.

Another approximation, which is common in single-screw extrusion, includes unrolling the channels to form a planar configuration

(Figure 4.9(b)). It should be noted that the channel-width to screw-diameter ratio is often greater in twin-screw extrusion, which decreases the validity of this assumption.

In fact, the degree of simplification required depends on the type of approach desired. If we seek to fundamentally understand some specific flow mechanisms in a limited part of a system, then we require sophisticated numerical techniques. 3D, time-dependent flow approaches were developed to study, in detail, the conveying zone, the interpenetrating zone and the mixing effects in a kneading block [YAN 92, KAJ 96, MAN 98, BRA 00, YOS 00, MOU 00, ISH 00, ISH 01, FUN 02, KAL 07]. On the contrary, if we aim to establish an overall model for this process, taking into account the different functional zones, a simplified approach needs to be used.

4.3. Solid conveying zone

In the previous chapter, we saw that in single-screw extrusion the conveying of pellets in the feed zone occurs due to the friction forces between the compacted solid, the screw and the barrel (Chapter 3, section 3.3). In twin-screw extrusion, the rotation of two interpenetrated screws works as a positive displacement pump, which pushes the solid downstream, regardless of the boundary friction conditions. This is particularly the case for "C-chamber" configurations (Figure 4.2(a)), that is to say, with a restricted passage from one screw to the other. It is therefore possible to starve feed the screws without affecting the process, and to efficiently convey materials that would pose problems in single-screw extrusion (e.g. formulation of highly lubricated PVC and PVC in powdered form which are extruded, almost exclusively, using counter-rotating twin-screw extrusion). For this reason, twin-screw extruders are generally fed by a volume or weight metering system, which allows a given feed rate to be set.

The solid material usually remains in the interpenetration zone, or falls into the lower part of the channel due to gravity, where it is pushed by the trailing flights and moves in the axial direction at a speed equal to the translation of the screw flights, i.e. NB (where N is the rotation speed per second), as shown in Figure 4.10.

If the material sticks to the screws, it will only move forward with each rotation, as it is passed from one screw to the other. The speed is, therefore, slower and can be calculated with: $NB(1-1/2n)$.

In fact, the solids conveying zone does not pose practical problems and hence does not need to be studied in detail. This is why few publications have been dedicated to this zone [CAR 93, POT 96a].

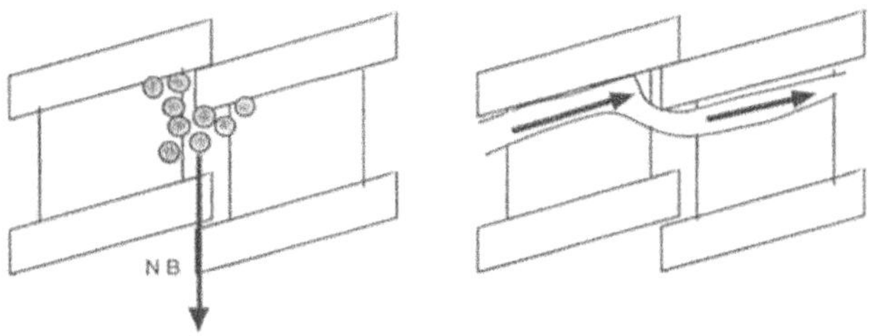

Figure 4.10. *The two conveying styles observed in the conveying zone*

The only challenge is to calculate the maximum feed rate of the machine under certain given conditions. Indeed, flow rate and rotation speed are independent of each other, but there are limiting parameters, for example, related to the overfilling of the hopper or the torque limits of the machine (leading to blocking).

For a given profile, at a fixed speed N, the feed rate can, generally, be increased up to a maximum value, Q_{max}, before reaching the machine limits (torque or power). Experimentally, we found that the flow rate, Q_{max}, increases with the pitch of the screw elements [NOE 92]. This is why screw elements with a large pitch are chosen in the feed zone. Q_{max}/N represents the overall maximum filling ratio of the machine. It is expressed in g/rpm and also reflects the intake capacity of the feed zone. To increase the maximum feed rate of a given machine, the screws must rotate as quickly as possible. This is why the maximum rotation speeds have increased from 300 to 600 and now to 1,200 rpm over the last 10 years.

4.4. Melting zone

Unlike in single-screw extrusion, studies on the melting mechanisms in co-rotating twin-screw extrusion are both recent and more limited. Most of the models are still under development and, unlike with single-screw extrusion, there is no consensus on these models.

4.4.1. *Experimental studies*

All the studies show that the melting mechanism in twin-screw extrusion is completely different from that observed in single-screw extrusion. Instead of the formation of a liquid film upon contact with the barrel, and the development of a melt pool at the rear of the channel, which is clearly separated from the solid phase, in twin-screw extrusion there is usually a succession of the following stages [GOG 98, VER 01b]:

– It all begins with the surface melting of pellets in contact with the barrel and the aggregation of these pellets, wetted by the molten polymer. At this point, the channel is not full, the pressure is zero and the pellets still have their initial shape.

– The complete filling of the channel occurs several C-chambers upstream of the restrictive element causing melting. This involves pressurization of the pellets, as well as significant plastic deformation. The space between the pellets is then occupied by the molten polymer.

– During the subsequent stages, we see a gradual decrease in the size of the solid pellets, which are randomly distributed within the molten polymer. A clear separation of solid and liquid is never observed, but rather there is a transition from an initially concentrated solid suspension to an increasingly more dilute suspension until the solid completely disappears.

Figure 4.11. *Melting zone in a left-handed screw element*

Figure 4.12 illustrates this melting process inside a left-handed screw element. This process depends on the geometry of the elements where the melting takes place (right- and left-handed screw elements, kneading elements) and the operating conditions (filling ratio, rotation speed and flow rate). Figure 4.13 illustrates the melting in a block of kneading elements and shows the general characteristics of the mechanisms described above.

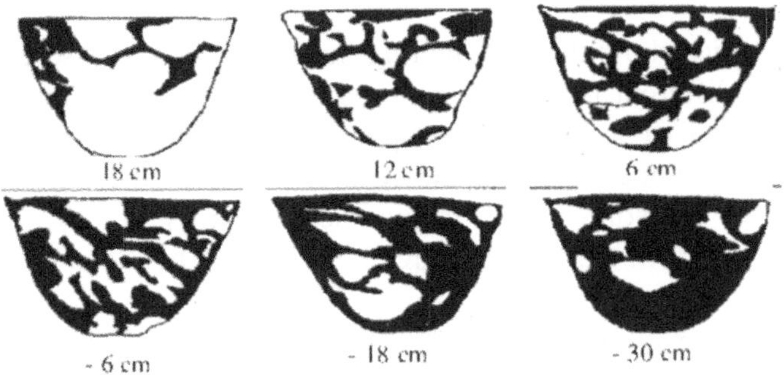

Figure 4.12. *Evolution of the solid/molten mixture during melting inside a left-handed screw element; the solid is in white, the liquid is in black; the distances are given according to the unrolled channel, the positive values correspond to the right-handed element preceding the left-handed one and the negative values correspond to the left-handed element [VER 01b]*

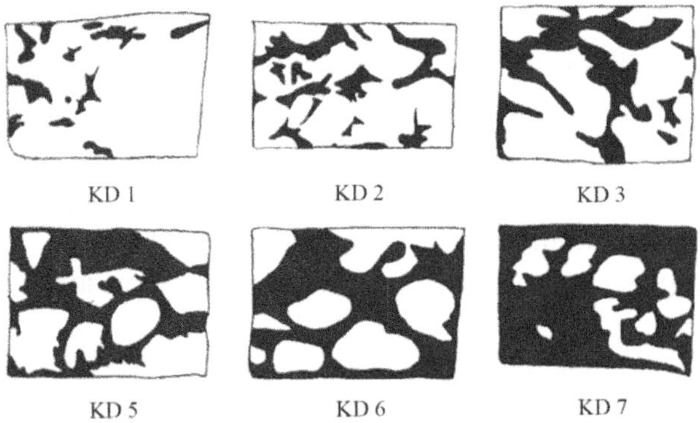

Figure 4.13. *Evolution of the solid/molten mixture during melting within the block of kneading discs; the solid is in white, the liquid is in black; the sections are from all discs, numbered from 1 to 8 from upstream to downstream [VER 01b]*

The melting mechanism can be quantified from the above-mentioned observations by analyzing the images and monitoring the evolution of the solid portion along the flow. Figure 4.14 shows this evolution for both cases presented above. We found that, under certain conditions, melting does not end, neither at the exit of a left-handed element (20% unmelted) nor at that of the block of eight kneading discs, staggered by 90° (35% unmelted). This quantification reveals that the different processing parameters affect the melting mechanisms.

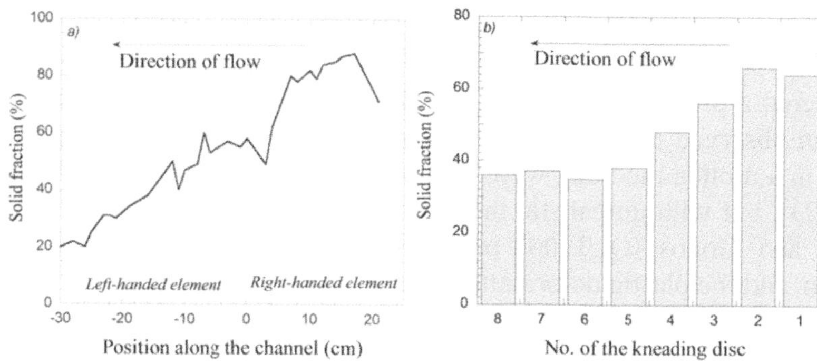

Figure 4.14. *Evolution of the solid fraction: a) along a left-handed screw element and b) along a block of kneading discs [VER 01b]*

A systematic study performed on a pilot scale machine revealed the following results [VER 01b]:

– A decrease in the feed rate at a constant rotation speed causes quicker melting. The same goes for an increase in screw speed at a constant flow rate [ZHU 99]. In both cases, this result can be attributed to a noticeable increase in the specific energy supplied to the material.

– The staggering angle of the kneading discs does not seem to greatly influence the melting speed. However, by modifying the filled length upstream of the block, it allows an advanced startup and thus more efficient melting. The literature states that a neutral staggering (90°) is most effective [CHA 95].

– The size of the pellets, or more specifically the size of the pellets relative to that of the channel, may play a role [TOD 93, GOG 98]. Indeed, our results show that the melting initiation mechanism can be different; but once the solid/liquid dispersion stage is reached, their evolution becomes similar.

– The size of the machine is an important parameter. From experiments on a laboratory machine, and an industrial-size machine, we have shown that, for a similar screw profile, the melting conditions were also similar. Thus, it seems that the mechanism mentioned so far can be generalized to all conditions of the co-rotating twin-screw extrusion process.

4.4.2. *Theoretical models*

The first theoretical approach to melting in twin-screw extrusion was derived by adapting the Tadmor model, developed for single-screw extrusion (Chapter 3, section 3.4) [POT 96b, BAW 98]. The derived models considered a preferential solid/liquid arrangement, which, as we have seen, was not observed experimentally. More realistic mechanisms, dispersion of solids in a molten medium, were also presented [HUA 93, ZHU 01, LIU 01, GUO 03], but with unrealistic theoretical basis. Gogos *et al.* [GOG 98] and Quian and Gogos [QUI 00] proposed to incorporate the interparticular friction, and the plastic deformation energy into the melting model; however, understanding of these phenomena is still incomplete and, therefore, difficult to apply.

To explain the experimental observations, first Noé [NOE 92] and then Todd [TOD 93] proposed to consider the mixture of solid and molten polymers as a highly filled medium, with a very high equivalent viscosity. This idea, which we present below [VER 01b], was developed further by Bleiman *et al.* [BLE 02].

A 1D model was applied to screw elements and kneading elements in a similar way (see Figure 4.15). We do not consider the initial instants of melting, but we start when the highly concentrated solid/liquid suspension has already formed. We start with an initial solid fraction Φ_0 and assume that the solid particles are undeformable spheres, as shown below. At this point, we have to be cautious of the fact that it is a rough estimate.

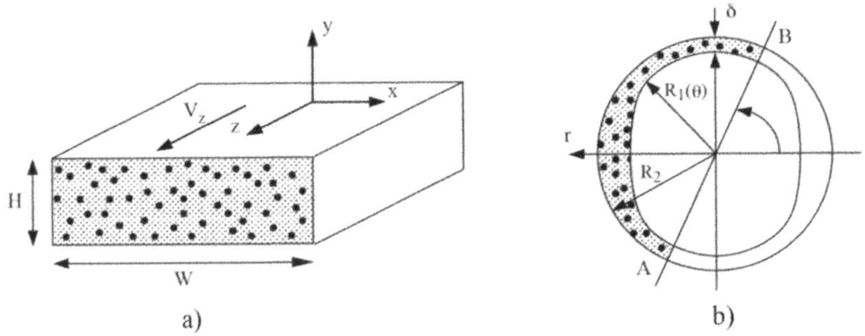

Figure 4.15. *A schematic description of the solid/liquid mixture: a) in the screw channel and b) in a kneading disc [VER 01b]*

The equivalent viscosity of this medium is given by the Maron–Pierce or Krieger–Dougherty equation, in the general form [AGA 14]:

$$\eta = \eta_0 \left(1 - \alpha\Phi\right)^{\beta} \quad [4.11]$$

where η_0 is the matrix viscosity, α is the inverse of the maximum solid fraction, corresponding to the maximum compaction in the solid phase, and β is a negative parameter, which depends on the particles' geometry and their interactions. By modifying α and β, it is possible to adjust the viscosity of the suspension.

From the 1D flow equations of a viscous fluid in a screw channel (section 4.5.1.1), or around a kneading disc (section 4.5.2.1), the model aims at calculating the power dissipated by deformation in this very viscous medium. This calculated power is then introduced into a local thermal balance equation, which enables the increase in temperature of the medium to be evaluated, in conjunction with heat transfer calculations involving the barrel and the solid particles:

$$\rho_l C_{pl} Q \Delta T_l = \dot{W} + h_T (T_F - T_l) S + n h_s (T_s - T_l) 4\pi R^2 \quad [4.12]$$

where ρ_l is the density of the molten polymer, C_{pl} is the specific heat capacity, T_f is the barrel temperature, T_s is the temperature of solid particles of radius R, S is the exchange surface with the barrel, h_T and h_s are the heat transfer coefficients between the molten polymer and barrel and solid particles, respectively, and n is the number of particles in the given volume.

Next, we calculate the decrease in the radius of the solid particles, assuming that at each time step Δt heat transfer from the surrounding liquid causes the melting of a peripheral layer with a thickness of ΔR:

$$\frac{\Delta R}{\Delta t} = \frac{h_s (T_s - T_l)}{\rho_s \left(C_{ps}(T_m - T_s) + \Delta H\right)} \quad [4.13]$$

Finally, we assume that the internal temperature of the solid particles increases due to heat transfer from the surface at temperature T_m. If T_{s0} is the starting temperature of the pellets, the temperature at instant t is given by [CAR 59]:

$$T_s = T_{s0} + \frac{T_m - T_{so}}{R}\left(6\sqrt{\frac{at}{\pi}} - \frac{3at}{R} + 12\sqrt{at}\sum_{j=1}^{\infty} \text{ierfc}\left(\frac{jR}{\sqrt{at}}\right)\right) \qquad [4.14]$$

where a is the heat diffusivity and "ierfc" is the function defined below:

$$\text{ierfc}(x) = \frac{1}{\sqrt{\pi}}\exp(-x^2) - x\left(1 - \frac{2}{\sqrt{\pi}}\int_0^x \exp(-u^2)du\right) \qquad [4.15]$$

The solution of equations [4.12–4.14], coupled with those for flow, allows the evolution in the pressure, the temperature of the solid, the temperature of the liquid and the radii of the pellets (the solid fraction) over time to be calculated along the length of the screw. Melting is considered to have been completed when the temperature of the solid is equal to T_m, or when the solid fraction is equal to zero.

Figure 4.16 shows an example of the results. The values of different parameters are consistent with those obtained from experimental measurements, within an order of magnitude.

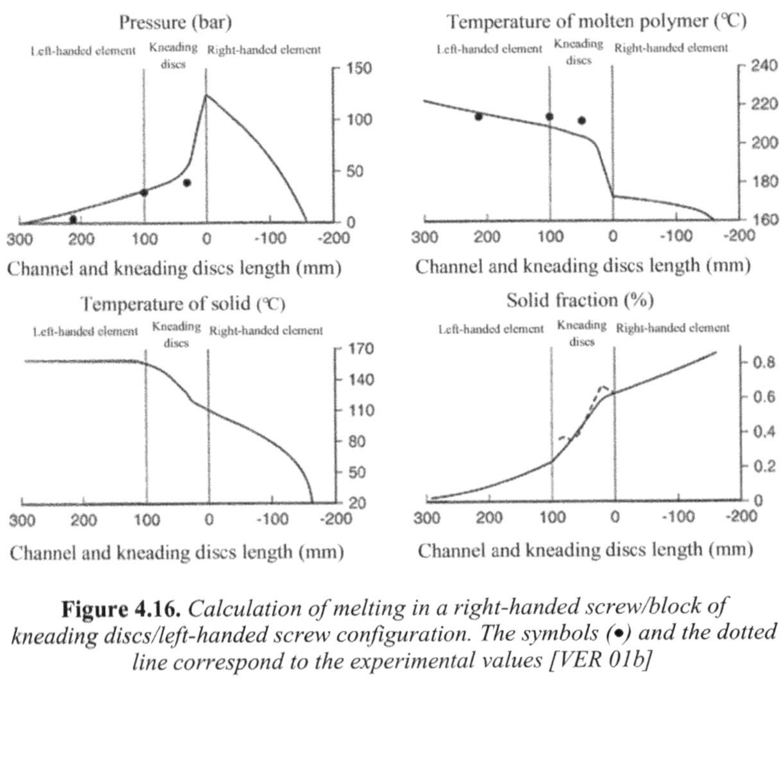

Figure 4.16. *Calculation of melting in a right-handed screw/block of kneading discs/left-handed screw configuration. The symbols (•) and the dotted line correspond to the experimental values [VER 01b]*

4.5. Flow in the molten state

Modeling flow in the molten state will vary according to the considered element and can be addressed in a simplified way depending on the objectives.

4.5.1. *Right- and left-handed screw element*

Let us consider the geometry in Figure 4.17, where, as in Figure 4.9b, the portions of the channel corresponding to the C-chambers have been unrolled.

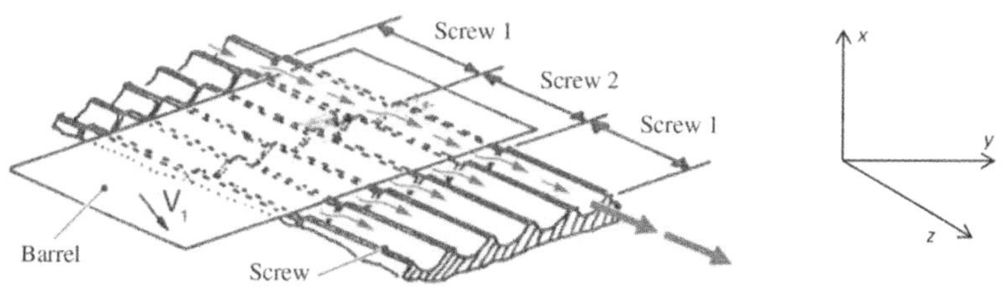

Figure 4.17. *The unrolled geometry of the screw channel and the reference frame used*

For a viscous fluid, the flow is governed by the following equations, which leads to Stokes equations if the behavior is Newtonian and isothermal:

$$\frac{\partial p}{\partial x} = 2\frac{\partial}{\partial x}\left(\eta\frac{\partial u}{\partial x}\right) + \frac{\partial}{\partial y}\left[\eta\left(\frac{\partial u}{\partial y} + \frac{\partial v}{\partial x}\right)\right] + \frac{\partial}{\partial z}\left[\eta\left(\frac{\partial u}{\partial z} + \frac{\partial w}{\partial x}\right)\right] \qquad [4.16]$$

$$\frac{\partial p}{\partial y} = \frac{\partial}{\partial x}\left[\eta\left(\frac{\partial u}{\partial y} + \frac{\partial v}{\partial x}\right)\right] + 2\frac{\partial}{\partial y}\left(\eta\frac{\partial v}{\partial y}\right) + \frac{\partial}{\partial z}\left[\eta\left(\frac{\partial v}{\partial z} + \frac{\partial w}{\partial y}\right)\right] \qquad [4.17]$$

$$\frac{\partial p}{\partial z} = \frac{\partial}{\partial x}\left[\eta\left(\frac{\partial u}{\partial z} + \frac{\partial w}{\partial x}\right)\right] + \frac{\partial}{\partial y}\left[\eta\left(\frac{\partial v}{\partial z} + \frac{\partial w}{\partial y}\right)\right] + 2\frac{\partial}{\partial z}\left(\eta\frac{\partial w}{\partial z}\right) \qquad [4.18]$$

$$\frac{\partial u}{\partial x} + \frac{\partial v}{\partial y} + \frac{\partial w}{\partial z} = 0 \qquad [4.19]$$

4.5.1.1. *One-dimensional analysis*

The simplest analyses, such as those in the metering zone of a single-screw extruder, involve considering a 1D flow along the z-direction, with a unique velocity component $w(x)$ [WYM 75, BOO 80, MEI 88a, TAY 89, POT 90]. We, therefore, obtain a relation of the following type:

$$Q_{ch} = Q_c - Q_p \quad [4.20]$$

where the flow rate along a channel results from the superposition of a shear flow rate Q_c and a pressure flow rate Q_p.

This analysis applies to right- and left-handed screw elements. In fact, as shown in Figure 4.18, the only difference between these elements is the direction of the velocity component V_{1z} at the barrel level.

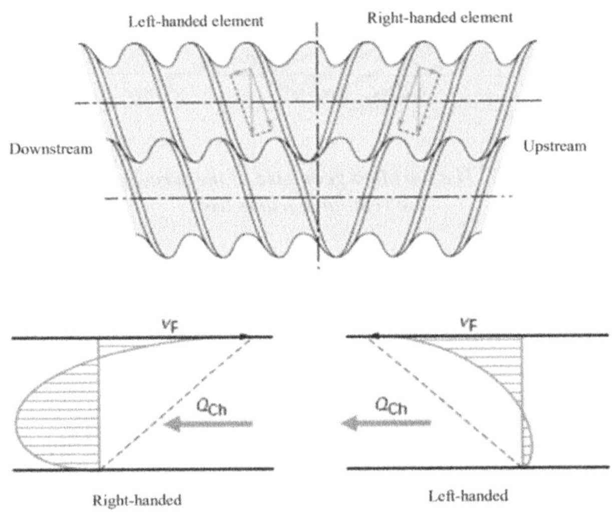

Figure 4.18. *Difference in flow between right- and left-handed screw elements [VER 01a]*

In the case of a right-handed element, the relative movement of the barrel drags the flow in the downstream direction, and the pressure gradient can be either positive or negative, similar to that in the metering zone of a single-screw extruder. In the case of a left-handed element, the moving barrel pushes the material upstream and the pressure gradient must be negative, and has to reach a critical value for the polymer to flow downstream.

Let us consider a C-chamber of length L_C, width W and depth H. For a right-handed element, the volumetric flow rate is expressed as the sum of the shear flow and the pressure flow (isothermal Newtonian behavior):

$$Q_{ch} = W\frac{V_{1z}}{2}H - \frac{WH^3}{12\eta}\frac{\Delta p_d}{L_C} \qquad [4.21]$$

This C-chamber, once filled, is capable of creating a pressure increase Δp_d equal to:

$$\Delta p_d = \frac{12\eta L_C}{WH^3}\left(\frac{WHV_{1z}}{2} - Q_{Ch}\right) \qquad [4.22]$$

If it is not filled, the pressure is zero and the material only occupies a part of the channel width, defined by:

$$W^* = \frac{2Q_{ch}}{V_{1z}H} \qquad [4.23]$$

NOTE 4.1. – If the actual flow rate Q_{ch} is greater than the shear flow rate, this C-chamber, although the element is right-handed, will only operate by consuming a pressure drop as follows:

$$\Delta p_d = \frac{12\eta L_C}{WH^3}\left(Q_{Ch} - \frac{WHV_{1z}}{2}\right) \qquad [4.24]$$

For a left-handed element with the same geometry, we obtain:

$$Q_{ch} = -W\frac{V_{1z}}{2}H + \frac{WH^3}{12\eta}\frac{\Delta p_i}{L_C} \qquad [4.25]$$

The flow rate Q_{ch} in this left-handed element requires a pressure drop Δp_i equal to:

$$\Delta p_i = \frac{12\eta L_C}{WH^3}\left(\frac{WHV_{1z}}{2} + Q_{Ch}\right) \qquad [4.26]$$

where Δp_i is greater than Δp_d: for a length L_C of the left-handed element, the length of the upstream right-handed element must be much greater to generate the pressure required to flow through the left-handed element (Figure 4.19). This length L_C' is:

$$L_C' = L_C \frac{V_{1z} WH + 2Q_{Ch}}{V_{1z} WH - 2Q_{Ch}} \qquad [4.27]$$

So far, the interpenetration zone between the two screws has not been taken into account. In this zone, the local geometry and the kinematics are complex (the screw walls move in opposite directions). Generally, this is simplified by considering the flow as a simple Poiseuille flow (therefore, requiring a pressure drop Δp_{zi}) in a geometry with reduced width W_{zi} and length L_{zi}:

$$Q_{ch} = \frac{W_{zi} H^3}{12\eta} \frac{\Delta p_{zi}}{L_{zi}} \qquad [4.28]$$

with $W_{zi} = W - e$ and $L_{zi} = \psi D_e$.

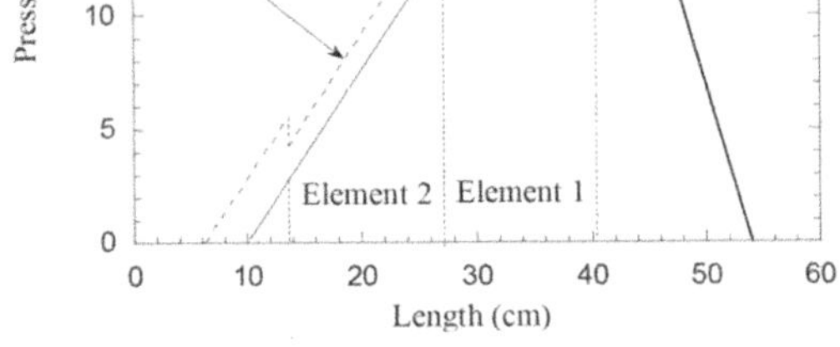

Figure 4.19. *Right- and left-handed elements in series, with and without considering the interpenetration zone*

NOTE 4.2. – By changing the sign, we found that the calculation for a left-handed element (equation [4.26]) is exactly the same as that for a right-handed element (equation [4.22]), except that the pressure gradient has to be positive and the flow rate has to be negative.

This simple approach is then used to construct, for a given type of element, the characteristic curves, describing the evolution of flow rate Q_{ch} as a function of the pressure gradient $\Delta p / L$. Figure 4.20 shows this type of curve for different rotation speeds. The sign agreement, mentioned above, is adopted, which allows the right- and left-handed elements to be plotted on the same graph. We found that a left-handed element always operates by decreasing the pressure, and that the pressure drop is always much greater when the flow rate and the screw speed are high. A right-handed element will usually generate an increase in pressure, which is much greater at high speeds. A drop in pressure (negative pressure gradient) is necessary beyond the limit shear flow rate, at which the Y-axis is intersected.

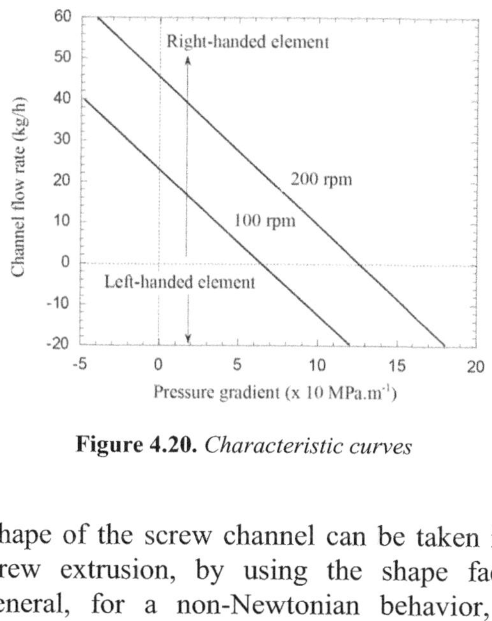

Figure 4.20. *Characteristic curves*

The specific shape of the screw channel can be taken into account, just like in single-screw extrusion, by using the shape factors F_d and F_p [BOO 80]. In general, for a non-Newtonian behavior, we assume an equivalent Newtonian viscosity, calculated using the generalized shear rate, taking into account the transverse flow component.

4.5.1.2. *Two-dimensional analysis*

The general set of equations [4.16–4.19] can be simplified by ignoring the velocity dependence in the *z*-direction:

$$\frac{\partial p}{\partial x} = 2\frac{\partial}{\partial x}\left(\eta\frac{\partial u}{\partial x}\right) + \frac{\partial}{\partial y}\left[\eta\left(\frac{\partial u}{\partial y} + \frac{\partial v}{\partial x}\right)\right] \qquad [4.29]$$

$$\frac{\partial p}{\partial y}=\frac{\partial}{\partial x}\left[\eta\left(\frac{\partial u}{\partial y}+\frac{\partial v}{\partial x}\right)\right]+2\frac{\partial}{\partial y}\left(\eta\frac{\partial v}{\partial y}\right) \quad [4.30]$$

$$\frac{\partial p}{\partial z}=\frac{\partial}{\partial x}\left(\eta\frac{\partial w}{\partial x}\right)+\frac{\partial}{\partial y}\left(\eta\frac{\partial w}{\partial y}\right) \quad [4.31]$$

$$\frac{\partial u}{\partial x}+\frac{\partial v}{\partial y}=0 \quad [4.32]$$

This system is identical to that obtained for the metering zone of a single-screw extruder. We have already seen that with Newtonian fluids, equations [4.29], [4.30] and [4.32], on the one hand, and [4.31], on the other hand, can be solved separately when addressing 2D transverse and longitudinal flows, respectively [DEN 80, WAN 89]. For example, the results presented below were obtained by solving uncoupled longitudinal and transverse flows using the finite element method in a C-chamber geometry [NOÉ 92]. Figure 4.21 shows the isovelocity plots in the screw channel (longitudinal flow) for a right-handed (a) and a left-handed element (b), under the same flow conditions. We found that the velocity profiles are very different: in the right-handed element, the velocity evolves gradually between the bottom of the channel and the barrel, whereas in the left-handed element the flow is split into two parts: one part flowing upstream and the other part flowing downstream. The vertical cross-section of the channel axis (Figure 4.22) shows the velocity profiles very similar to those obtained using the 1D approach, which was presented previously (see Figure 4.18).

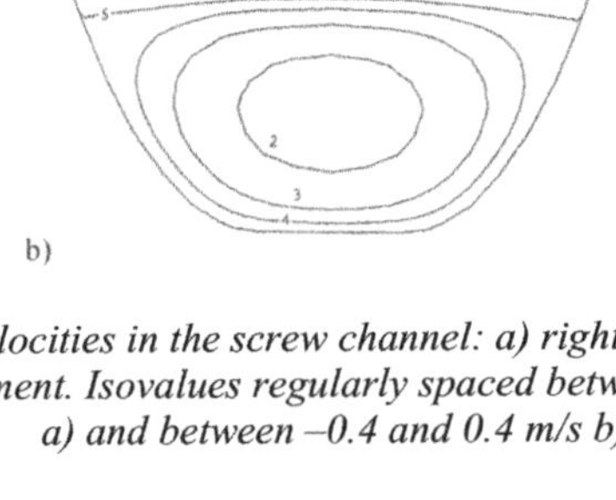

Figure 4.21. *Isovelocities in the screw channel: a) right-handed element and b) left-handed element. Isovalues regularly spaced between –0.1 and 0.4 m/s a) and between –0.4 and 0.4 m/s b)*

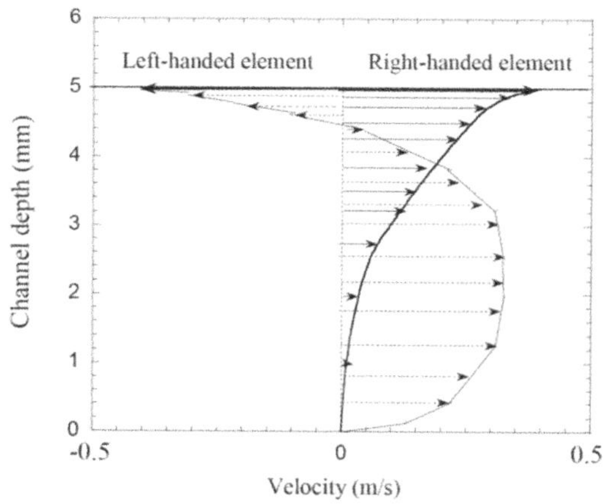

Figure 4.22. *Velocity profiles in the median plane of the screw channel for a right- and a left-handed element*

These calculations allow the characteristic curves to be plotted by linking the pressure gradient to the flow rate. Generally, these curves are dimensionless to remove the impact of velocity [WAN 89]:

$$Q^* = \frac{4Q}{\pi D_e^3 N \cos\theta} \qquad [4.33]$$

$$\text{grad}\, P^* = \frac{1}{K(2\pi N \cos\theta)^m} \text{grad} P \qquad [4.34]$$

Therefore, only two parameters will intervene in the plotting of these curves: the power-law index m (for a power-law behavior) and the pitch (linked to the angle θ). Figure 4.23 provides an example of a dimensionless curve, showing the influence of the pitch. With regard to Figure 4.20, nonlinear evolutions are obtained due to the non-Newtonian behavior of the material.

In the case of non-Newtonian fluids, the longitudinal and transverse flow equations are coupled by means of viscosity, which depends on the generalized shear rate:

$$\dot{\gamma}^2 = \left(\frac{\partial u}{\partial x}\right)^2 + \left(\frac{\partial v}{\partial y}\right)^2 + \frac{1}{2}\left[\left(\frac{\partial w}{\partial x}\right)^2 + \left(\frac{\partial w}{\partial y}\right)^2\right] + \frac{1}{2}\left(\frac{\partial u}{\partial y} + \frac{\partial v}{\partial x}\right)^2 \qquad [4.35]$$

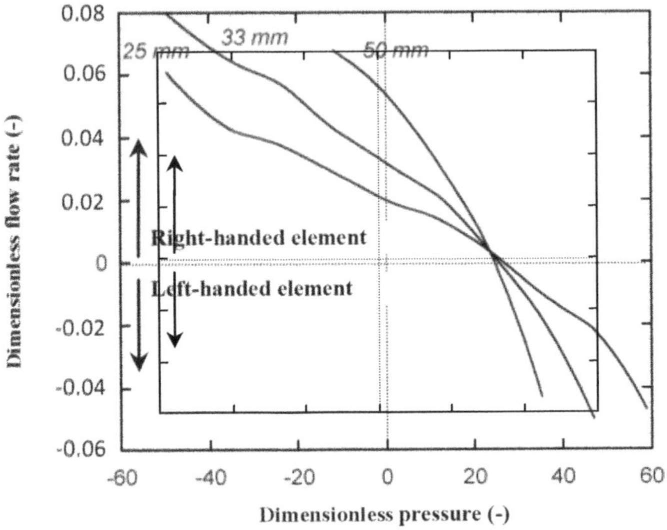

Figure 4.23. *Dimensionless characteristic curves for different pitches*

In this case, an iterative method can be used [KAL 88; WAN 89]:

– First, we solve the longitudinal flow (equation [4.31]), only considering the terms in w in expression [4.35].

– We then solve the transverse flow (equations [4.29], [4.30] and [4.32]) with the terms in w calculated previously.

– We then solve, the longitudinal flow again, using the new value for $\bar{\dot{\gamma}}$ calculated with all the components and so on.

4.5.1.3. *Three-dimensional analysis*

Alongside the simplified approach we have just discussed, 3D models were developed in the late 1990s. The main purpose of these models is to study the actual geometry and the kinematics (two screws rotating in a stationary barrel). Of course, these models are complex from a numerical point of view and require significant computing resources. Despite the progress made in solving meshing problems, and improving the numerical techniques to calculate flows, they are still the privilege of specialists.

The main challenge is the absence of symmetry, already mentioned above, which requires a special treatment of the moving screws. Early methods [KAJ 96, MAN 98, BRA 00] involved choosing a certain number of relative positions of the screws and the barrel (for example, every 30°), and meshing the space occupied by the polymer. This thus implied as many meshes as the number of considered positions. A different approach, called the fictitious domain method, was introduced by Avalosse [AVA 96] and Bertrand *et al.* [BER 97]. It involves using a unique global mesh, on which the position and movement of the screws are set [GAL 02, ALS 04]. In our case, we have used an immersed domain method, where the mesh of the surface of the screws is introduced into a global mesh representing the internal volume of the barrel [VAL 07]. The flow equations are solved for the whole domain (polymer and metal) using a multiphasic approach [VAL 09]. Figure 4.24 illustrates this immersion technique.

Figure 4.25 provides an example of the results obtained by the Ximex© software, marketed in France by SCC (www.scconsultants.com), in the case of an extruder with a diameter of 50 mm. The area under study was composed of a left-handed element 50 mm in length with a 25 mm pitch, preceded by right-handed screw elements of an identical geometry. The study included the pressure field, and the completely filled zone, as the flow rate was increased from 10 to 40 kg/h, at a constant screw speed (200 rpm). It was found that the pressure at the left-handed/right-handed junction increases and, as a result, the filled length also increases.

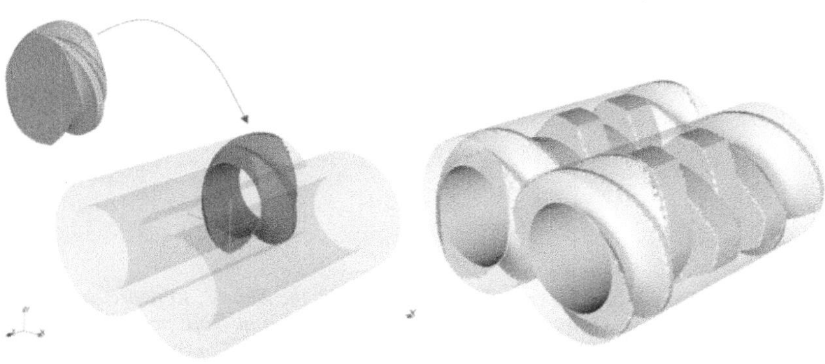

a) b)

Figure 4.24. *a) Immersion of a screw element into the cylinder mesh and b) final geometry of the studied configuration*

When the screw speed was increased from 100 to 300 rpm at a constant flow rate (20 kg/h), we found that the pressure increases and the filled length decreases more significantly (Figure 4.26).

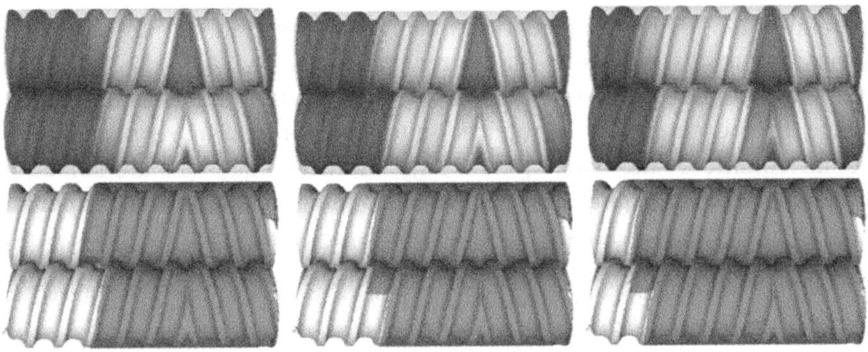

Figure 4.25. *Example of a 3D calculation obtained by the Ximex© software. Influence of flow rate at 200 rpm, from left to right: 10, 20 and 40 kg/h. Top: pressure fields from 0 (in blue) to 10 MPa (in red); Bottom: filled length [DUR 12]. For a color version of this figure, see www.iste.co.uk/lafleur/polymer.zip*

Figure 4.26. *Example of a 3D calculation obtained by the Ximex© software. Influence of screw speed at 20 kg/h, from left to right: 100, 200 and 300 rpm. Top: pressure fields from 0 (in blue) to 10 MPa (in red); Bottom: filled lengths [DUR 12]. For a color version of this figure, see www.iste.co.uk/lafleur/polymer.zip*

4.5.2. *Mixing elements*

In the majority of industrial applications, particular elements are incorporated into the screw profile that favor mixing or dispersion. These elements can be of different types:

– kneading elements, with two or three tips (Figure 4.4);

– eccentric discs (staggered relative to each other, similar to the kneading elements);

– left-handed elements whose sides are pierced with axial openings (Figure 4.27);

– mixing discs (cogwheels, used to separate and recombine the flow).

Originally, left-handed elements with axial opening were, essentially, used for food applications [TAY 88, BAR 92]. Nowadays, they are being increasingly used for thermoplastic polymers, with axial openings of varied shapes and sizes. In the following, we will only discuss the most frequently used elements for compounding applications, that is to say, the kneading elements (with two or three tips). Just as for flows in the screw elements, different approaches are encountered, 1D or 2D flows in unrolled geometries and 3D flows in real geometries.

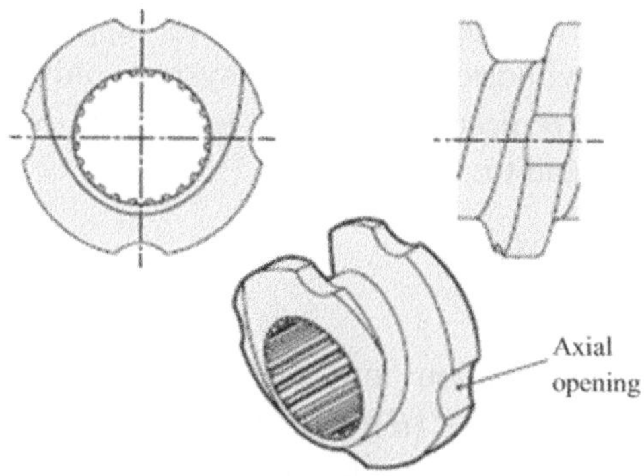

Figure 4.27. *Example of a left-handed element with axial opening (single flight) [VER 01a]*

A study by Werner [WER 76] was the first to shed light on how material flowed in a block of mixing elements. In these elements without helicity, the conveying is rather peculiar: first, the circumferential flow about an isolated element creates a peripheral pressure profile, characterized by a peak located just upstream of the tip (Figure 4.28). Since the elements are staggered by a certain angle relative to each other, this causes a shift in tips locations and, as a result, a shift in pressure peaks, which will create an axial pressure gradient in the direction perpendicular to the disc. It is this axial pressure gradient that causes the axial flow.

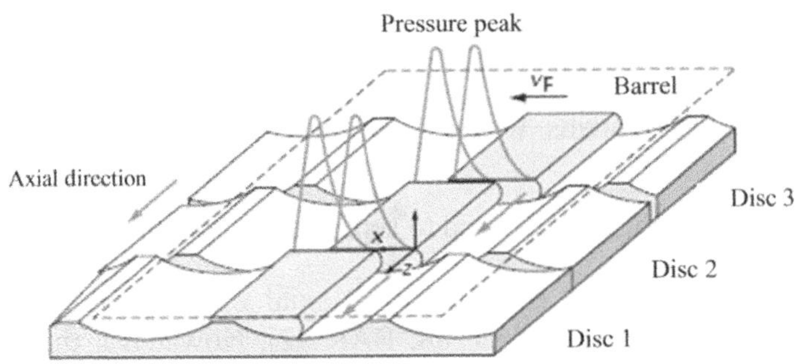

Figure 4.28. *Conveying of material in a block of kneading discs [WER 76]*

4.5.2.1. *One-dimensional analysis*

If we consider a fixed and isolated kneading disc, with the barrel rotating around it, the circumferential flow is equivalent to that of a hydrodynamic bearing (Figure 4.29). If we apply the lubrication approximation [AGA 14], we can consider, in a cylindrical coordinate system, only one velocity vector component $v(r)$ in the θ direction. For an isothermal Newtonian fluid, the momentum equations lead to:

$$\frac{1}{\eta}\frac{dp}{d\theta} = r\frac{d}{dr}\left(\frac{1}{r}\frac{d\,rv(r)}{\partial r}\right) \qquad [4.36]$$

By integrating this equation into the conventional non-slip boundary conditions $v(R_2) = \Omega R_2$ and $v(R_1(\theta)) = 0$, where $R_1 = R_2 - H(\theta)$, with $H(\theta)$ being defined by equation [4.8], we obtain the circumferential velocity profile:

$$v(r) = \frac{\Omega R_2^{\,2}}{R_2^{\,2} - R_1^{\,2}}\frac{r^2 - R_1^{\,2}}{r} - \frac{1}{2\eta}\frac{dp}{d\theta}\frac{1}{R_2^{\,2} - R_1^{\,2}} \cdot \left[r\left(R_1^{\,2}\ln\frac{r}{R_1} + R_2^{\,2}\ln\frac{R_2}{r} - \frac{R_1^{\,2}R_2^{\,2}}{r}\ln\frac{R_2}{R_1}\right)\right] \qquad [4.37]$$

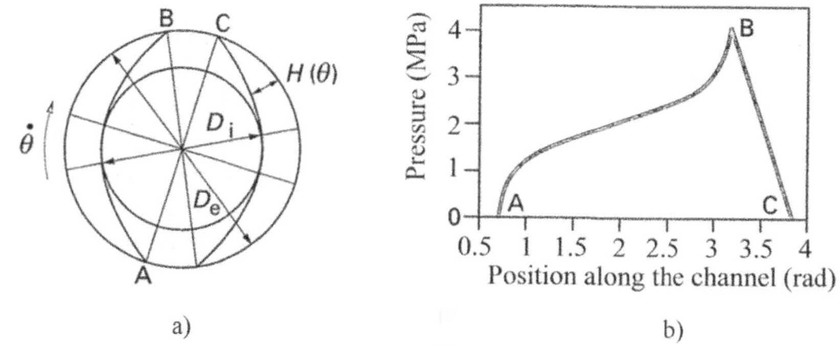

Figure 4.29. *a) Circumferential flow about a kneader and b) corresponding pressure profile [VER 01a]*

Integrating equation [4.37] from R_1 to R_2 leads to a circumferential flow rate per unit width Q_θ / E (where E is the thickness of the kneading disc), which gives the following pressure gradient expression:

$$\frac{dp}{d\theta} = \frac{\dfrac{Q_\theta}{E} + \dfrac{\pi\Omega}{30} R_2^{\,2} \left(\dfrac{R_1^{\,2}}{R_2^{\,2} - R_1^{\,2}} \ln \dfrac{R_2}{R_1} - \dfrac{1}{2} \right)}{\dfrac{1}{8\eta} \left[\dfrac{4 R_1^{\,2} R_2^{\,2}}{R_2^{\,2} - R_1^{\,2}} \left(\ln \dfrac{R_2}{R_1} \right)^2 - \left(R_2^{\,2} - R_1^{\,2} \right) \right]} \qquad [4.38]$$

Now, to obtain the pressure profile $p(\theta)$, equation [4.38] is numerically integrated along the kneading disc, by imposing, for reasons of symmetry, that the pressure at points A and C is the same.

This provides the value of the flow rate, Q_θ, around the kneading disc. This flow rate is independent of the pressure in the block and the flow rate in the extruder, which crosses the kneading disc in the axial flow direction. The pressure profile is shown in Figure 4.29(b).

As indicated in the analysis by Werner [WER 76], the maximum pressure is located just upstream of the tip of the disc. We can show that this pressure profile is not sensitive to the flow rate, but it strongly depends on the clearance δ_f between the tip of the disc and the barrel.

4.5.2.2. *Two-dimensional analysis*

For a more realistic approach, we can use the finite volume method by considering the lubrication approximation in a flat geometry [SZY 87, SZY 88] or a cylindrical geometry [NOÉ 92]. These methods, which are simple to implement, already provide an adequate description of the flows.

Figure 4.30 shows the division of a block of four kneading discs into finite volumes. The kneading discs are considered to be fixed, the barrel as mobile and the interpenetration zone has been disregarded. It is assumed that the flow is isothermal and the polymer has an equivalent Newtonian behavior.

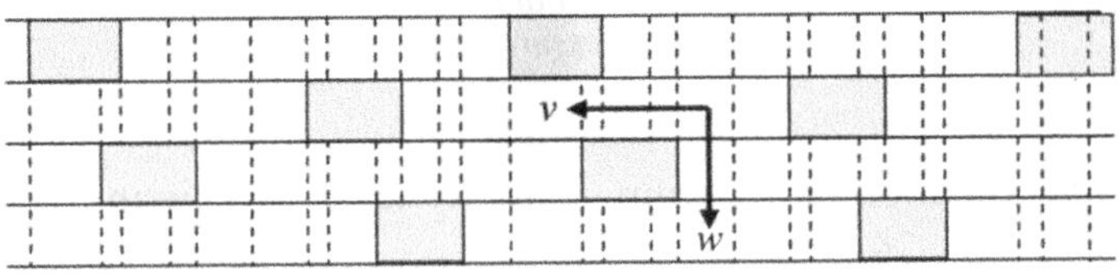

Figure 4.30. *Division of the kneading block into finite volumes*

If we apply the lubrication approximation, we consider only two velocity components v and w, both of which depend on z. Solving the Stokes equations allows us to obtain, for each elementary volume in the plane (θ, z), referenced by the node (i, j), the expressions of local flow rates according to θ and z:

$$Q_\theta(i,j) = \alpha(i,j) + \beta(i,j)\Delta p_\theta(i,j) \tag{4.39}$$

$$Q_z(i,j) = \delta(i,j) + \Delta p_z(i,j) \tag{4.40}$$

with:

$$\alpha(i,j) = EV_{1z}R_2\left[\frac{1}{2} - \frac{R_1^2(i,j)\ \ln\dfrac{R_2}{R_1(i,j)}}{R_2^2 - R_1^2(i,j)}\right] \tag{4.41}$$

$$\beta(i,j) = \frac{E}{\eta\Delta\theta(i,j)}\frac{R_2^2 - R_1^2(i,j)}{4} - \frac{R_1^2(i,j)R_2^2\left[\ln\dfrac{R_2}{R_1(i,j)}\right]^2}{R_2^2 - R_1^2(i,j)} \tag{4.42}$$

$$\delta(i,j)=\frac{\Delta\theta(i,j)}{8\eta E}\left[R_2^{\,4}-R_1^{\,4}(i,j)-\frac{\left(R_2^{\,2}-R_1^{\,2}(i,j)\right)^2}{\ln\dfrac{R_2}{R_1(i,j)}}\right] \qquad [4.43]$$

where Δp_z and Δp_θ are the differences in pressure between neighboring nodes, according to the respective directions z and θ; E is the thickness of the element, $R_1(i,j)$ is the radius of the kneading disc for a given volume element and $\Delta\theta(i,j)$ is the angle associated with this volume element. For each volume element, we obtain a mass balance:

$$Q_\theta(i-1,j)+Q_z(i,j-1)=Q_\theta(i,j)+Q_z(i,j) \qquad [4.44]$$

By substituting Q_θ and Q_z by their expressions [4.39] and [4.40] in equation [4.44], we obtain a set of equations with the pressure as the unknown; by solving this, the pressure field in the kneading block is obtained. Figure 4.31 gives an example of pressure fields, calculated for the two kneading blocks, staggered by 55° in the direct direction (Figure 4.31(a)) and in the inverse direction (Figure 4.31(b)).

Under identical processing conditions (200 rpm and 30 kg/h), the pressure between the entry and exit of the block increases by approximately 1.2 MPa for the direct staggering, whereas it decreases by 2.2 MPa for the inverse staggering. The shape of the isobars shows that the material flows through a pseudo-channel, defined by the tips of the kneading discs. The change in the staggering direction of the tips is equivalent to the change in the sign of the pseudo-channel pitch. It can be concluded that a direct kneading block behaves like a right-handed screw element, whereas an inverse kneading bloc shows a behavior like that of a left-handed element.

This behavior is explained more clearly in Figure 4.32, where the evolution of the pressure gradient is plotted in dimensionless quantities (equations [4.33] and [4.34], with $\cos\theta=1$), as a function of the staggering angle, for various flow rates. Except for the case where the flow rate $Q^*=0$ (no flow across the kneading block), we see that the curves are not symmetrical: the pressure drop created by an inverse block is always greater than the increase in pressure created by the same block with direct staggering. This pressurization capacity increases with the staggering angle up to a value of approximately 45°, then decreases due to increasingly significant leakage flows between the disc tips. It becomes zero for a much smaller staggering angle when the flow rate increases (up to 60° for a flow rate of 94 kg/h at 200 rpm shown in Figure 4.32). Beyond this value and regardless of the staggering, the kneading block consumes the pressure, the maximum being when the staggering angle is approximately 45°.

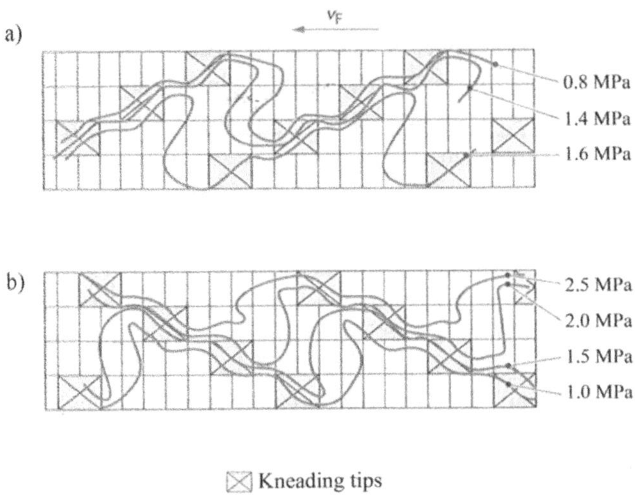

Figure 4.31. *The pressure field for two types of staggering: a) direct and b) inverse (four two-tips kneading discs) [VER 01a]*

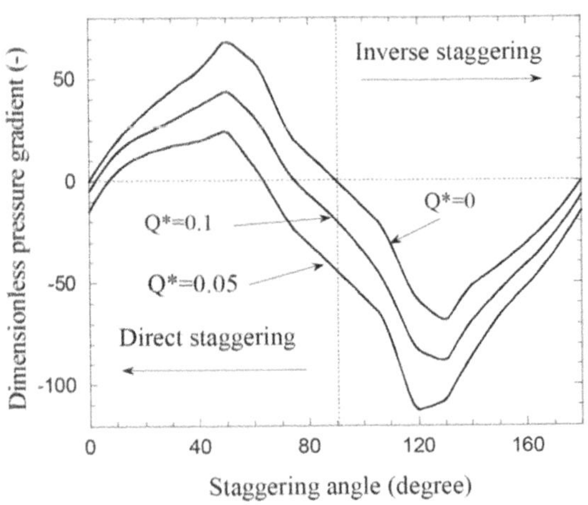

Figure 4.32. *Evolution of the dimensionless pressure gradient with the staggering angle*

The thickness of the kneading disc E and the clearance δ_f between the tip and the barrel are also important parameters that influence the flow conditions; in general, a decrease in these parameters tends to increase the pressure gradients.

4.5.2.3. *Three-dimensional analysis*

With regard to kneading elements, 3D models [YAN 92, CHE 97, ALS 04, BRA 04], based on the methods described in section 4.5.1.3, were also developed. These methods generally validate the previous simplifications (main peripheral flow, similar velocity and pressure fields), but allow the local mixing mechanisms to be treated in greater detail, posing a significant calculation challenge. As we have previously done for flows in screw elements, here we present some examples obtained using the immersed domains method and the Ximex© software [VAL 09].

Figure 4.33 shows pressure fields obtained for four different configurations of a block of kneading elements. We consider a section of an extruder, composed of a left-handed screw element preceded by a block of kneading discs and a right-handed screw element. The staggering angles are 0°, 30°, 60° and 90°, respectively. We have chosen a flow rate of 20 kg/h and a screw speed of 200 rpm as the processing conditions. Figure 4.33 shows that the pressure field varies depending on the configuration. In all cases, in the block of kneading discs, the pressure in a plane perpendicular to the screw axes reaches a maximum before the tip of the discs, as shown earlier with the 1D and 2D methods.

For a zero staggering angle (Figure 4.33(a)), the pressure varies substantially in the plane perpendicular to the screw axes. With an angle of 30° (Figure 4.33(b)), the tips of the discs create a well-defined pseudo-channel and the maximum pressure in the block is reached at the entry of the left-handed screw element. The staggering angles of 60° and particularly 90° generate a much more uniform pressure field, with the pressure only slightly varying within the block.

Now, if we consider the kneading discs staggered by 30°, but narrower, we will find that, by comparing Figures 4.34(a) and 4.33(b), this essentially modifies the angle of the pseudo-channel defined by the tips of the discs, that is to say the pitch of this pseudo-channel. As stated above, the decrease in thickness also leads to an increase in the pressure gradient. However, if the staggering direction is modified from 30° to –30°, then we will find that, by comparing Figures 4.34(b) and 4.33(b), this radically changes the flow conditions. As already seen for 2D models (Figure 4.32), the kneading block behaves like a left-handed element: it consumes the pressure drop, and the maximum pressure is now reached at the block entry.

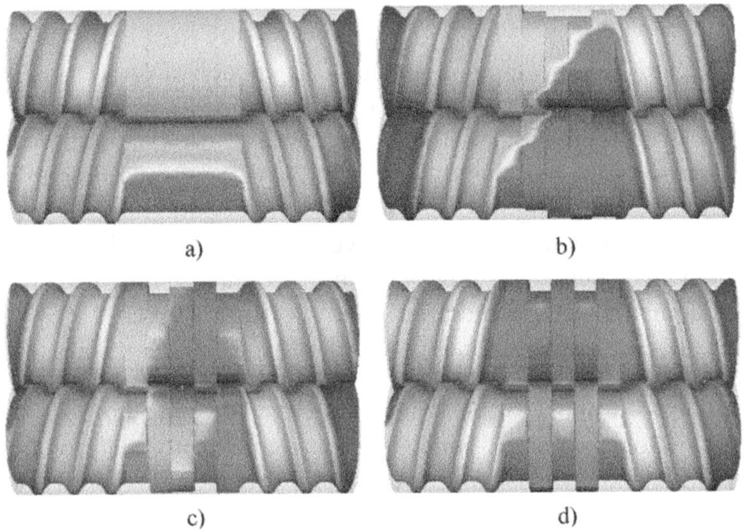

Figure 4.33. *3D pressure fields obtained by the Ximex© software from 0 (in blue) to 10 MPa (in red). Influence of the staggering angle a) 0, b) 30°, c) 60° and d) 90° (200 rpm, 20 kg/h) [DUR 12]. For a color version of this figure, see www.iste.co.uk/lafleur/polymer.zip*

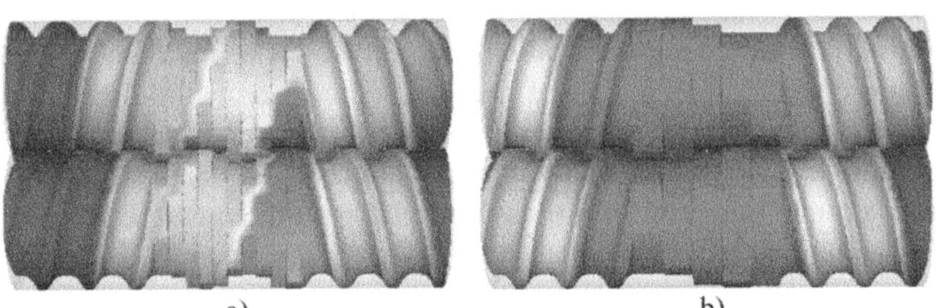

Figure 4.34. *Examples of 3D pressure fields obtained by the Ximex© software from 0 (in blue) to 10 MPa (in red: a) influence of thickness of kneading discs and b) influence of the staggering direction (200 rpm, 20 kg/h) [DUR 12]. For a color version of this figure, see www.iste.co.uk/lafleur/polymer.zip*

The 3D calculation, despite challenges during implementation, provides interesting data, particularly for quantifying the mixing capacity of different configurations of kneading elements.

4.5.3. *Heat transfer*

So far, we have only considered isothermal flows. Heat transfer is also very important in twin-screw extrusion, and it must be considered. However, just like in single-screw extrusion, a realistic method must either be limited to a 1D estimate of the average temperature or follow the evolution of the local temperature along the current trajectories, that is to say it must use a true 3D method.

As shown previously for single-screw extrusion, the evolution in average temperature is directly linked to the heat transfer coefficient between the barrel and the polymer. The heat transfer condition on the screws is generally adiabatic, except for the case where they are thermally regulated, which is the case for large machines. However, the choice of the heat transfer coefficient between the barrel and the polymer is always challenging. In the literature, we find values from 360 $W.m^{-2}.K^{-1}$ [FUK 94] to 20,000 $W.m^{-2}.K^{-1}$ [WHI 94], most of which are poorly justified. It would appear that the experimental values are between 300 and 900 $W.m^{-2}.K^{-1}$ and are much greater for screws with large diameters [LAR 88]. Based on the study of co-rotating twin-screw extruders, Todd [TOD 88] proposed an equation similar to those used in forced convection, where the polymer acts as the fluid.

The Nüsselt number Nu (which is a dimensionless transfer coefficient) is given by:

$$Nu = \frac{h_T D_e}{k} = 0.94 \left(\frac{D_e^2 N \rho}{\eta} \right)^{0.28} \left(\frac{c_p \eta}{k} \right)^{0.33} \left(\frac{\eta}{\eta_p} \right)^{0.14} \qquad [4.45]$$

where D_e is the screw diameter, N is the rotation speed (expressed in rad/s), η is the average viscosity and η_p is the viscosity at the wall, which is calculated for the maximum shear rate.

This expression typically involves the dimensionless Reynolds ($Re = D_e^2 N\rho / \eta$) and Prandtl ($Pr = c_p\eta / k$) numbers. It allows a local transfer coefficient to be defined, which evolves along the screws, depending on the changing values of η and η_p. However, in the majority of the cases we have considered, experimental results suggest that the transfer coefficient determined from the Todd equation tends to underestimate the actual value.

Domschke and Justus [DOM 93] also proposed a theoretical approach for heat transfer in twin-screw extrusion, based on the existence of a "boundary layer" in the clearance between the screws and the barrel, similar to that used by Barrès *et al.* [BAR 91] to deal with heat flow in the solid conveying zone. The result is that the heat transfer coefficient depends on this clearance (a small value significantly increases heat transfer), the rotation speed (particularly with large diameters) and the screw diameter (mainly at high velocities). Industrial machines, between 50 and 160 mm in diameter, would have heat transfer coefficients from 100 to 1,200 $W.m^{-2}.K^{-1}$.

4.5.4. *Residence time distribution*

The residence time distribution (RTD) is essential in twin-screw extrusion, where in most cases, and especially in reactive extrusion (section 4.7.2), the product quality directly depends on the time spent in the machine. Therefore, it is important to know how this RTD depends on the processing parameters and screw profile, and if possible, how to estimate it, as simply as possible.

In flows as complex as those found in twin-screw extrusion, not all particles follow the same trajectory, nor do they spend the same amount of time in the machine. Therefore, there is an RTD, a concept introduced by Danckwerts in 1953 [DAN 53]. By definition, $E(t) = f(t)\, dt$ is the number of particles residing in the machine for the period of time between t and $t + dt$. We also define a cumulative RTD by:

$$F(t) = \int_0^t f(t)dt \qquad [4.46]$$

where $F(t)$ is the number of particles with a residence time less than or equal to t. By definition, $F(t)$ tends toward 1 when t tends toward infinity. The average residence time $\bar{t}$ is given by the first moment of the distribution:

$$\bar{t} = \int_{t_{min}}^{\infty} t\, f(t)\, dt \qquad [4.47]$$

where t_{min} is the minimum residence time.

If the velocity field of a flow is known, we can calculate the corresponding RTD. For example, the RTD of the Poiseuille flow of a Newtonian fluid in a pipe is expressed by:

$$f(t) = \frac{2t_{min}^2}{t^3} \qquad [4.48]$$

where t_{min}, in this case, is the residence time on the axis where the velocity is at its maximum.

The bibliography provides numerous studies regarding RTD measurements in twin-screw extrusion. This includes measurement of electrical conductivity [TOD 75], γ radiation [WOL 86], ultraviolet (UV) fluorescence [CHE 93], colorimetry [ALT 86], techniques such as Raman spectroscopy [WAR 96] and infrared or UV [CAS 91] spectroscopy. We used the last technique mentioned above to obtain the results presented below [POU 03b].

Figure 4.35 shows the general shape of an RTD (function $E(t)$) for different flow rates and screw rotation speeds. When the flow rate increases at a constant screw speed (Figure 4.35(a)), the RTD moves toward shorter times and becomes narrower. When the speed increases at a constant flow rate (Figure 4.35b), we essentially observe a shift in the curves toward even shorter times, without any noticeable change in the shape of the distribution curve. This effect is increasingly reduced as the speed increases, with almost no difference between 300 and 600 rpm, in this case. Of course, the average residence time follows these general trends, that is to say it decreases when the flow rate or the screw speed increases.

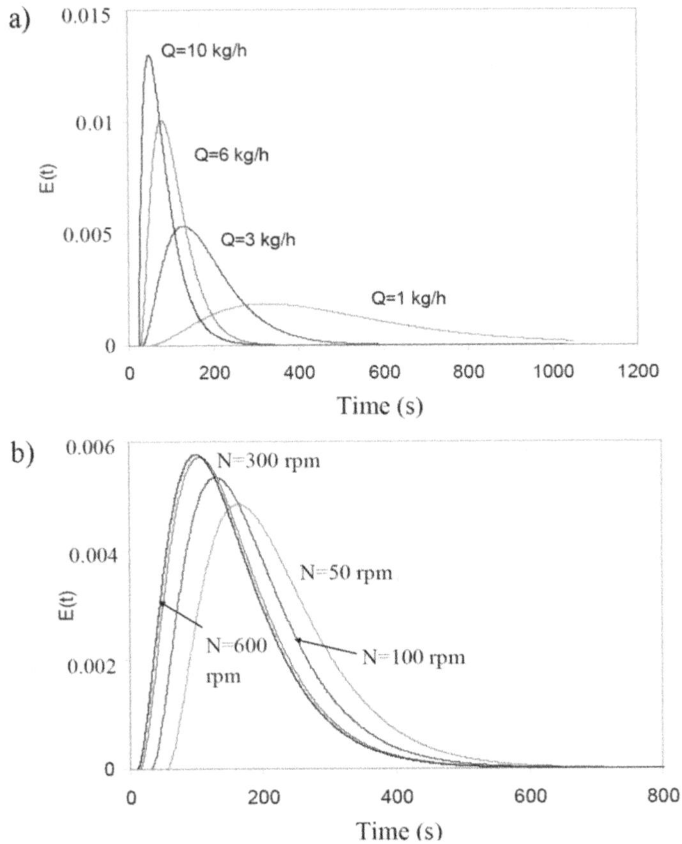

Figure 4.35. *Influence of a) flow rate and b) rotation speed on the residence time distribution [POU 03a]*

To interpret these results, it is necessary to understand where and how the material spends time in the extruder. As mentioned previously, and as shown in Figure 4.3, there is a succession of completely and partially filled zones along the screw profile.

In the completely filled zones, the average residence time $\bar{t}$ is defined by:

$$\bar{t}_r = \frac{\mathcal{V}}{Q} \qquad [4.49]$$

where $\mathcal{V}$ is the volume occupied by the polymer and Q is the mass flow rate. $\mathcal{V}$ can easily be calculated using the expressions presented in section 4.2.2.

In the partially filled zones (essentially the conveying screw elements), the pressure is equal to the atmospheric pressure, and the flow is a simple shear flow between the screw and the barrel. The average residence time is, therefore, equal to the distance covered, divided by the average speed, i.e. [POU 03a]:

$$\bar{t}_{nr} = \frac{120\,L}{NB\cos^2\theta} \quad [4.50]$$

where L is the axial length of the element considered, B is its pitch and θ is the thread angle.

To obtain the average residence time, we simply need to add up expressions [4.49] and [4.50] for all zones, i.e.:

$$\bar{t} = \sum_{\text{restrictive elements}} \frac{\mathcal{V}_i}{Q} + \sum_{\text{conveying elements}} \frac{120\,L_i}{N\,B_i\cos^2\theta_i} \quad [4.51]$$

We found that the average time $\bar{t}$ depends on both the rotation speed N and the flow rate Q, as shown in Figure 4.35. More specifically, $\bar{t}$ is inversely proportional to these two parameters. Since the average time spent in the filled elements is much longer than the time spent in the conveying elements, the effect of a variation in the flow rate is stronger than a variation in speed, which can be clearly seen in Figure 4.35. Finally, in equation [4.51], we found that the viscosity of the polymer does not affect the average residence time: at a constant flow rate and screw speed, a very fluid polymer has the same average residence time as a very viscous one, which is far from being intuitive.

From equation [4.51], we can also calculate the overall filling ratio of the extruder, which can be considered as the filled volume divided by the free volume, i.e.:

$$\tau_r = \frac{Q\bar{t}}{\sum_{elements} \mathcal{V}_i} \quad [4.52]$$

By substituting $\overline{t}$ with its expression, we obtain:

$$\tau_r = \frac{\sum_{\text{restrictive elements}} \mathcal{V}_i + \sum_{\text{conveying elements}} \frac{120\, L_i}{B_i \cos^2 \theta_i} \frac{Q}{N}}{\sum_{\text{elements}} \mathcal{V}_i} \qquad [4.53]$$

For a given screw geometry, the obtained filling ratio is a linear function of the *Q*/*N* ratio, which can always be experimentally verified in twin-screw extrusion.

4.6. An overall model of twin-screw extrusion

Just like in the case of single-screw extrusion, overall models of twin-screw extrusion were developed by connecting the flows in the conveying and mixing zones in series. Among the softwares currently sold are Akroco-Twin Screw, which was developed from studies undertaken at the University of Akron by Professor White and his team [WHI 87, WAN 89, WHI 94, CHE 94], and Sigma, which was developed at the University of Paderborn by Professor Potente and his team [POT 94]. Here, we describe the LUDOVIC© software in detail (Logiciel pour l'Utilisation des DOubles VIs Corotatives) [VER 98], which was developed by CEMEF and INRA, from initial studies on the extrusion cooking of cereal products [TAY 92, DEL 93].

4.6.1. *General description*

The initial aim was to obtain a global model of the twin-screw extrusion process that is predictive, and to rapidly calculate the main flow parameters (pressure, temperature, residence time, shear rate, strain, etc.), in an industrial configuration. For this purpose, we have developed a simplified method, using 1D slab approximation and calculating mean values. Obviously, this method has been validated by many sets of experimental results [CAR 00] and, as we will see later, through comparisons with the results from 3D calculations. LUDOVIC© is organized into modules, with each module corresponding to a particular type of element: filled right-handed screw element under pressure, partially-filled right-handed screw element at atmospheric pressure, left-handed screw element, kneading elements (with one, two or three tips), left-handed screw element with axial opening, etc.

The calculation may include a melting model, such as the one presented in section 4.4.2, but most of the time it is only concerned with the molten state. We must, therefore, form hypotheses about where and how melting occurs. We will base this on the experimental study in section 4.4.1, assuming that the melting occurs in the first restrictive element (left-handed screw element or block of kneading discs) encountered by the pellets along the screw axis. This melting, which in reality occurs very rapidly, is assumed to be instantaneous. We also assume that, at the point where the pressure begins to increase, upstream of this first restrictive element, the temperature of the polymer is equal to its melting temperature (or glass transition temperature for an amorphous product).

In right- and left-handed screw elements, we employ a 1D approach, taking into account the shape factors F_d and F_p, which are functions of the channel width/height ratio (see section 3.5.2, Chapter 3). To validate this approach, we have compared the results obtained with the results from 3D calculations, presented above. In the case of a right/left-handed sequence, we have compared the maximum pressure at the left-handed element entry, the total specific mechanical energy (SME) and the completely filled upstream length for different processing conditions. Figure 4.36 shows that the results are, overall, very satisfactory. The conclusion is that a 1D model is capable of taking into account the general flow characteristics along the screw elements.

For a block of kneading elements, we compared the 2D analysis presented in section 4.5.2.2 to the 3D approach in the case of a right-handed screw/kneading block/left-handed screw configuration. Here we will compare the maximum pressure and the specific energy, as a function of the staggering angle of the elements.

As explained before, the pressure value is maximal at the entry of the left-handed screw element when the discs are staggered in a direct direction. However, this occurs at the block entry when the discs are neutral or staggered in an inverse direction. Figure 4.37 shows that the maximum pressure is correctly estimated by the 2D calculation, both in absolute values and depending on staggering. The 2D calculation, however, does not foresee the effect of the angle when the staggering is direct: the pressure is, in fact, directly linked to the flow in the left-handed element, which does not alter with a change in angle of the discs preceding it. Here, the 3D calculation is more accurate, and the differences are relatively small (around 15%).

For the specific energy, the simplified calculation gives a constant value, regardless of the staggering angle, whereas the 3D calculation shows a very distinct effect. This is explained by the fact that, in the simplified model, the peripheral flow around the kneading disc is dominant. However, the order of magnitude is correctly estimated (47 kWh/t in 2D; from 36 to 51 kWh/t in 3D). In summary, the simplified approach is capable of providing highly relevant results, despite some understandable differences considering the flow complexity.

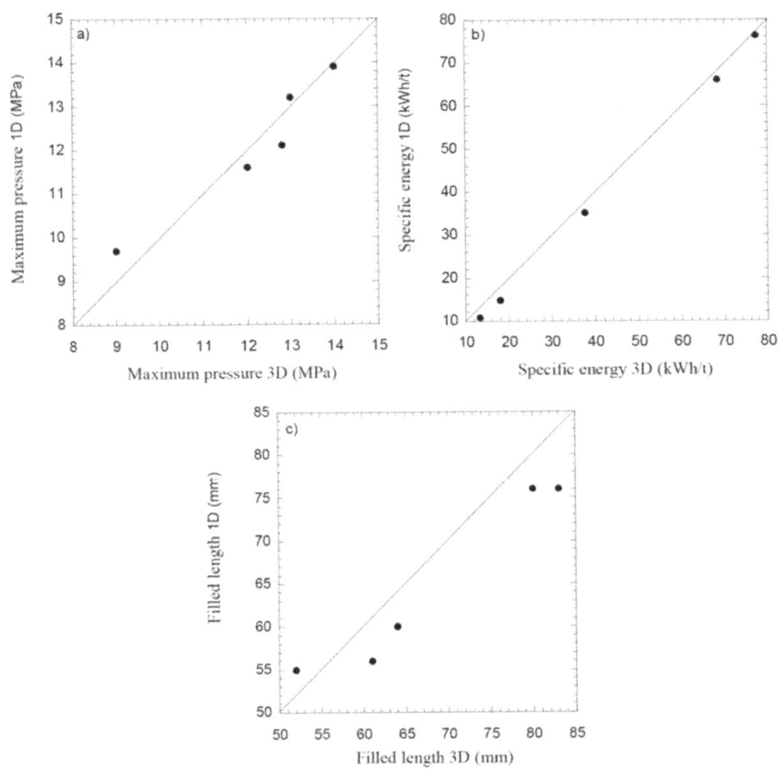

Figure 4.36. *Comparison between 1D (Ludovic©) and 3D (Ximex©) methods for a) maximum pressure, b) specific energy and c) filled length*

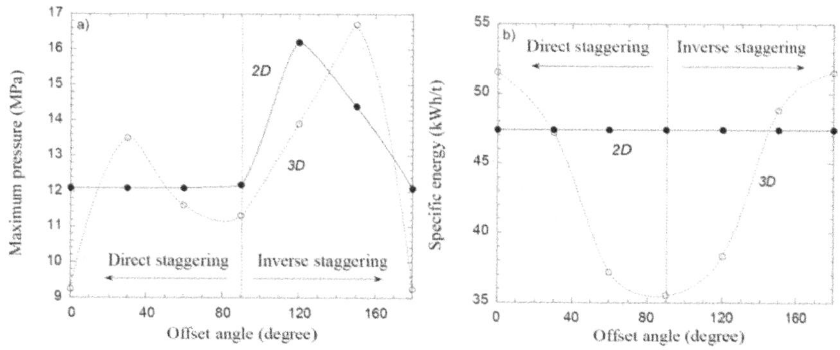

Figure 4.37. *Comparison between 2D (Ludovic©) and 3D (Ximex©) methods for a) maximum pressure and b) specific energy*

The evolution of temperature between two successive elements is calculated using the thermal balance equation (section 1.3, Chapter 1), and is presented in terms of the average temperature:

$$\rho c_p Q_{ch} \Delta \bar{T} = h_f \left(T_f - \bar{T}\right) + h_v \left(T_v - \bar{T}\right) + \dot{W} \qquad [4.54]$$

where h_f and h_v are the heat transfer coefficients between the polymer and the barrel at temperature T_f, and between the polymer and the screws at temperature T_v, respectively. $\dot{W}$ is the power dissipated by the longitudinal and transverse flows.

The challenge lies in estimating the transfer coefficients. When the screws are not thermally regulated, the conditions are adiabatic $(h_v = 0)$. To estimate heat transfer between the polymer and the barrel, Todd's equation can be used as a first approximation (equation [4.44]).

4.6.2. *Calculation algorithm*

As the screw is only partially filled, we cannot know *a priori* the point where pressurization begins. It is therefore necessary to start from the die and go back upstream step by step. However, since the final temperature is an unknown of the problem, an iterative procedure must be performed:

– A starting product temperature is chosen at the exit of the die. This choice is arbitrary and, once the calculation has converged, the final result will not depend on this initial choice. However, the number of iterations will be much lower when this value is close to the solution.

– We then calculate the pressure drop corresponding to the flow in the die, which gives us the value for the pressure to be created upstream.

– The evolution in pressure and temperature is calculated, element by element, working back upstream until the pressure becomes zero. This means that the filling of the screw becomes incomplete.

– We then verify whether there is a restrictive element (left-handed screw element or block of kneading elements) upstream on the screw axis.

– If yes, the calculation is continued upstream, taking into account that the screw is now only partially filled (zero pressure).

– Otherwise, we compare the temperature calculated at the point where we compare the local temperature, calculated at the point where the pressure becomes zero, and the melting temperature T_m. If they are identical, the calculation is considered complete. If not, the final temperature is adjusted according to the difference at the point of convergence and the calculation is reiterated until convergence. Usually, a few dozen iterations are sufficient to obtain a solution.

4.6.3. *Residence time distribution*

To calculate the RTD in a twin-screw extruder, most authors use a "chemical engineering" approach, where they describe the extruder as an assembly of ideal reactors whose RTD is known [OBE 94, GAO 99]. This type of approach is implemented in LUDOVIC© [POU 03a].

As shown in section 4.5.4, it is easy to calculate the average residence times in the different sections of the extruder. The idea is to replace particular geometric configurations (for example, a filled and an unfilled conveying element, a block of kneading elements, etc.) with ideal chemical reactors, for which we can calculate the RTD $E(t)$ from the simple average residence time $\bar{t}$. For example, for a plug flow reactor:

$$E(t) = \delta(t - \bar{t}) \quad [4.55]$$

where δ is the Dirac function; for a continuous stirred-tank reactor (CSTR):

$$E(t)=\frac{1}{\bar{t}}\exp\left(-\frac{t}{\bar{t}}\right) \qquad [4.56]$$

If we consider n-cascaded CSTRs, the RTD is written as:

$$E(t)=\left(\frac{n}{\bar{t}}\right)^{n} t^{n-1}\frac{\exp\left(-\frac{nt}{\bar{t}}\right)}{(n-1)!} \qquad [4.57]$$

For a tubular reactor (Poiseuille flow in a tube), with a power law fluid, we obtain:

$$E(t)=\frac{2m}{3m+1}\left[1-\frac{m+1}{3m+1}\frac{\bar{t}}{t}\right]^{\frac{m-1}{m+1}}\frac{\bar{t}^{2}}{t^{3}} \text{ for } t> t_0=\frac{m+1}{3m+1}\bar{t} \qquad [4.58]$$

where t_0 is the minimum residence time and m is the power-law index.

Then, using numerous experimental results, the sections of the machine are associated with types of ideal reactors. For example, the conveying zones are associated with plug flow reactors, mixing zones with cascaded CSTRs, whose number depends on the number of discs, and the die is associated with a tubular reactor [POU 01a, POU 03b]. For each zone, once the mean residence time is calculated using equation [4.49] or [4.50], we calculate the local RTD $E_i(t)$ according to the type of reactor utilized (equations [4.55]–[4.58]). The total RTD is then calculated as a convolution product of the local RTD:

$$E(t)=E_1(t)\otimes E_2(t)\ldots\otimes E_n(t) \qquad [4.59]$$

For example, Figure 4.38 shows that this technique, which does not require any adjustable parameter, accurately describes the RTD in twin-screw extrusion, as well as the effect of the processing parameters such as the rotation speed and flow rate.

4.6.4. *Example of results using the LUDOVIC© software*

To illustrate the use of LUDOVIC©, we will study a polypropylene flow in the extruder whose profile is given in Figure 4.39. This extruder has a diameter of 50 mm.

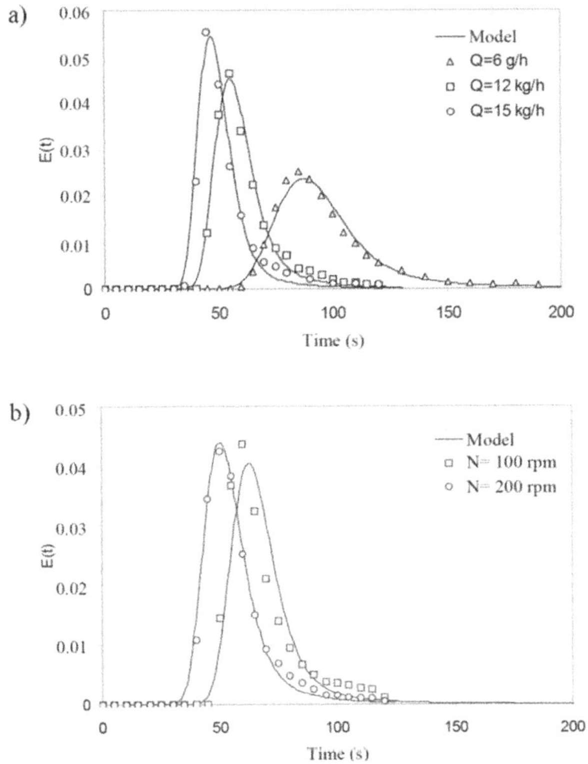

Figure 4.38. *A comparison of experimental and calculated RTDs: a) effect of flow rate and b) effect of rotation speed [POU 03b]*

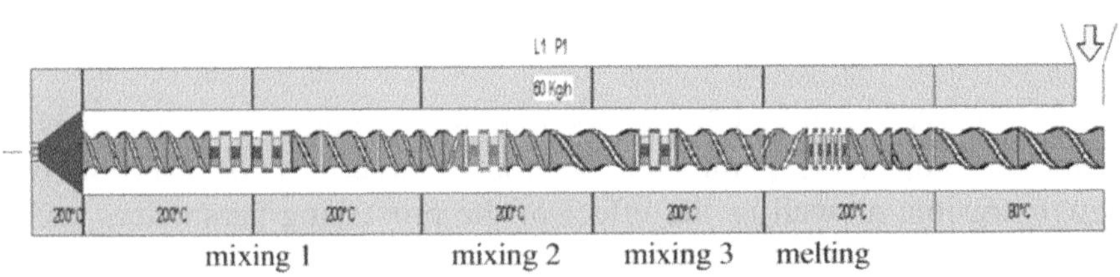

Figure 4.39. *Extruder screw profile*

The profile includes a melting zone, composed of kneading elements at 90° followed by a left-handed screw element, and three mixing zones, one of them including a left-handed screw element. The total length of the extruder is 1,200 mm.

Figure 4.40 shows the evolution of pressure and temperature along the screws for a flow rate of 60 kg/h and a screw speed of 200 rpm. We observe that the screw is only partially filled. The pressurized zones correspond to the restrictive elements (melting and mixing zones), and to the die. The highest pressure (11 MPa) is found after the melting zone, where the polymer is the coolest and, therefore, the most viscous.

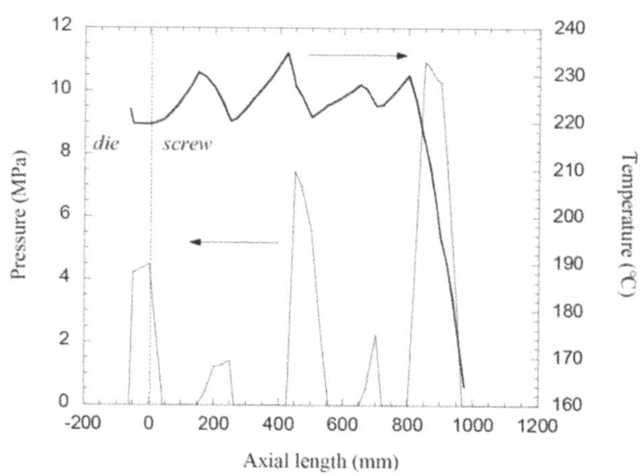

Figure 4.40. *Evolution of pressure and temperature along the screws (Nu = 40)*

We found that the temperature increases rapidly due to viscous dissipation in this zone. Afterward, as the shear rate is low along the unfilled conveying zones, the heat transfer tends to bring the temperature back to the regulation temperature value (here, 200°C). Figure 4.40 shows an increase in temperature from 10°C to 20°C in each high-shear zone, followed by a fairly rapid return to equilibrium temperature.

This calculation was performed using a Nüsselt number of 40, corresponding to a relatively high heat transfer coefficient, of approximately 1,600 W/m^2/°C. Figure 4.41 shows the impact of the heat transfer coefficient on temperature evolution: when it increases from 20 to 60 (i.e. the coefficient h_f from 800 to 2,400 W/m^2/°C), the final temperature decreases from 240°C to 216°C and the maximum temperatures reached (in the second mixing zone where the left-handed screw element is situated) drop from 248°C to 223°C.

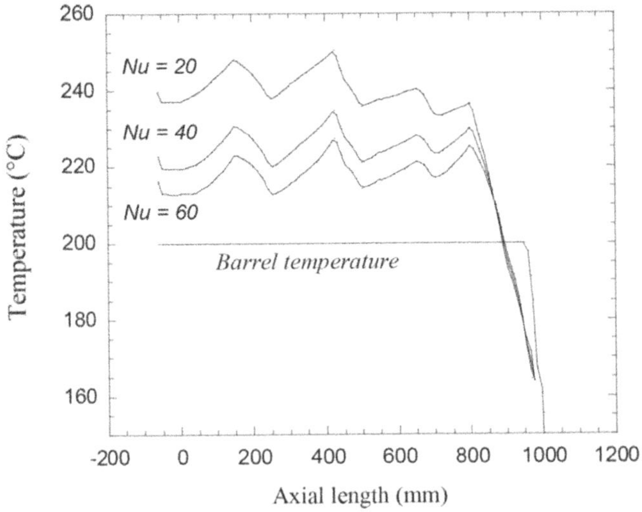

Figure 4.41. *Temperature evolution along the screws for three different Nüsselt numbers*

Figure 4.42(a) shows the pressure evolution when the flow rate is increased, maintaining a constant speed (Nu = 40): when it increases from 60 to 80 kg/h, the maximum pressure increases only slightly (from 11 to 12 MPa), whereas the zones situated upstream of the restrictive elements are filled more. When the flow rate increases to 90 kg/h, the maximum pressure becomes 13 MPa, but as a result, the filling ratio is considerably increased, and pushes the melting zone upstream. This is where the machine limits are, and without increasing the speed, any further increase in the flow rate would cause a blockage. However, in Figure 4.42(b), we clearly see that these changes barely influence the temperature profile. Under the chosen heat transfer conditions, we see a slight decrease in the final temperature along the screws (only 1.2°C), when the flow rate is increased from 60 to 90 kg/h.

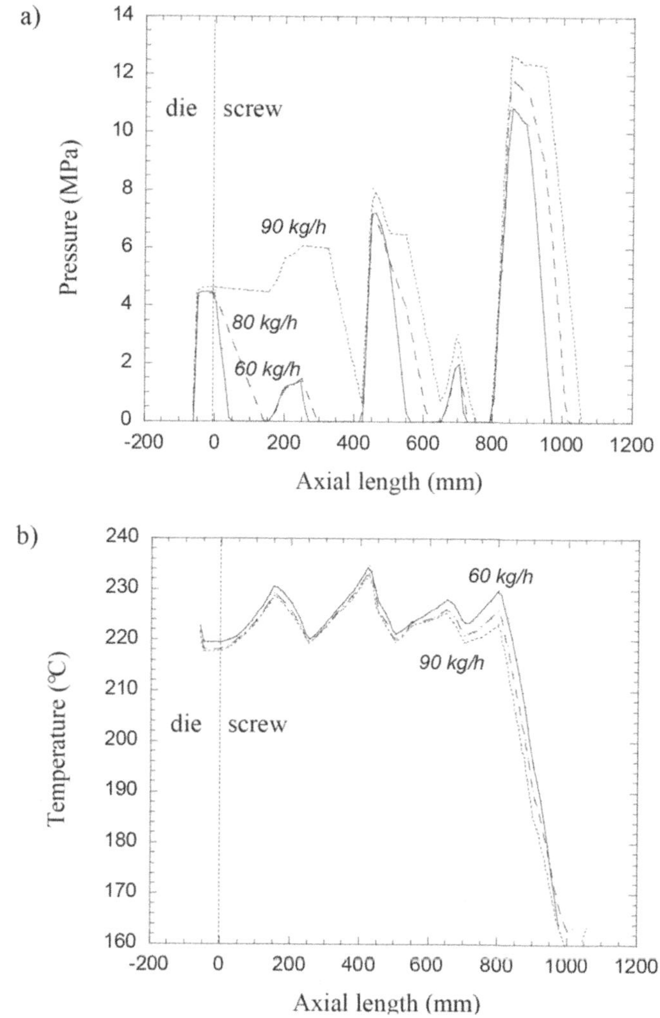

Figure 4.42. *Influence of the flow rate at 200 rpm on a) the pressure profile and b) the temperature profile*

Let us now study the effect of rotation speed at a constant flow rate. In Figure 4.43(a), we see that the pressure is hardly affected and only the filling ratio increases when the rotation speed decreases. However, the effect on the temperature profile is large (Figure 4.43b). In fact, due to the viscous dissipation, which cannot be completely compensated for by controlling the barrel temperature, the temperature is much greater at high speed. In this case, for a barrel temperature of 200°C, the final temperature of the polymer increases from 215 to 223 and then to 238°C, when the speed increases from 150 to 200 and then to 300 rpm. Therefore it is practically possible for some materials to have a temperature 40°C greater than the barrel temperature; this can undoubtedly lead to degradation.

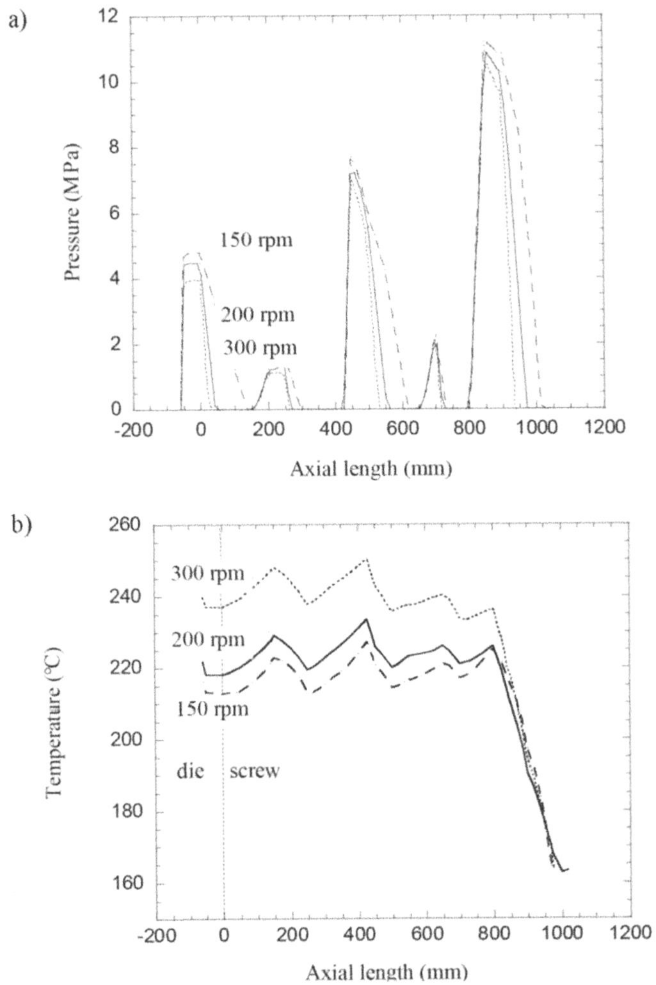

Figure 4.43. *Effect of screw speed at 60 kg/h on a) the pressure profile and b) the temperature profile*

4.7. Compounding application and production of complex materials

In this section, we will present some more realistic examples of twin-screw extrusion applications, including *compounding*, reactive extrusion or the preparation of polymer blends and nanocomposites. For each example, after introducing the theoretical concepts and their challenges, we will show how the use of simulation software allows us to better understand the process and to optimize both the screw profile and the processing conditions.

4.7.1. *Compounding and mixing*

One of the main applications of twin-screw extrusion is *compounding*, which is the mixing of a solid phase with a molten phase, and it is used to obtain a filled material or composite with enhanced properties (mechanical properties, impact resistance, heat or electrical conductivity, fire resistance, etc.). We will see how mixing is defined and how it can be implemented in an extrusion process.

4.7.1.1. *Different mixing types*

There are different types of mixing [MAN 09]. As shown in Figure 4.44, it is conventional to distinguish between two types of mixing:

– distributive mixing, which homogenizes the spatial distribution of the particles within the matrix;

– dispersive mixing, which reduces the size of the particles.

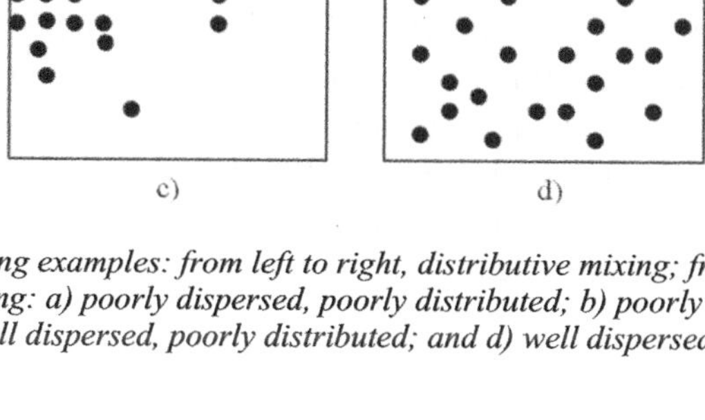

Figure 4.44. *Mixing examples: from left to right, distributive mixing; from top to bottom, dispersive mixing: a) poorly dispersed, poorly distributed; b) poorly dispersed, well distributed; c) well dispersed, poorly distributed; and d) well dispersed, well distributed*

Distributive mixing is controlled by the flow kinematics. The complex trajectories followed by the particles allow them to be distributed homogenously within the matrix. Dispersive mixing is associated with the stresses applied to the particles, which break them and reduce their size. In an industrial compounding operation, these two types of mixing generally occur sequentially: first, the agglomerates or aggregates are reduced in size and become as fine as possible, and then these particles are homogenized within the polymer matrix. It is therefore important to understand how different screw or kneading block configurations can contribute to these actions, and what their best positions along the screw axis are.

4.7.1.2. *Distributive mixing*

One way for theoretically characterizing the distributive mixing is to perform particle tracking, that is to say following a certain number of material points during the flow calculation. From their successive positions, we can define mixing indices, which represent the ability of the flow to correctly distribute these particles. Of course, this type of characterization can only be implemented using an accurate 3D calculation [AVA 96, MAN 98, ALS 04, BRA 04]. The results that we present, hereafter, were obtained using the code Ximex©, developed at CEMEF and described above [DUR 12].

Let us consider, for example, a configuration made up of right-handed screw element/kneading discs at 90°/left-handed screw element. The 3D calculation (pressure field example in Figure 4.45(a)) allows us to obtain the velocity field at any position and for all time steps. Next, we introduce many fictitious particles at the beginning of the flow (here 20,000), which do not interact with the fluid, and we follow their trajectories over time. Figure 4.45(b) shows these particles at different times.

By knowing the positions at each time t, we can calculate the mixing indices. There are many of these indices mentioned in the literature. One of the simplest is the length stretch [AVA 96]. If two particles are separated by dX at time t and by dx at time $t + dt$, then the length stretch λ is the ratio of dx/dX. Averaged over all the particles, the greater this value is, the more efficient distributive mixing becomes. We can also define the Lyapounov exponent [AGA 09], which characterizes the limit of length stretch at infinite time for initially close particles:

$$\Gamma(\vec{X},\vec{M}) = \lim_{\substack{t\to\infty \\ |d\vec{X}|\to 0}} \left[\frac{1}{t} \ln\left(\frac{|d\vec{x}|}{|d\vec{X}|} \right) \right] \qquad [4.60]$$

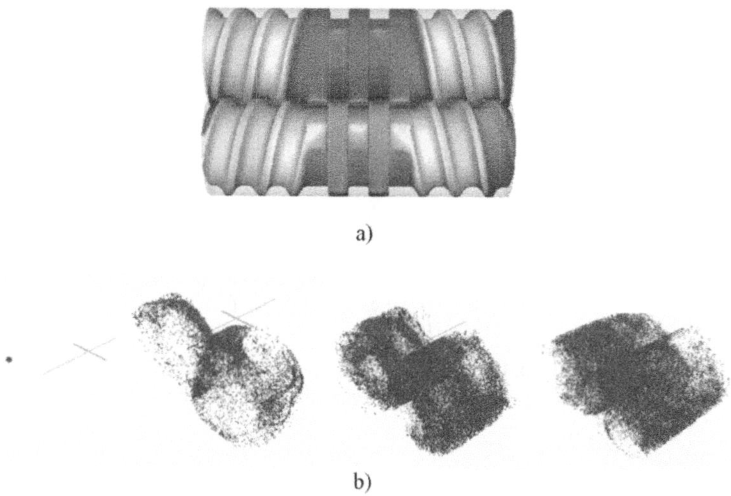

Figure 4.45. *a) 3D pressure field in the studied configuration and b) position of the 20,000 particles at varying successive time steps [DUR 12]. For a color version of this figure, see www.iste.co.uk/lafleur/polymer.zip*

If this exponent is positive, both particles will exponentially separate from each other, which guarantees efficient mixing. Otherwise, they will diverge linearly and mixing will be poor.

This characterization of distributive mixing can be used to compare the mixing ability of various screw configurations. However, it is important to note that these calculations are time-consuming and complex, they do not always converge and often lack the accuracy needed [DUR 12]. They must, therefore, be carefully considered. Nevertheless, we will provide an illustrative example where we compare the length stretch evolution over time, under identical conditions, of two blocks of kneading discs at 30°, differing only by their width (see Figure 4.46). This figure shows that the largest elements are the most efficient. However, it is worth noting that this observation is made for a certain rotation speed and flow rate; it can be different under different conditions.

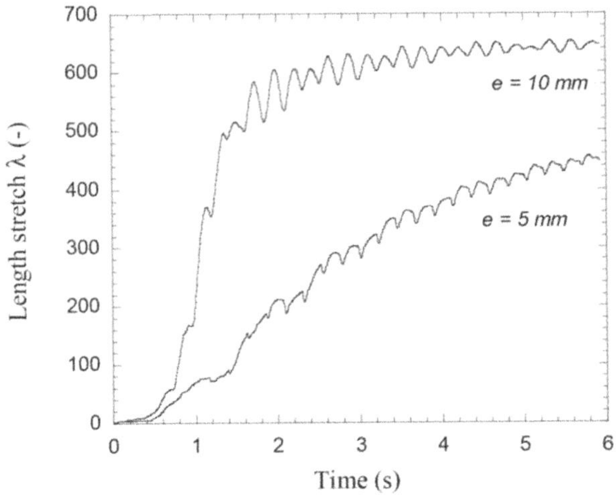

Figure 4.46. *Evolution of length stretch over time for kneading discs at 30°, with thicknesses of 5 and 10 mm*

Particle tracking requires a 3D flow calculation. For a 1D approach, such as that used with the LUDOVIC© software, it is no longer possible to apply the same technique. To comparatively estimate the distributive mixing, the cumulative strain is usually considered, i.e. the integral of $\dot{\gamma}t$ along the screw axis. The variance in RTD, calculated as shown in section 4.6.3, may also be of interest. The greater the cumulative strain or variance is, the better the distributive mixing becomes.

4.7.1.3. *Dispersive mixing*

Distributive mixing can be approached in a very general way, but dispersive mixing requires a specific approach for each type of particle. It is understandable that a black carbon aggregate, lamellar clay or a bundle of glass fibers will not equally dissociate. Generally speaking, to break up an agglomerate we must apply a force higher than the internal cohesion force holding it together, which depends on the organization, the structure and the bonding forces of the constituent elementary particles. For most mineral particles including calcium carbonate, silica, carbon black, etc., there are two main mechanisms [RWE 90, RWE 92]:

– if the hydrodynamic stress is high, the agglomerate usually breaks into large fragments;

– if the hydrodynamic stress is low, we observe an erosion phenomenon, where the elementary particles are pulled one by one from the agglomerate surface.

Therefore, it is necessary to define the laws of size evolution, describing the rupture and erosion mechanisms. Numerous publications related to these phenomena are available in the literature. Here, we present a model as an example, which was successfully applied to the dispersive mixing of calcium carbonate in a polypropylene matrix [LOZ 03].

Let us define $f(v, t) = v\, n(v)$, where $n(v)$ is the distribution function of the particles of volume v. The time evolution of $f(v)$ is given by:

$$\frac{\partial f(v,t)}{\partial t} = \frac{\partial}{\partial v}\big(f(v)V(v,f)\big) + \frac{\partial}{\partial v}\big(f(v)W(v,f)\big) \qquad [4.61]$$

The term $V(v, f)$ represents the contribution from erosion and $W(v, f)$ represents that from the breakage and agglomeration processes. Note that equation [4.61] is a continuity equation corresponding to:

$$\frac{dv}{dt} = \alpha_1 V(v,f) + \alpha_2 W(v,f) \qquad [4.62]$$

where α_1 and α_2 are the normalized weights.

Equation [4.61] implies that $\int_{v_0}^{v_{max}} f(v)dv$, which represents the total volume of the particles, remains constant over time.

The quantity $\dfrac{f(v,t)}{\int_{v_0}^{v_{max}} f(v,t)\,dv}$ is the volume fraction of particles with the volume v, where v_0 is the volume of the elementary particle and v_{max} is the volume of the particle composed of all the elementary particles.

The algorithm to calculate the time evolution of $f(v)$ is as follows:

– we assume that $f(v, t)$ is known;

– we calculate dv/dt (by evaluating the functions of the differential equation 4.62);

– we choose the values $v_0<v_1<v_2<v_3<....<...v_{max}$ and solve the differential equation [4.62] at time Δt for the initial conditions v_1, v_2, v_3,... etc., obtaining: $v_1(t+\Delta t)$, $v_2(t+\Delta t)$, $v_3(t+\Delta t)$, etc.;

– the distribution function is updated:

$$f(v,t) = f(v^{\leftarrow(\Delta t)},t)\,\frac{dv(f(v,t);v^{\leftarrow(\Delta t)})}{dv(f(v,t);v)} \quad [4.63]$$

where $v^{\leftarrow(\Delta t)}$ is the initial volume which, after Δt, moves toward v;

– we go back to the first step, replacing $f(v, t)$ by $f(v, t+\Delta t)$;

– by repeating the following stages, we successively calculate $f(v, t+2\Delta t)$, $f(v, t+3\Delta t)$, etc.

The erosion term $V(v,f)$ is given by [LOZ 02]:

$$V(v,f) = \upsilon(v) - \frac{1}{f(v)}\int_{v_0}^{v_{max}} f(w)\frac{\upsilon(w)}{w}dw \quad [4.64]$$

where $\upsilon(v)$ is the expression proposed by Bohin *et al.* [BOH 96a, BOH 96b]. For dense agglomerates:

$$\upsilon(v) = -\frac{3^{4/3}k_{e1}v^{4/3}}{(4\pi)^{1/3}} \quad [4.65]$$

and for loosely compacted agglomerates:

$$\upsilon(v) = -\frac{3^{4/3}k_{e2}v^{4/3}}{(4\pi)^{1/3}} + \frac{3^{2/3}k_{e3}v^{2/3}}{(4\pi)^{-1/3}} \quad [4.66]$$

where k_{e1}, k_{e2} and k_{e3} are given by:

$$k_{e1} = K(\frac{5}{2}\pi\mu\dot{\gamma}\beta - \kappa) \quad [4.67]$$

$$k_{e2} = K\frac{5}{2}\pi\mu\dot{\gamma}\beta \quad [4.68]$$

$$k_{e3} = \kappa K \quad [4.69]$$

where K is a parameter of the model.

The rupture term $W(v, f)$ is given by [LOZ 02]:

$$W(v,f) = \frac{1}{f(v)}\zeta(v) \quad [4.70]$$

with:

$$\varsigma(v) = \int_{v_0}^{v_{\max}} dw \int_0^1 w\varphi(v^{(\varsigma)}, w^{(\varsigma)})\,d\varsigma \quad [4.71]$$

$$v^{(\varsigma)} = v - \varsigma w \quad [4.72]$$

$$w^{(\varsigma)} = v + (1-\varsigma)w \quad [4.73]$$

$$\phi(v,w) = (v/\widehat{w})k_1(w/v_0)^m[\frac{1}{(2\pi^{0.5})\sigma(w)}][\exp(\frac{-(v-\widehat{w})^2}{2\sigma(w)^2})]f(w) \quad [4.74]$$

$$\widehat{w} = \frac{w}{2 + k_2(w/v_0)^p} \quad [4.75]$$

$$\sigma(w) = k_3\widehat{w}(w/v_0)^q \quad [4.76]$$

where k_1, k_2, k_3, m, p and q are the parameters introduced by Pandya and Spielman [PAN 82].

NOTE 4.3. – To implement this model, the different coefficients must first be identified. This was accomplished by conducting experiments in simple flows [LOZ 02]. Then, as shown in Figure 4.47, samples were taken along the extruder and the size distributions of the agglomerates were measured.

To simulate this, the results for local shear rate and residence time obtained by the LUDOVIC© software are used. Using this information, we can calculate the evolution of the size distribution by solving equation [4.62], as explained above. Figure 4.48 compares the calculated and measured distributions, at one point along the extruder with identical weights for erosion and rupture ($\alpha_1 = \alpha_2 = 0.4$). We found that there is a satisfactory agreement between the two.

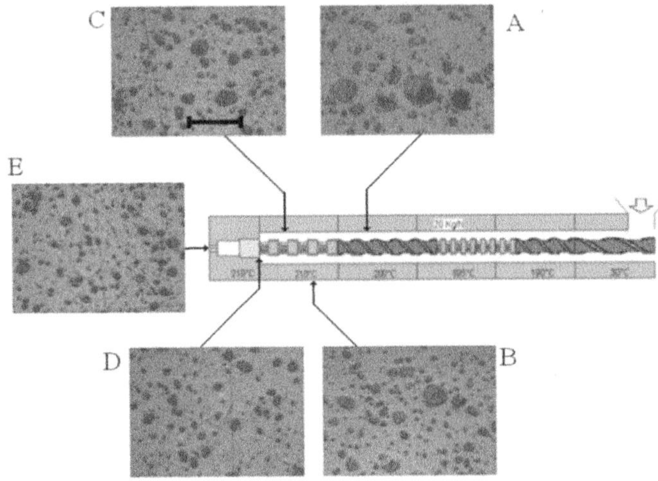

Figure 4.47. *Evolution of the dispersion of calcium carbonate along the extruder (scale: 500 µm) [LOZ 03]*

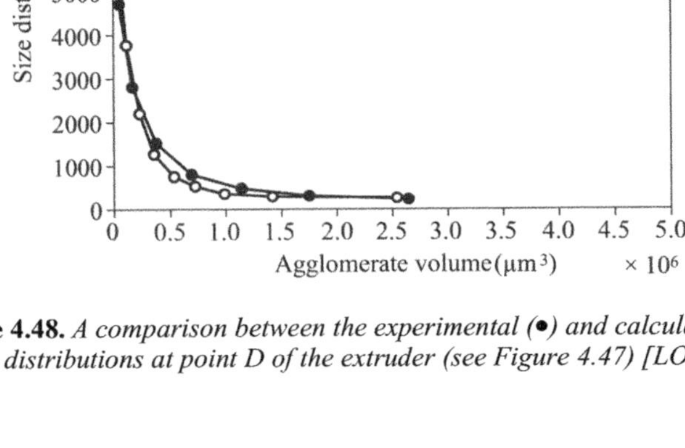

Figure 4.48. *A comparison between the experimental (•) and calculated (○) size distributions at point D of the extruder (see Figure 4.47) [LOZ 03]*

We have just seen one application of the LUDOVIC© software in the case of the dispersion of micro-scale mineral particles. It was also used in the compounding of glass fiber composites. Again, by combining the flow calculation with a model of rigid fiber breakage, it was possible to predict the evolution of the distribution of fiber lengths during the extrusion process [INC 11, VIL 13, DUR 13].

4.7.2. *Reactive extrusion*

Another important application of twin-screw extrusion involves using the extruder as a continuous chemical reactor. This is called reactive extrusion. Polymerization (or polycondensation), chemical modifications (e.g. maleic anhydride grafting), controlled degradation (in particular for polypropylenes), dynamic vulcanization (production of thermoplastic elastomers), etc., can be performed [XAN 92, BER 03]. Compared with a conventional batch reactor, its main aim is to have a continuous and very flexible process (modular geometry, independence of processing parameters, sequence of operations, etc.), to work without a solvent, and to be able to use very viscous components. However, the residence times are limited, the cooling capacities are lower and it is often difficult to disperse a reactive liquid in a very viscous medium. Therefore, not all reactions are likely to be used in reactive extrusion, since they should be fast and only slightly exothermic.

4.7.2.1. *Modeling principles*

The development and optimization of a reactive extrusion process can be greatly facilitated by modeling [VER 06]. However, this approach is complicated by the many interactions involved in the process. For example, as shown in Figure 4.49, modifying a single process parameter (e.g. rotation speed) affects all the other parameters. It is therefore almost impossible for the operator to predict what these mutual effects are. Only a theoretical model can help us to understand this process.

Implementing such a model requires three modules:

– a thermomechanical calculation of the flow along the screws: here, we will use the LUDOVIC© software;

– a kinetic module, which calculates the reaction rate as a function of time and temperature;

– a rheokinetic module, where necessary, which describes the effect of the reaction rate on the rheological properties.

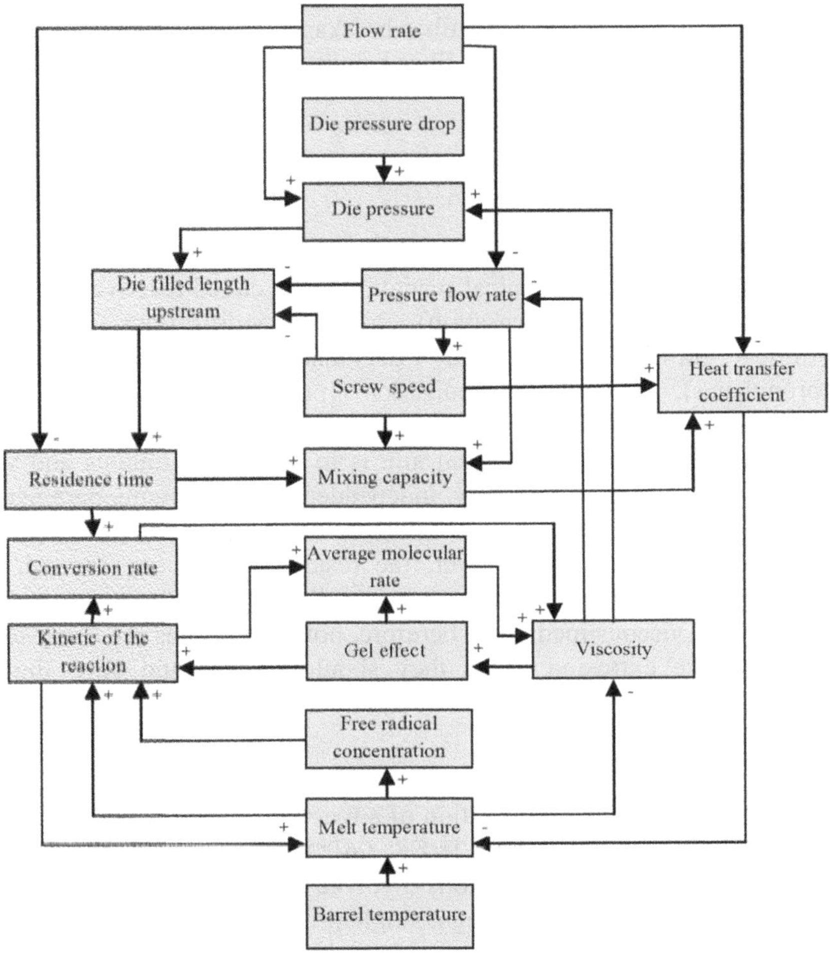

Figure 4.49. *A diagram showing the interactions between the parameters in the case of PMMA polymerization in a twin-screw extruder; the signs + or – indicate an increase or a decrease in the considered parameter [JON 95]*

The last module is only necessary when the studied reaction significantly modifies the rheological behavior. For example, during polymerization, we change from the viscosity of a monomer ($\approx 10^{-3}$ Pa.s) to that of a polymer ($\approx 10^3$ Pa.s) over a short length. This is also the case for controlled degradation of polypropylene, as we will see later. However, a grafting operation or an exchange reaction, which only slightly modifies the structure of the chain, will not require the use of such a module.

The complexity of the kinetic module depends on the studied reaction. For example, if we consider the polymerization of ε-caprolactone, the conversion rate C is simply given by an analytical expression [GIM 00]:

$$C = 1 - \exp\left(-K_0 \exp\left(-\frac{E_a}{RT}\right)[I_0]^{\alpha_c}\, t\right) \quad [4.77]$$

where K_0 is the kinetic parameter, E_a is the activation energy of the reaction, $[I_0]$ is the initiator concentration and α_c is the partial order of the reaction with regard to the initiator. On the other hand, if we study the degradation of polypropylene by peroxide, the kinetic scheme must take into account several reactions, such as the dissociation of the peroxide, β-scission, intermolecular transfer, heat degradation and termination [TZO 88]. A balance of chemical species (peroxides, free radicals, macroradical polymers and polymer chains) leads us to a complex system of eight equations of partial derivatives, which must be solved at each time step using a numerical method, for example the Runge–Kutta method [KRE 94].

For the rheokinetic module, the laws can also be rather complicated. In the controlled degradation of polypropylene, the Carreau–Yasuda law is used (Chapter 1, section 1.2.2), whose parameters (η_0, a, λ, m and E_a) all depend on the average molecular weight of the polymer [BER 00, BER 01]. In the polymerization of ε-caprolactone, Gimenez *et al.* [GIM 00] have shown the need to consider two regimes:

– below a critical molecular weight, a Rouse regime with a Newtonian behavior:

$$\eta = K\, a_T\, M_w^{1,2}\, C \quad [4.78]$$

where C is the conversion rate defined by equation [4.77];

– above this critical molecular weight, an entangled regime, with a viscosity obeying the Carreau–Yasuda law where η_0 and λ depend on the average molecular weight.

4.7.2.2. *Example of an uncoupled calculation*

We begin with the simplest case, when the rheological behavior does not evolve with chemical kinetics. In this case, an *a priori* calculation is performed using LUDOVIC© to obtain the evolution of local residence time and temperature along the screws. Then the kinetic module is used to calculate the reaction rate.

Figure 4.50 shows the transesterification reaction of an ethylene vinyl acetate (EVA) copolymer, which results in the substitution of the vinyl acetate groups with OH groups [BER 98b]. Under these conditions, we obtain a conversion rate of 38% at the machine exit (it is a maximum of 65% at equilibrium). We found that it mainly evolves along the mixing zones, where the residence time is large and the temperature is high due to viscous heating.

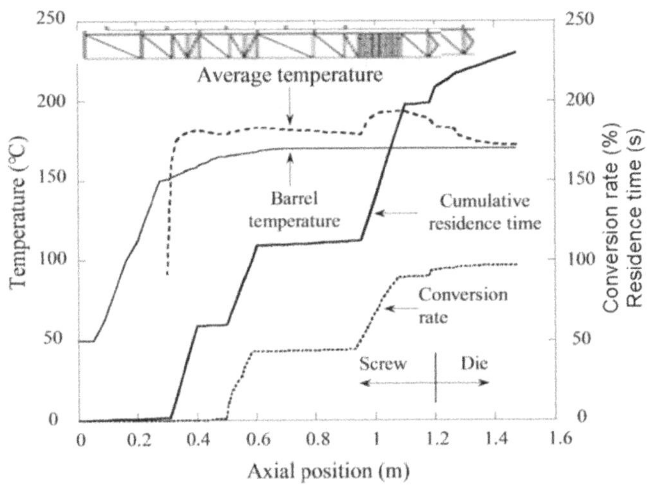

Figure 4.50. *Esterification of EVA; evolution of temperature, cumulative residence time and conversion rate along the screws [BER 98b]*

If the feed rate is modified, we find that this barely changes the temperature profile (Figure 4.42(b)), but significantly decreases the residence time (Figure 4.35(a)). Therefore, a decrease in the conversion rate is expected. However, if the rotation speed is increased, the residence time decreases (Figure 4.34b), but the temperature increases (Figure 4.43(b)). Both parameters controlling the reaction vary in opposite ways and, without a model, it is not possible to know which one predominates. A calculation provides this answer and shows that, in this case, an increase in speed from 50 to 150 rpm allows the conversion to increase from 27 to 37% (Figure 4.51). Therefore, the increase in temperature takes precedence over the decrease in residence time, but this is by no means generalizable. The opposite effect could be seen with different screw profiles or processing conditions.

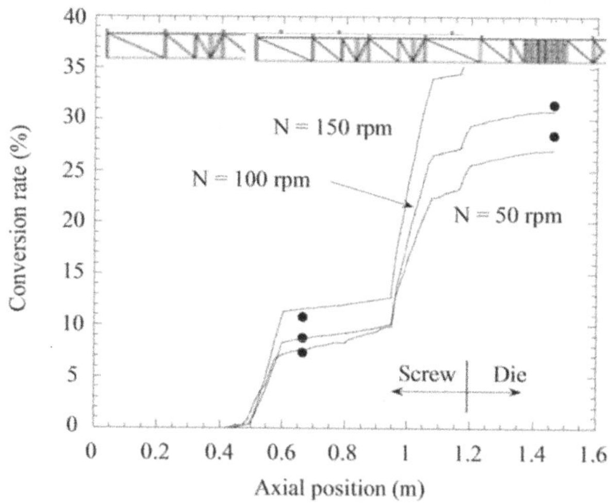

Figure 4.51. *Esterification of EVA: evolution of the conversion rates with screw speed. Symbols (•) indicate the experimental points and lines (—) indicate the theoretical model [BER 98b]*

Figure 4.51 also demonstrates that the comparison of the model to experimental measurements, performed using infrared spectrometry, shows excellent agreement, without any adjustable parameters. This is, thus, a predictive and reliable tool, which can subsequently be used to optimize the process, or to extrapolate it to an industrial scale.

4.7.2.3. *Examples of a coupled calculation*

The cases that involve a modification in the rheological behavior as the reaction progresses are somewhat more complex. The coupled calculation is carried out as follows [BER 98a]:

– A calculation is performed by the LUDOVIC© software using the initial rheological behavior (zero conversion rate). This gives the first estimates of the filled zones, the local residence times and the temperatures. This calculation is carried out from downstream to upstream.

– From these estimated parameters, and now starting upstream (at the introduction of the reactants), we calculate the coupling between the flow, the reaction and the viscosity step by step.

– The time t_i and the temperature T_i are known in element i:

- using the kinetic equation, the increment in the conversion in element i is calculated;

- using the rheokinetic module and the new conversion rate, the corresponding viscosity is calculated;

- with this new viscosity, the flow conditions in element $i + 1$ are calculated along with the new residence time and temperature.

Thus, we progress element by element until the end of the flow. In Figure 4.52, we have shown the continuous polymerization of ε-caprolactone [POU 01b]. We found that the model is capable of following the evolution of the viscosity from the monomer (10^{-4} Pa.s) to the polymer (10^{3} Pa.s), an increase of seven decades. Under tested conditions, the conversion rate is almost 100% at the die exit.

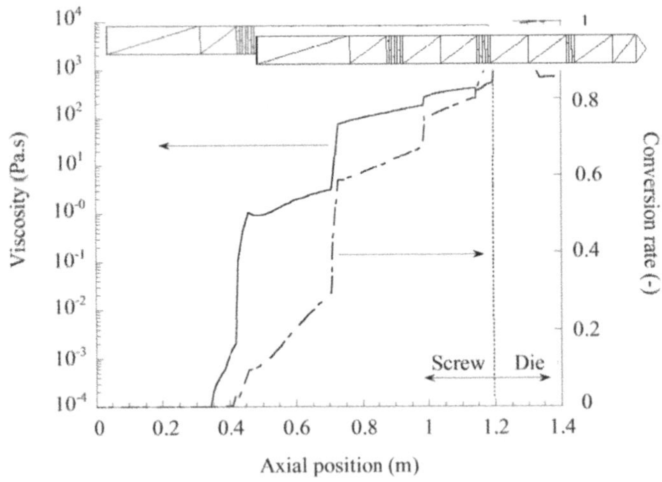

Figure 4.52. *Polymerization of ε-caprolactone: evolution of the conversion rate and viscosity along the screws [POU 01b]*

Another interesting case is that of the controlled degradation of polypropylene (PP) using peroxides. At the exit of the polymerization reactor, PP has a high molar mass, with viscous and elastic properties that make it poorly suitable for processing. This is why it is necessary to reduce the molar mass and adjust the viscosity to make it easy to use in processes such as extrusion or injection. This operation is carried out in a twin-screw extruder, using peroxides to break the long PP chains. These peroxides are generally introduced after the melting zone and they decompose with heat. As a result, they give rise to free radicals, which then attack and cut the chains by hydrogen abstraction and β-scission. Using the kinetic and rheokinetic relationships, presented previously, we can calculate the evolution of the molar mass along the screws as a function of the amount of peroxide (Figure 4.53). In this model, we have only one adjustable parameter, which is the efficiency of the peroxide, that is to say, the number of radicals out of the total radicals which actually lead to β-scission.

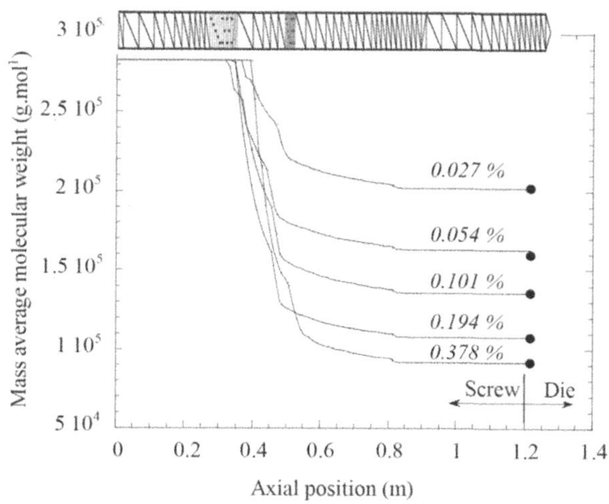

Figure 4.53. *Controlled degradation of polypropylene: evolution of mass average molecular weight along the screws as a function of the peroxide content [BER 00]*

This efficiency was experimentally estimated and it mainly varies with peroxide content [BER 00]. To validate the theoretical model, we took samples along the screws [BER 06]. Their analysis confirmed that the rapid decrease in molar mass when the peroxide is introduced (Figure 4.53) is real (Figure 4.54). In addition, we found that the model correctly takes account of the effects of the screw rotation speed and the barrel temperature.

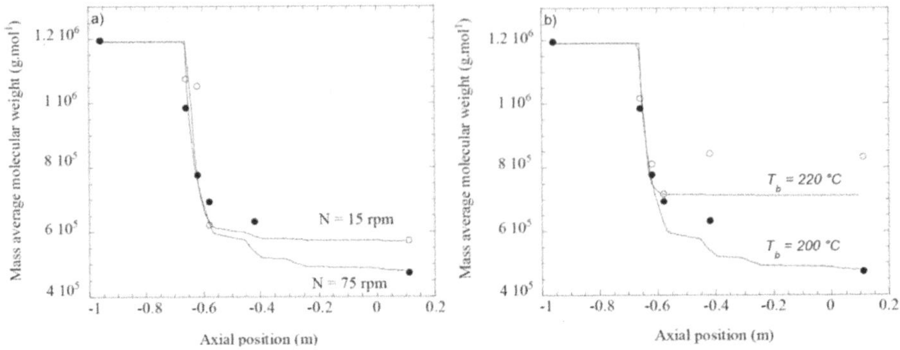

Figure 4.54. *Controlled degradation of polypropylene: evolution of mass average molecular weight along the screws for a) various rotation speeds and b) temperatures of the barrel. Symbols (•, ○) are experimental points and lines (—) are the theoretical model [BER 06]*

Besides the examples we have just presented, we have successfully applied this modeling approach to other reactions, such as the extension of polyamide 12 chains with a coupling agent [CHA 02], starch cationization [BER 07a, BER 07b] and the synthesis of nanoparticles using the sol–gel method [BAH 11]. By considering the breakage and erosion mechanisms discussed in the previous section as kinetic equations, it was even possible to address the dispersive mixing of particles, using these reactive extrusion models [BER 02].

4.7.3. *Polymer blends*

A particular type of compounding in twin-screw extrusion, which was developed in the 1990s, involved producing polymer blends. Instead of a particle–polymer mixture, here we focus on a polymer–polymer mixture. Just like in coextrusion, this allows us to combine the properties of the different polymers mixed together and even, in some cases, achieve synergy, that is to say, better properties than those predicted by a simple mixing rule [UTR 02]. For immiscible polymers, this process induces a particular morphology, which will then define the properties. Depending on the proportions, and the rheological and interfacial properties of each phase, the blend of both polymers A and B may give the following morphologies:

– phase A nodules dispersed in matrix B;

– often complex co-continuous morphology;

– phase B nodules dispersed in matrix A.

Here, we will only focus on the simplest nodular morphology without mentioning the use of a compatibilizer, which, in practice, greatly reduces the size of the nodules and improves the properties. First, we present the basic mechanisms controlling the evolution of this nodular morphology and then show how these evolutions can be calculated along a twin-screw extruder.

4.7.3.1. *Basic mechanisms*

Let us consider polymer A dispersed as nodules within polymer B. The two basic mechanisms that will cause the morphology to evolve are breakup, which decreases the size of the nodules, and coalescence, which has the opposite effect.

Deformation and breakup mechanisms of an isolated Newtonian drop, sheared within a Newtonian matrix, were studied by Taylor in 1932 [TAY 32]. It was shown that the deformation of the drop depends on the shear forces, which tend to deform it, and the surface tension forces, which tend to restore its spherical shape. The ratio of these two forces is called the capillary number or the Weber number:

$$Ca = \frac{\eta_m \dot{\gamma} R}{\sigma_{AB}} \qquad [4.79]$$

where η_m is the matrix viscosity, R is the nodule radius and σ_{AB} is the surface tension. Different scenarios of deformation and rupture of the drop are therefore established, depending on the value of the critical capillary number Ca^*, which is experimentally defined [MEI 88b, HUN 95, DEL 96]:

– if $Ca < Ca^*$, the drop is deformed, but does not break,

– if $Ca^* < Ca < 2\ Ca^*$, the drop can be split into two identical drops, provided that the shear time is greater than the rupture time, defined by:

$$t_b = \frac{Ca}{\dot{\gamma}} f\left(\frac{\eta_d}{\eta_m}\right) \qquad [4.80]$$

where $f(\eta_d / \eta_m)$ is the function of the ratio between the viscosity of the dispersed phase and that of the matrix [GRA 82].

– If $Ca > 2\ Ca^*$, the drop follows the affine deformation of the matrix and is converted into a fiber. For an initial diameter equal to D_0, the fiber will have a diameter of $D(t) = \frac{D_0}{\sqrt{\dot{\gamma}t}}$ and a length $L(t) = D_0\ \dot{\gamma}t$. This fiber can then be further broken up by two mechanisms:

– when the diameter reaches a critical value d^*, unique to each system and defined from experimental observations;

– after the shear terminates, by Rayleigh instability [LOR 92].

This instability, the development of which is shown in Figure 4.55, causes the fiber to break into main and satellite drops, by amplification of a small variation in diameter over time. The equation of this mechanism was proposed by Tomotika [TOM 35, TOM 36].

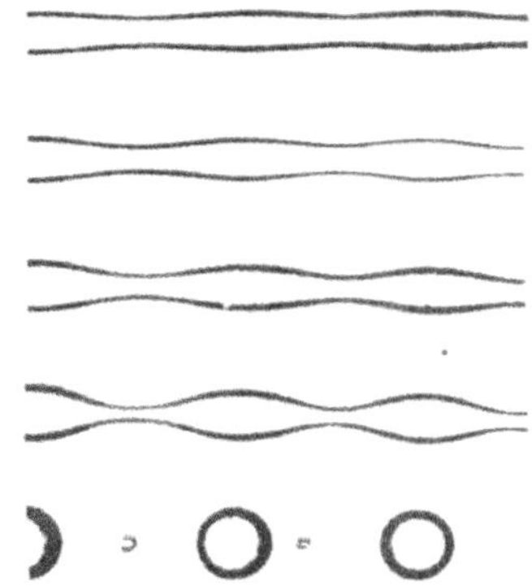

Figure 4.55. *Breakage of the fiber by Rayleigh instability*

The value of the critical capillary number is the underlying factor in these rupture mechanisms because it determines the evolution in size. All the estimates, found in the literature, are based on observations from laboratory experiments, or industrial mixing systems. In all the cases, Ca^* is a simple function of the viscosity ratio η_d/η_m. In Figure 4.56, the curves for simple shear and uniaxial elongation were determined [GRA 82]. In simple shear, the rupture is no longer possible when η_d/η_m is greater than 4. However, for elongation, breakage is always possible, regardless of the viscosity ratio.

Moreover, the Ca^* values are smaller, which means that the drops break more easily with elongation than with shear.

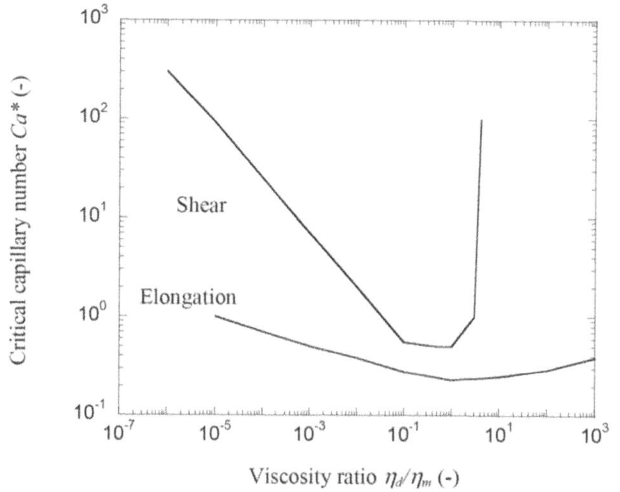

Figure 4.56. *Evolution of the critical capillary number with viscosity ratio [GRA 82]*

Another common definition of the critical capillary number Ca^* is based on observations made at the twin-screw extruder exit, where the flows involve both shear and elongation [WU 87]. In this case, Ca^* is defined by (Figure 4.57):

$$Ca^* = 4\left(\frac{\eta_d}{\eta_m}\right)^{\pm 0{,}84} \qquad [4.81]$$

The sign of the exponent is positive if the viscosity ratio is greater than 1 and it is negative in the opposite case. As with uniaxial elongation, rupture is possible regardless of the viscosity ratio, but it is easier if this ratio is equal to 1.

The other basic mechanism of morphology evolution is coalescence, when two nodules collide to create a single nodule of a larger size. In order to make this possible, the two nodules must remain in contact for the time required to remove the thin matrix film that separates them (Figure 4.58).

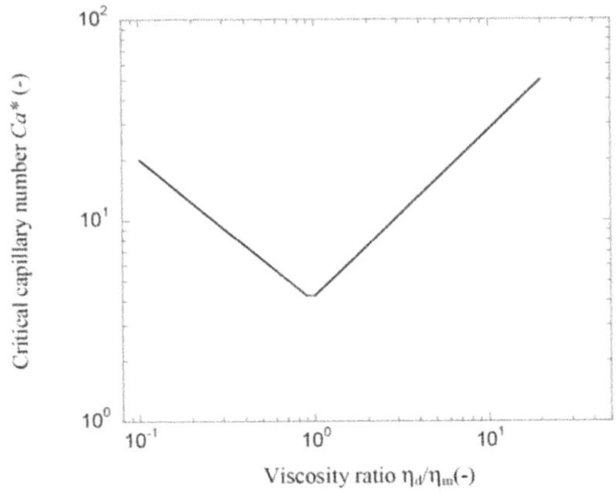

Figure 4.57. *Evolution of the critical capillary number with the viscosity ratio [WU 87]*

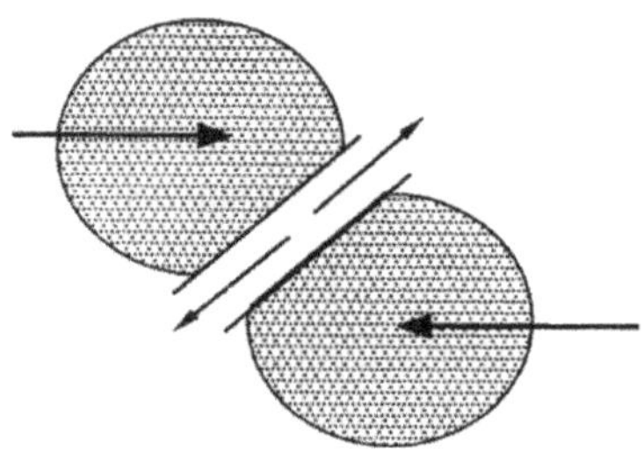

Figure 4.58. *Coalescence mechanism of two nodules*

We can model the coalescence of two spherical nodules of radius R by successively calculating [ELM 86, CHE 91] the following:

– *the collision frequency*:

$$C = 8\frac{\dot{\gamma}\varphi}{\pi} \quad [4.82]$$

where φ is the mass fraction of the dispersed phase;

– *the collision probability:*

$$P_{col} = \exp\left(\frac{-\pi}{8\dot{\gamma}\varphi\, t_{loc}}\right) \quad [4.83]$$

where t_{loc} is the local residence time;

– *the probability of removing the liquid film*, which depends on the "mobility" of this film, linked to the viscosity ratio [CHE 91]. For example, for mobile interfaces ($\eta_d/\eta_m << 1$):

$$P_{exp} = \exp\left[-\frac{3}{2} Ca \ln\frac{R}{h^*}\right] \qquad [4.84]$$

where h^* is the minimum thickness of the film, below which it will rupture. h^* is defined by:

$$h^* = \left(\frac{10^{-20} R}{8\pi\sigma_{AB}}\right)^{\frac{1}{3}} \qquad [4.85]$$

– *the coalescence probability* is defined as the product of the collision (equation [4.83]) and removal (equation [4.84]) probabilities. If coalescence is possible, the new radius of the nodule is calculated by conservation of volume.

4.7.3.2. *Modeling along the twin-screw extruder and an example of the results*

To calculate the morphology of nodules in a blend of incompatible polymers along a twin-screw extruder, we must define the scenarios that describe the potential mechanisms according to the local flow conditions [DEL 95].

In many situations, the most significant reduction in size in the dispersed phase occurs in the melting zone, where we move from a size of several millimeters (that of pellets) to a size of several microns. However, as we have seen previously, melting mechanisms are still poorly understood, even for a single polymer. We start the morphology calculation after this zone by introducing a parameter which is the initial diameter of the nodules, D_0. We will then consider "strong" and "weak" flow zones. The "strong" flow zones correspond to the filled zones under pressure. In these zones, depending on the local conditions, rupture or fibrillation of the drop, as well as coalescence, may occur. In the "weak" zones, which are the partially filled conveying zones, rupture by Rayleigh instability of fibers, which were previously formed in "strong" zones, and coalescence occur.

An example of the results is presented in Figure 4.59. It shows the evolution along the screws of the average diameter of the dispersed phase, which varies from 0.53 to 0.1 μm, depending on the local flow conditions. This variation is controlled by the ratio of the local capillary number to the critical capillary number (Figure 4.59(b)), the ratio of rupture time to local residence time (Figure 4.59(c)) and the coalescence probability (Figure 4.59(d)). Just after melting, the diameter remains constant, since it is in a "weak" zone, and the coalescence probability is zero. In the pressurized zone upstream of the kneading block, fibrillation and immediate rupture occur, since the critical diameter is very quickly reached. These nodules are then too small to be fibrillated again. They do not coalesce, since the probability is still close to zero, and they can only be split into two when the rupture time is reached. Thus, the diameter progressively decreases until stabilizing in a second "weak" zone (before the pressure increases upstream of the die). When the pressure starts to increase, there is, once again, a possibility of rupture into two, after which the capillary number is too small to induce new modifications. The only mechanism that is still possible is coalescence, with a probability of approximately 50%, which can be seen as an increase in size toward the end of the flow.

This type of model allows us to follow and understand the evolutions of the average size of a nodule depending on the geometry of the extruder and the processing conditions. For example, Figure 4.60 shows that the process conditions can be chosen to control the final morphology, characterized by particle size reduction as the screw speed increases, or as the flow rate decreases. However, note that the speed and the flow rate influence the melting mechanism and, therefore, could affect the initial size of the dispersed phase, which we have considered in this case to be constant.

The previous examples illustrated the calculation of the mean diameter. It is relatively simple to generalize this calculation to a size distribution, by considering different size categories with different evolution regimes. In the same machine, at the same location, some categories may break into two whereas others fibrillate or coalesce. We, thus, obtain an evolution of the size distribution along the extruder, which allows us a more realistic description of the morphology.

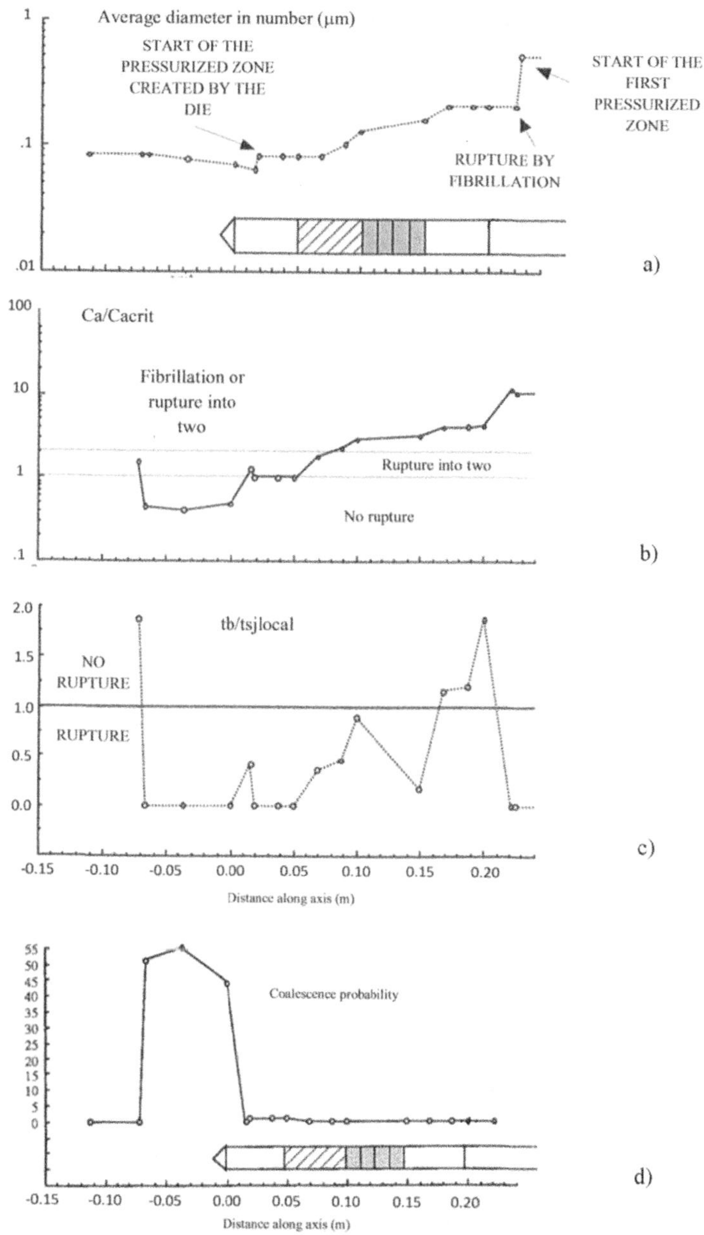

Figure 4.59. *An example of calculation of the morphology of a polymer blend along a twin-screw extruder [DEL 95]. Evolutions of a) average diameter, b) Ca/Ca* ratio, c) t_b/t_s ratio and d) coalescence probability*

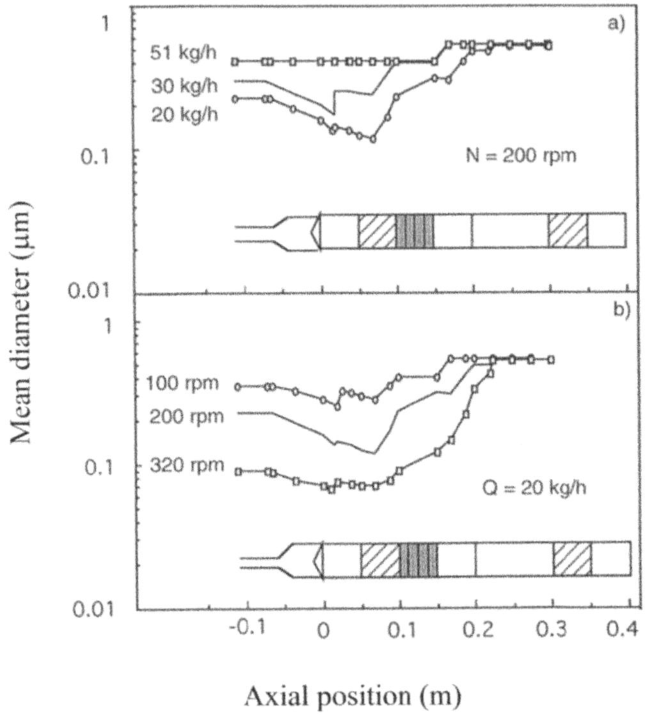

Figure 4.60. *An example of calculation of the morphology of a polymer blend along a twin-screw extruder: a) influence of flow rate and b) influence of rotation speed [DEL 96]*

4.7.4. *Production of clay-based nanocomposites*

In this section, we will discuss the production of nanocomposites based on polypropylene and lamellar clay [LER 06a, DOM 12a]. These materials were developed at the beginning of the 1980s by a research group at Toyota, who showed that by dispersing a lamellar clay in polyamide, it is possible to significantly improve the mechanical properties, with very low filler content, approximately 5% [OKA 87, FUK 87].

The ability to significantly reduce the required filler content is mainly due to the large specific surface area of these clays. These clays are composed of 1 nm thick layers, stacked in tactoids of several dozen layers, associated as aggregates, and agglomerates of several dozen microns (see Figure 4.61).

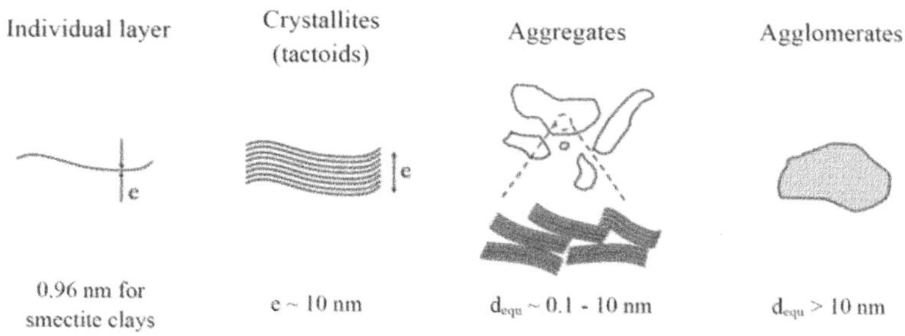

Figure 4.61. *The different structural levels of lamellar clay [DOM 12a]*

When mixed with a molten matrix, it is essential to separate the agglomerates present in the initial clay into individual sheets in order to obtain adequate properties. This is relatively easy in the case of polar polymers such as polyamide, which has a high affinity for clay, but it is much more difficult for non-polar polymers, such as polyolefins. In this case considered in this section, there are different structures (Figure 4.62). If the clay remains in the form of agglomerates or even tactoids (Figure 4.62(a)), we obtain a microcomposite, which does not have enhanced properties due to the low filler content. Improvements in the properties start to become visible when the matrix chains intercalate between the clay layers (Figure 4. 62(b)), thereby increasing the interlamellar distance. This is called an intercalated structure. Finally, if all the layers are individually dispersed (Figure 4.62(c)), the structure is said to be exfoliated. It is this structure that gives the best properties. In reality, for polyolefins, this rarely occurs and the most common structure obtained is a mixed one, intercalated and partially exfoliated.

Nanocomposites can be prepared by *in situ* polymerization or solution blending, but the most industrial and commonly used method is melt blending by twin-screw extrusion [VER 08]. The flow conditions along the machine allow the agglomerates to disperse and the tactoids, subsequently, to intercalate, and progressively exfoliate. However, to facilitate the dispersion of clay in a non-polar matrix, it is essential to use organomodified clays and add a coupling agent, also known as a compatibilizer. Organomodified clays are natural clays (usually from montmorillonite), which have been modified by surfactants to make them more compatible with polymers. Coupling agents, typically maleic anhydride grafted polymers, are required to intercalate and exfoliate clay in non-polar matrices [LER 05].

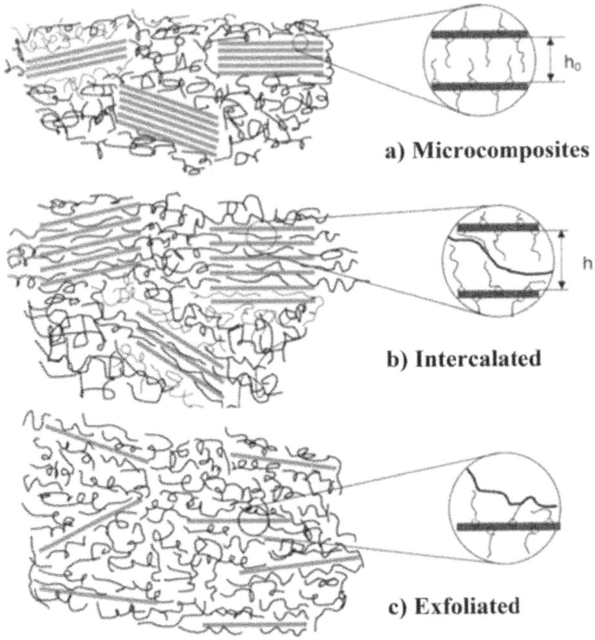

Figure 4.62. *The main structures of lamellar clay-based nanocomposites [KRI 01]*

As there are many levels of dispersion, ranging from agglomerates of several micrometers to exfoliated sheets, it is necessary to quantify these different levels and use complementary techniques, which provide us with different information:

– scanning electron microscopy (SEM) allows the large agglomerates to be visualized and to quantify them using, for example, the area ratio (area of the agglomerates divided by the total area analyzed);

– X-ray diffraction allows us to measure the interlamellar distance (distance between the sheets in a tactoid), and to quantify the level of intercalation;

– transmission electron microscopy (TEM) allows the structure to be viewed at the nanoscale. This can confirm or invalidate the exfoliated nature of the nanocomposite;

– small amplitude oscillatory shear rheometry allows us to identify the interactions between the exfoliated sheets. In fact, these interactions, which are much greater when there are more sheets, result in the appearance of a plateau on the elastic modulus G' at low frequencies, or, by a significant increase in viscosity (Figure 4.63). We have shown that this behavior can be described by the Carreau–Yasuda law, which incorporates a yield stress:

$$\eta^* = \frac{\sigma_0}{\omega} + \eta_0\left(1+(\lambda\dot{\gamma})^a\right)^{\frac{m-1}{a}} \qquad [4.86]$$

The yield stress value σ_0 allows the level of exfoliation to be quantified [LER 05, VER 11].

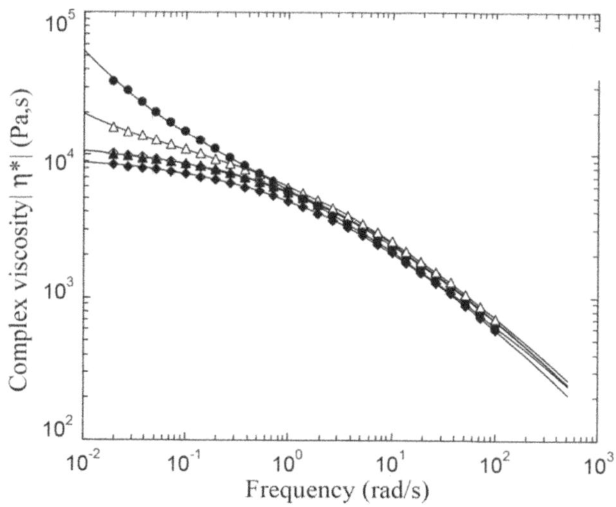

Figure 4.63. *Viscosity curves of polypropylene/clay nanocomposites at different dispersion levels. The continuous lines represent the Carreau–Yasuda model with a yield stress [LER 05]*

As with the preparation of blends or compounds, the extrusion conditions improve and determine the structure of a nanocomposite. In the following examples, we will focus on the dispersion of an organomodified clay in a polypropylene matrix, compatibilized with maleic anhydride grafted polypropylene (PP-g-MA).

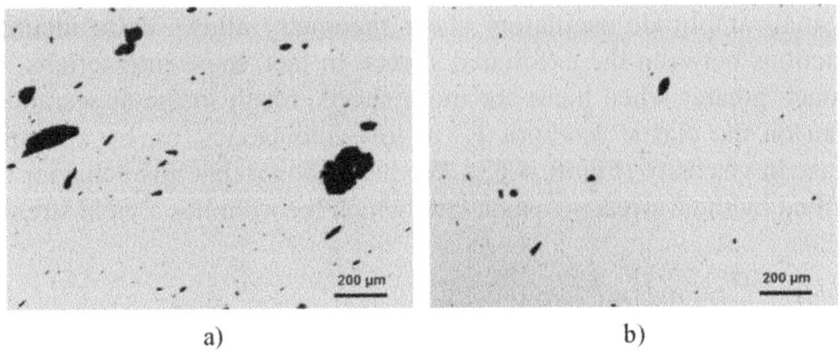

Figure 4.64. *SEM images of nanocomposites prepared at a) 100 and b) 700 rpm [DOM 12b]*

Figure 4.64 shows the effect of the screw rotation speed on the micro-sized agglomerates. We see a significant decrease in the number and size of these agglomerates with screw speed.

Under the same conditions, the melt yield stress, and, therefore, the level of exfoliation, increases up to 500 rpm before stabilizing (Figure 4.65). This improvement in exfoliation with screw speed is confirmed by TEM observations (Figure 4.66).

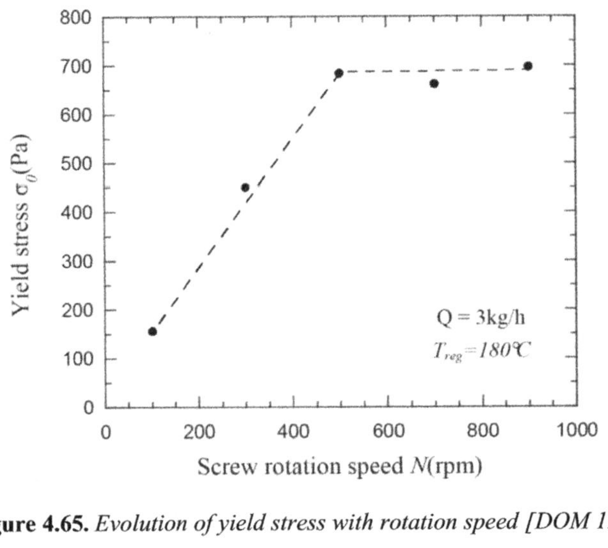

Figure 4.65. *Evolution of yield stress with rotation speed [DOM 12b]*

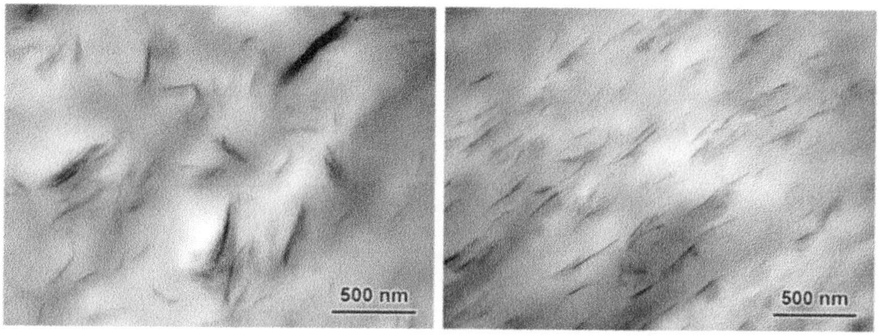

a) b)

Figure 4.66. *TEM images of nanocomposites prepared at a) 100 and b) 700 rpm [DOM 12b]*

An increase in feed rate or barrel regulation temperature leads to a poorer dispersion, with larger and much more numerous agglomerates and reduced exfoliation [DOM 12b].

Therefore, the processing conditions have a significant impact on different structural levels of nanocomposites. These results can be summarized by plotting descriptive parameters of the structure (area ratio for agglomerates and yield stress for exfoliation) as a function of the SME (Figure 4.67). This allows master curves to be obtained, confirming that the dispersion at different scales is controlled by an energetic parameter [DOM 12b].

The screw profile has an important role [LER 06b, LER 07, VER 08]. However, its actual influence is still not well understood, since the structure obtained, in addition to SME, seems to also depend on the thermomechanical history, which can vary considerably from one profile to another. To better understand the role played by the geometry, we have taken samples along the screw axis: once the extruder in the steady-state regime is stopped, it is rapidly cooled, the barrel is opened and samples are taken, which are then analyzed as described above (rheometry, image analysis, X-ray diffraction, etc.).

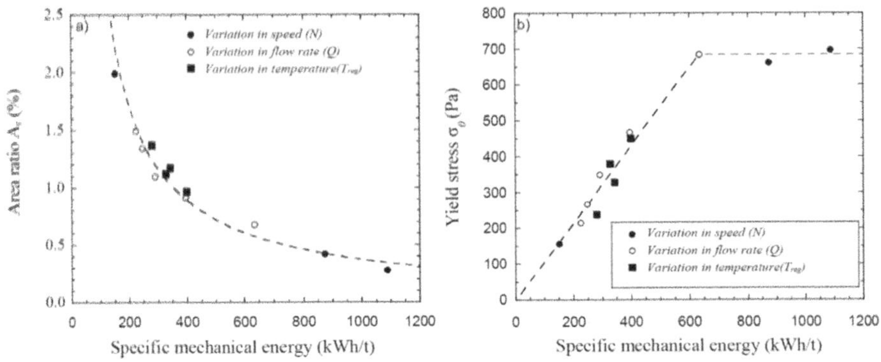

Figure 4.67. *a) Evolution of the area ratio and b) evolution of yield stress with specific mechanical energy [DOM 12b]*

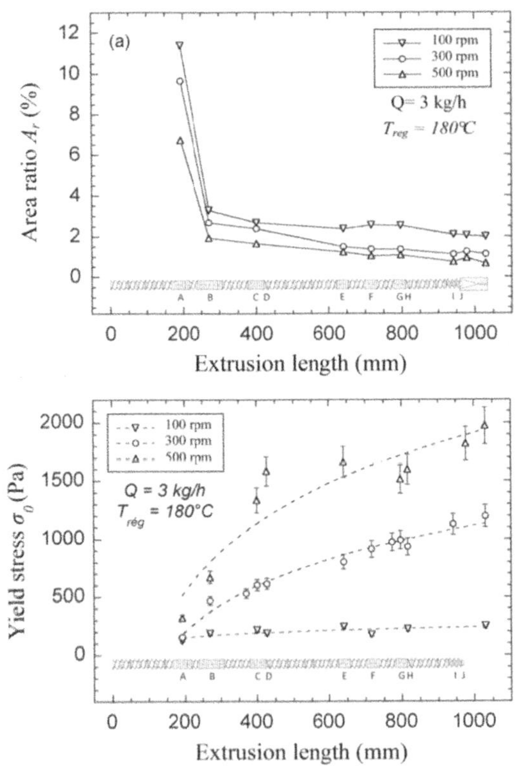

Figure 4.68. *a) Evolution of the area ratio and b) evolution of the yield stress along the screw axis for different rotation speeds [DOM 13]*

Figure 4.68 shows that the area ratio, and therefore the size of the agglomerates, decreases rapidly during the melting, as we have already seen, for example, for incompatible polymer blends. The subsequent changes are more limited. However, the yield stress, and, therefore, the level of exfoliation, progressively increases after melting up until the die, with a significant effect of the rotation speed.

If we now plot the yield stress as a function of the area ratio for the samples in Figure 4.68, we are able to distinguish the two different zones along the extruder (Figure 4.69):

– in zones A and B (see position of points in Figure 4.68), there is a significant decrease in the area ratio, accompanied by a slight increase in the yield stress, which indicates the predominance of agglomerate rupture and erosion mechanisms over exfoliation;

– from zone B until the die exit, there is simultaneously a large increase in the yield stress and a slight decrease in the area ratio, which means that the clay is predominantly dispersed by the progressive exfoliation of sheets along this section of the screw profile.

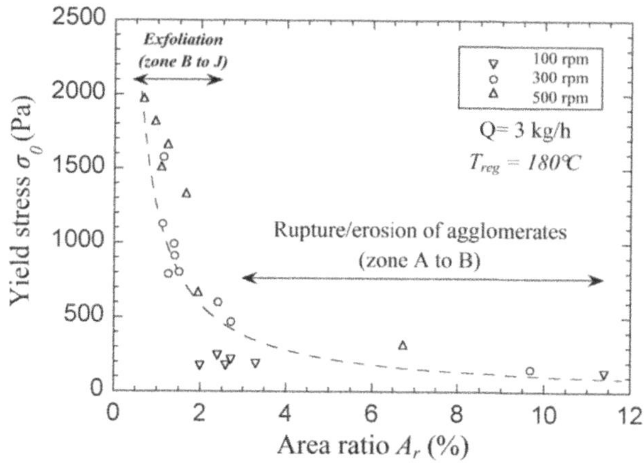

Figure 4.69. *Evolution of the yield stress versus the area ratio along the screw profile for different rotation speeds [DOM 13]*

From the results in Figure 4.67, it is possible to establish correlations between the area ratio, the yield stress and the SME. Introducing these correlations into the LUDOVIC© software was sufficient to enable us to predict the theoretical evolution of the nanocomposite structure, all along the screw axis. As shown in Figure 4.70, the general trend is correctly described for the different tested rotation speeds [DOM 13].

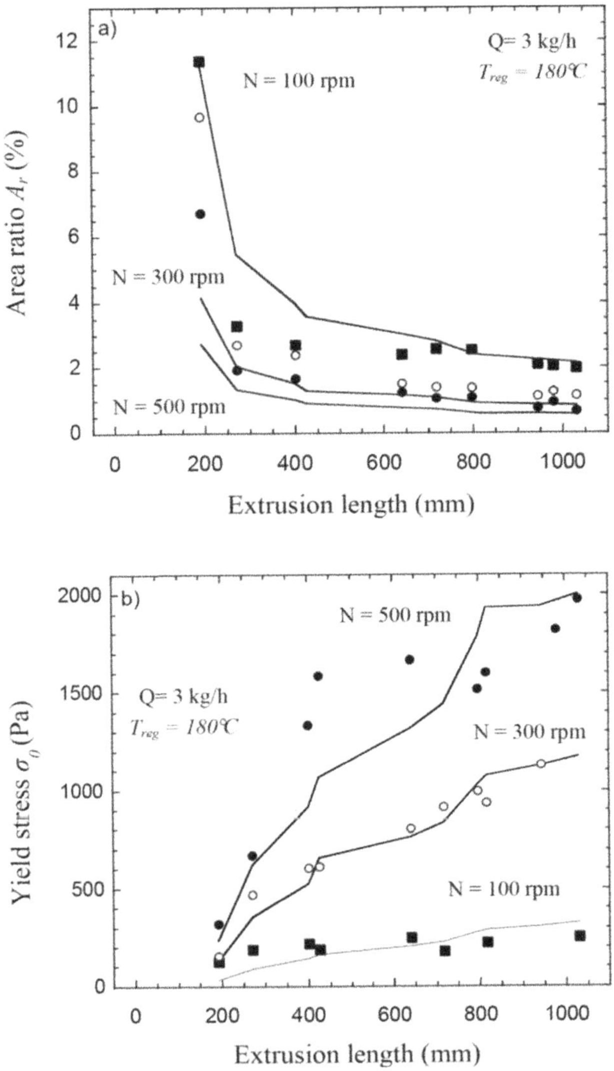

Figure 4.70. *a) Evolution of the area ratio and b) evolution of the yield stress along the screw profile for different rotation speeds. Comparison of theoretical prediction (—) and experimental results (○, ●, ■) [DOM 13]*

4.7.5. *Optimization and scale-up*

For complex issues, such as those frequently found in industrial twin-screw extrusion applications, modeling is a great help because it allows us to simultaneously understand what is happening during the process and suggest improvements. A recurring issue is scale-up, that is to say, moving from a laboratory extruder, where the flow rate is only several kg/h, to an industrial extruder, with a diameter 10 times greater, and a flow rate that can reach several tons/h. This scale-up can be accomplished by trial-and-error. However, this technique is very time-consuming and costly in terms of machine and personnel resources, as well as raw material consumption. It is, thus, quite beneficial to use a modeling approach. In the following section, we give some examples of using software such as LUDOVIC© to solve these optimization and scale-up problems.

First, the computer code can be used to directly optimize a process by trial-and-error, as we would do experimentally, but in a much faster and less expensive way. If the numerical solution can be obtained quickly enough, then more sophisticated optimization techniques can be used. For example, we have already used genetic algorithms to optimize extrusion conditions in the case of a predefined screw profile [GAS 02]. These algorithms imitate the natural evolution processes, with an initial population of individuals that evolves with different operations (selection, hybridization, mutation, etc.) over successive generations. To run this optimization process, three modules are required:

– *an objective function*, which quantifies the performance of each individual (i.e. each calculated extrusion condition) compared to the desired performance;

– *a calculation module*, here LUDOVIC©, which estimates the value of the objective function for each individual in a population;

– *an optimization algorithm*, which controls the evolution of a population and leads it toward the objective. The initial population is randomly chosen (e.g. by varying the rotation speed, flow rate and barrel temperature). For each new generation, the algorithm selects the best individuals and reproduces them.

We can define an overall objective function, including several different and ultimately conflicting objectives with assigned weights, as follows:

$$Fo = \sum_i w_i f_i \qquad [4.87]$$

where w_i are the weights ($\sum_i w_i = 1$) and f_i are the functions which maximize, minimize or maintain a parameter in a certain range. Let us consider, as an example, the polymerization of ε-caprolactone by reactive extrusion, presented in section 4.7.2.3. Let us assume that we require a maximum flow rate, a conversion of 100%, the maximum temperature of the material along the screw to remain below 220°C and the consumed power to be less than 9.2 kW. Figure 4.71 shows the conversion rate obtained at 150 rpm, according to the respective values of Q/Q_{max} and T_b/T_{bmax} (maximum flow rate Q_{max} = 30 kg/h and maximum barrel temperature T_{bmax} = 210°C).

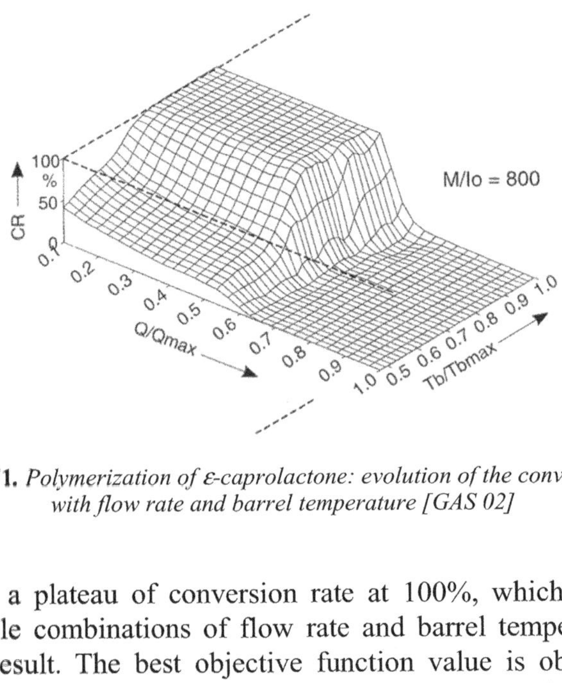

Figure 4.71. *Polymerization of ε-caprolactone: evolution of the conversion rate with flow rate and barrel temperature [GAS 02]*

We obtain a plateau of conversion rate at 100%, which indicates that several possible combinations of flow rate and barrel temperature lead us to the same result. The best objective function value is obtained here at 200 rpm and 26 kg/h, which gives a consumed power of 6.4 kW and a residence time of 32 s.

Utilizing more advanced optimization techniques (multi-objective evolutionary algorithms), we then studied the optimization of the screw design [GAS 05, GAS 1, TEI 11]. In this case, the optimizations carried out often resulted in many similar solutions; it was difficult to choose the best one among them. It is possible that some simplifications of the LUDOVIC© software, in particular the lack of an efficient melting model, prevent us from following through with these sophisticated techniques.

We are going to end this section with an example of optimization and scale-up in the case of starch cationization. Cationic starches are often used in the paper-making industry. Cationization involves substituting, on a starch chain, a hydroxyl group by quaternary ammonium. This modification can be made by reactive extrusion and the result is quantified in terms of efficiency (ratio of the obtained degree of substitution and the theoretical degree of substitution) [TAR 04]. First, we calculated on a laboratory machine (Clextral BC21, diameter 25 mm) the flow conditions for different screw designs, whose complexity increases from design 1 to design 5, by varying the feed rate from 2.5 to 40 kg/h at a screw speed of 400 rpm. Figure 4.72 shows the evolution of the efficiency of the reaction versus the screw design for different flow rates.

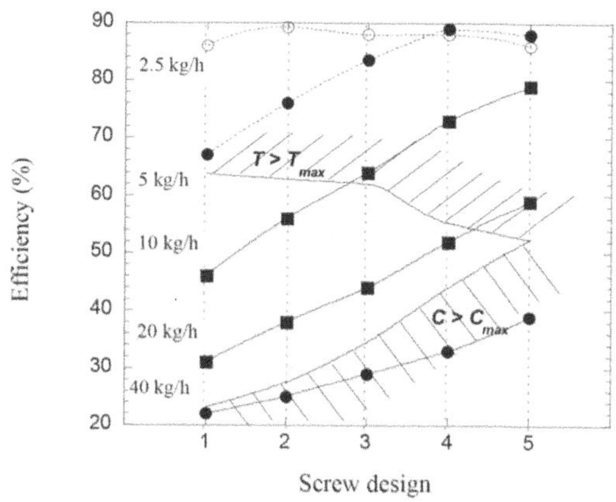

Figure 4.72. *Laboratory machine; variation in efficiency with screw profile and flow rate [BER 07b]*

We found that the efficiency is much higher when the flow rate is low (long residence time) and the design is restrictive (high temperature). If we want the temperature to remain below 165°C to prevent degradation, and the torque to remain less than C_{max} = 50 mN, the processability window is reduced. Depending on whether we are focusing on efficiency or flow rate, we can choose various points on the upper boundary: an efficiency of 64% at 5.5 kg/h with profile 1, 63% at 8.5 kg/h with profile 2, 62% at 11 kg/h with profile 3, 55% at 18 kg/h with profile 4 or, finally, 52% at 25 kg/h with profile 5. Here, it is up to the users to make a choice according to their priorities.

Now let us assume that we want to predict the processing conditions for this material using a much larger machine (Clextral BC72, diameter 82 mm). First, we will define profiles similar to those of the laboratory extruder, which is to say, with the same number of restrictive elements and identical length ratios. Then the flows in these new profiles are simulated in the same way, for a range of flow rates adapted to the size of the industrial machine, at its maximum speed, i.e., here, 360 rpm. The results are shown in Figure 4.73, with a torque limit C_{max} = 1,500 mN. We observed results similar to those for the laboratory extruder. As mentioned above, the temperature and the torque restrictions reduce the processability window, leading to the establishment of the optimal conditions:

– an efficiency of 60% at 155 kg/h with profile 1;

– or 58% at 225 kg/h with profile 2;

– or 58% at 310 kg/h with profile 3;

– or 49% at 560 kg/h with profile 4.

Again, the final choice is left to the users. However the calculation provides a realistic estimate of accessible flow rates and, therefore, helps predict the potential profitability of the industrial process.

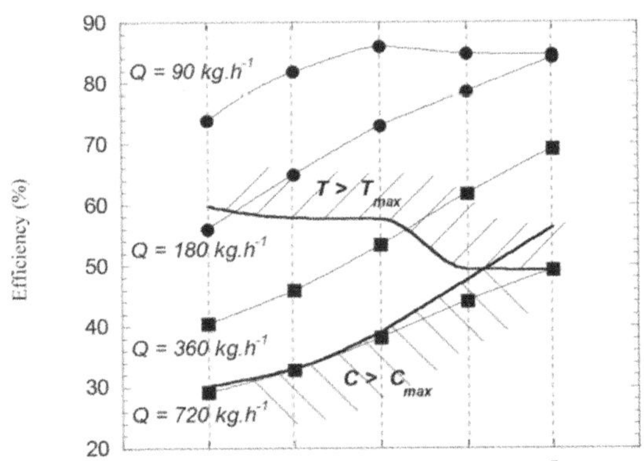

Figure 4.73. *For an industrial machine; variation in efficiency with screw profile and flow rate [BER 07b]*

4.8. Conclusion

The twin-screw extrusion process is a very interesting process, which, due to its high versatility, can deal with a variety of challenges to develop increasingly complex materials. Except for the melting process, whose mechanisms are still not completely understood, this process can be modeled both locally, with 3D models, and globally, with a software such as LUDOVIC©. This software can define the optimal processing conditions for a given screw design. It can also define a design adapted to the desired application and material. It also can help us to solve the complicated task of scale-up, similar to the case in single-screw extrusion.

We have shown through numerous examples that this software is an excellent starting point to address issues regarding the production of filled products, polymer blends, (nano)composites, reactive systems, etc. We have also had the opportunity to use it for specific developments in related fields, for example the extrusion-cooking of cereals [DEL 93, DEL 11], the development of gluten-based materials [RED 99] and, more recently, the dispersion of radioactive waste in bitumen [MOU 11].

4.9. Bibliography

[AGA 09] AGASSANT J.-F., POITOU A., VALETTE R., "A kinematic approach to distributive mixing", in MANAS-ZLOCZOWER I. (ed.), *Mixing and Compounding of Polymers*, Hanser, Munich, 2009.

[AGA 14] AGASSANT J.-F., AVENAS P., SERGENT J.-P., *et al.*, *Mise en forme des matières polymères. Approche thermomécanique de la plasturgie*, Tec & Doc, Lavoisier, Paris, 2014.

[ALS 04] ALSTEENS B., LEGAT V., AVALOSSE T., "Parametric study of the mixing efficiency in a kneading block section of a twin-screw extruder", *International Polymer Processing*, vol. 19, pp. 207–217, 2004.

[ALT 86] ALTOMARE R.E., GHOSSI P., "An analysis of residence time distribution in a twin screw cooking extruder", *Biotechnology Progress*, vol. 2, pp. 157–163, 1986.

[AVA 96] AVALOSSE T., "Numerical simulation of distributive mixing in 3-D flows", *Macromolecular Symposia*, vol. 112, pp. 91–98, 1996.

[BAH 11] BAHLOUL W., ODDES O., BOUNOR-LEGARÉ V., *et al.*, "Reactive extrusion processing of polypropylene/TiO_2 nanocomposites by in-situ synthesis of the nanofillers", *AIChE Journal*, vol. 57, pp. 2174–2184, 2011.

[BAR 91] BARRÈS C., VERGNES B., TAYEB J., "An improved thermal model for the solid conveying section of a twin-screw cooker-extruder", *Journal of Food Engineering*, vol. 15, pp. 167–185, 1991.

[BAR 92] BARRÈS C., VERGNES B., TAYEB J., "Computation of residence time and energy distribution in the reverse screw element of a twin-screw cooker-extruder", *Journal of Food Engineering*, vol. 16, pp. 215–237, 1992.

[BAW 95] BAWISKAR S., WHITE J.L., "Solids conveying and melting in a starve fed self-wiping co-rotating twin screw extruder", *International Polymer Processing*, vol. 10, pp. 105–110, 1995.

[BAW 98] BAWISKAR S., WHITE J.L., "Melting model for modular self wiping co-rotating twin screw extruders", *Polymer Engineering & Science*, vol. 38, pp. 727–740, 1998.

[BER 97] BERTRAND F., TANGUY P.A., THIBAULT F., "A three-dimensional fictitious domain method for incompressible fluid flow problems", *International Journal for Numerical Methods in Fluids*, vol. 25, pp. 719–736, 1997.

[BER 98a] BERZIN F., Etude expérimentale et modélisation d'une opération d'extrusion réactive, PhD Thesis, Paris School of Mines, 1998.

[BER 98b] BERZIN F., VERGNES B., "Transesterification of ethylene acetate copolymer in a twin screw extruder", *International Polymer Processing.*, vol. 13, pp. 13–22, 1998.

[BER 00] BERZIN F., VERGNES B., DUFOSSÉ P., *et al.*, "Modelling of peroxide initiated controlled degradation of polypropylene in a twin screw extruder", *Polymer Engineering & Science*, vol. 40, pp. 344–356, 2000.

[BER 01] BERZIN F., VERGNES B., DELAMARE L., "Rheological behavior of controlled-rheology polypropylenes obtained by peroxide promoted degradation during extrusion: comparison between homopolymer and copolymer", *Journal of Applied Polymer Science*, vol. 80, pp. 1243–1252, 2001.

[BER 02] BERZIN F., VERGNES B., LAFLEUR P.G., *et al.*, "A theoretical approach of solid filler dispersion in a twin screw extruder", *Polymer Engineering & Science*, vol. 42, pp. 473–481, 2002.

[BER 03] BERZIN F., HU G.H., "Procédés d'extrusion reactive", *Techniques de l'Ingénieur*, vol. AM3654, pp. 1–16, 2003.

[BER 06] BERZIN F., VERGNES B., CANEVAROLO S.V., *et al.*, "Evolution of peroxide-induced degradation of polypropylene along a twin-screw extruder: experimental data and theoretical predictions", *Journal of Applied Polymer Science*, vol. 99, pp. 2082–2090, 2006.

[BER 07a] BERZIN F., TARA A., TIGHZERT L., *et al.*, "Computation of starch cationization performances by twin screw extrusion", *Polymer Engineering & Science*, vol. 47, pp. 112–119, 2007.

[BER 07b] BERZIN F., TARA A., VERGNES B., "Optimization and scale-up of thermoplastic wheat starch cationization in a twin screw extruder", *Polymer Engineering & Science*, vol. 47, pp. 814–823, 2007.

[BLE 02] BLEIMAN P., BULTERS M., ELEMANS P., *et al.*, "Melting model for co-rotating twin screw extruder", *SPE ANTEC Technical Papers*, vol. 1, pp. 355–359, 2002.

[BOH 96a] BOHIN F., FEKE D.L., MANAS-ZLOCZOWER I., "Analysis of power requirements and dispersion quality in batch compounding using a dispersion model for single agglomerates", *Rubber Chemistry and Technology*, vol. 69, pp. 1–7, 1996.

[BOH 96b] BOHIN F., MANAS-ZLOCZOWER I., FEKE D.L., "Kinetics of dispersion for sparse agglomerates in simple shear flows: ppplication to silica agglomerates in silicone polymers", *Chemical Engineering Science*, vol. 51, pp. 5193–5204, 1996.

[BOO 78] BOOY M.L., "Geometry of fully wiped twin screw equipments", *Polymer Engineering & Science,* vol. 18, pp. 973–984, 1978.

[BOO 80] BOOY M.L., "Isothermal flow of viscous liquids in corotating twin screw devices", *Polymer Engineering & Science*, vol. 20, pp. 1220–1228, 1980.

[BRA 00] BRAVO V.L., HRYMAK A.N., WRIGHT J.D., "Numerical simulation of pressure and velocity profiles in kneading elements of a co-rotating twin screw extruder", *Polymer Engineering & Science*, vol. 40, pp. 525–541, 2000.

[BRA 04] BRAVO V.L., HRYMAK A.N., WRIGHT J.D., "Study of particle trajectories, residence times and flow behavior in kneading discs of intermeshing co-rotating twin-screw extruders", *Polymer Engineering & Science*, vol. 44, pp. 779–793, 2004.

[CAR 59] CARSLAW H.S., JAEGER J.C., *Conduction of Heat in Solids*, Oxford University Press, Oxford, 1959.

[CAR 93] CARROT C., GUILLET J., MAY J.F., *et al.*, "Modeling of the conveying of solid polymer in the feeding zone of intermeshing co-rotating twin screw extruders", *Polymer Engineering & Science*, vol. 33, pp. 700–708, 1993.

[CAR 00] CARNEIRO O.S., COVAS J.A., VERGNES B., "Experimental and theoretical study of the twin screw extrusion of polypropylene", *Journal of Applied Polymer Science*, vol. 78, pp. 1419–1430, 2000.

[CAS 91] CASSAGNAU P., MIJANGOS C., MICHEL A., "An ultraviolet method for the determination of the residence time distribution in a twin screw extruder", *Polymer Engineering & Science*, vol. 31, pp. 772–778, 1991.

[CHA 95] CHAN H.T., DUFRESNE D.A., "Kneading block melting study", *SPE ANTEC Technical Papers*, pp. 302–311, 1995.

[CHA 02] CHALAMET Y., TAHA M., BERZIN F., *et al.*, "Carboxyl terminated polyamide 12 chain extension by reactive extrusion using a dioxazoline coupling agent. Part II: effects of extrusion conditions", *Polymer Engineering & Science*, vol. 42, pp. 2317–2327, 2002.

[CHE 91] CHESTERS A.K., "The modeling of coalescence processes in fluid liquid dispersions – a review of current understanding", *Chemical Engineering Research and Design*, vol. 69, pp. 259–270, 1991.

[CHE 93] CHEN L.Q., PAN Z.Q., HU G.H., "Residence time distribution in screw extruders", *AIChE Journal*, vol. 39, pp. 1455–1464, 1993.

[CHE 94] CHEN Z., WHITE J.L., "Simulation of non-isothermal flow in twin screw extrusion", *International Polymer Processing*, vol. 9, pp. 310–318, 1994.

[CHE 97] CHENG H.F., MANAS-ZLOCZOWER I., "Study of mixing efficiency in kneading discs of co-rotating twin screw extruders", *Polymer Engineering & Science*, vol. 37, pp. 1082–1090, 1997.

[DAN 53] DANCKWERTS P.V., "Continuous flow systems: distribution of residence times", *Chemical Engineering Science*, vol. 2, pp. 1–13, 1953.

[DEL 93] DELLA VALLE G., BARRÈS C., PLEWA J., *et al.*, "Computer simulation of starchy products transformation by twin-screw extrusion", *Journal of Food Engineering*, vol. 19, pp. 1–31, 1993.

[DEL 95] DELAMARE L., Evolution morphologique d'un mélange de polymères incompatibles dans une extrudeuse bi-vis corotative, PhD Thesis, Paris School of Mines, 1995.

[DEL 96] DELAMARE L., VERGNES B., "Computation of the morphological changes of a polymer blend along a twin screw extruder", *Polymer Engineering & Science*, vol. 36, pp. 1685–1693, 1996.

[DEL 11] DELLA VALLE G., BERZIN F., VERGNES B., "Modelling of twin screw extrusion process for food products design and process optimization", in MASKAN M., ALTAN A. (eds), *Advances in Food Extrusion Technology*, CRC Press, Boca Raton, pp. 327–354, 2011.

[DEN 80] DENSON C.D., HWANG B.K., "The influence of the axial pressure gradient on flow rate for Newtonian liquids in a self-wiping co-rotating twin screw extruder", *Polymer Engineering & Science*, vol. 20, pp. 965–971, 1980.

[DOM 93] DOMSCHKE T., JUSTUS C., "Heat-transfer in single and twin-screw extruders", *International Polymer Processing*, vol. 8, pp. 294–307, 1993.

[DOM 12a] DOMENECH T., Structure et propriétés de nanocomposites polypropylène/argile lamellaire préparés par mélange à l'état fondu, PhD Thesis, Paris School of Mines, 2012.

[DOM 12b] DOMENECH T., BONNY S., PEUVREL-DISDIER E., *et al.*, "Influence of twin-screw processing conditions on structure and properties of polypropylene – organoclay nanocomposites", *International Polymer Processing*, vol. 27, pp. 517–526, 2012.

[DOM 13] DOMENECH T., PEUVREL-DISDIER E., VERGNES B., "The importance of specific mechanical energy during twin screw extrusion of organoclay based polypropylene nanocomposites", *Composites Science and Technology*, vol. 75, pp. 7–14, 2013.

[DUR 12] DURIN A., Modélisation du mélange de polymères chargés et de la casse de fibres rigides en extrusion bivis, PhD Thesis, Paris School of Mines, 2012.

[DUR 13] DURIN A., DE MICHELI P., VILLE J., *et al.*, "A matricial approach of fiber breakage in twin-screw extrusion of glass fibers reinforced thermoplastics", *Composites Part A.*, vol. 48, pp. 47–56, 2013.

[ELM 86] ELMENDORP J.J., VAN DER VEGT A.K., "A study on polymer blending microrheology. Part IV: the influence of coalescence on blend morphology origination", *Polymer Engineering & Science*, vol. 26, pp. 1332–1338, 1986.

[FUK 87] FUKUSHIMA Y., INAGAKI S., "Synthesis of an intercalated compound of montmorillonite and 6-polyamide", *Journal of Inclusion Phenomena*, vol. 5, pp. 473–482, 1987.

[FUK 94] FUKUOKA T., MIN K., "Numerical non-isothermal flow analysis of non-Newtonian fluid in a non-intermeshing counter-rotating twin screw extruder", *Polymer Engineering & Science*, vol. 34, pp. 1033–1046, 1994.

[FUN 02] FUNATSU K., KIHARA S.I., MIYAZAKI M., *et al.*, "3-D numerical analysis on the mixing performance for assemblies with filled zone of right-handed and left-handed double-flighted screws and kneading blocks in twin-screw extruders", *Polymer Engineering & Science*, vol. 42, pp. 707–723, 2002.

[GAL 02] GALAKTIONOV O.S., ANDERSON P.D., PETERS G.W.M., *et al.*, "Mapping approach for 3D laminar mixing simulations: application to industrial flows", *International Journal for Numerical Methods in Fluids*, vol. 40, pp. 345–351, 2002.

[GAO 99] GAO J., WALSH G.C., BIGIO D., *et al.*, "Residence-time distribution model for twin-screw extruders", *AIChE Journal*, vol. 45, pp. 2541–2549, 1999.

[GAS 02] GASPAR-CUNHA A., POULESQUEN A., VERGNES B., *et al.*, "Optimization of processing conditions for polymer twin-screw extrusion", *International Polymer Processing*, vol. 17, pp. 201–213, 2002.

[GAS 05] GASPAR-CUNHA A., VERGNES B., COVAS J.A., "Defining the configuration of co-rotating twin-screw extruders with multi-objective evolutionary algorithms", *Polymer Engineering & Science*, vol. 45, pp. 1169–1173, 2005.

[GAS 11] GASPAR-CUNHA A., COVAS J.A., VERGNES B., *et al.*, "Reactive extrusion-optimization of representative processes", in GASPAR-CUNHA A., COVAS J.A. (eds), *Optimization of Polymer Processing*, Nova Science Publishers, New York, pp. 115–143, 2011.

[GIM 00] GIMENEZ J., BOUDRIS M., CASSAGNAU P., *et al.*, "Control of bulk ε-caprolactone polymerization in a twin screw extruder", *Polymer Reaction Engineering*, vol. 8, pp. 135–157, 2000.

[GOG 98] GOGOS C.C., TADMOR Z., KIM M.H., "Melting phenomena and mechanisms in polymer processing equipment", *Advances in Polymer Technology*, vol. 17, pp. 285–305, 1998.

[GRA 82] GRACE H.P., "Dispersion phenomena in high-viscosity immiscible fluid systems and application of static mixers as dispersion devices in such systems", *Chemical Engineering Communications*, vol. 14, pp. 225–277, 1982.

[GUO 03] GUO Y., ZHU F., "Study of the different flow patterns in the melting section of a co-rotating twin screw extruder", *Polymer Engineering & Science*, vol. 43, pp. 306–316, 2003.

[HUA 93] HUANG H.X., PENG Y.C., "Theoretical modeling of dispersive melting mechanism of polymers in an extruder", *Advances in Polymer Technology*, vol. 12, pp. 343–352, 1993.

[HUN 95] HUNEAULT M.A., SHI Z.H., UTRACKI L.A., "Development of polymer blend morphology during compounding in a twin-screw extruder. Part IV: a new computational model with coalescence", *Polymer Engineering & Science*, vol. 35, pp. 115–127, 1995.

[INC 11] INCEOGLU F., VILLE J., GHAMRI N., *et al.*, "Correlation between processing conditions and fiber breakage during compounding of glass fiber-reinforced polyamide", *Polymer Composites*, vol. 32, pp. 1842–1850, 2011.

[ISH 00] ISHIKAWA T., KIHARA S.I., FUNATSU K., "3-D numerical simulations of nonisothermal flow in rotating twin screw extruders", *Polymer Engineering & Science*, vol. 40, pp. 357–364, 2000.

[ISH 01] ISHIKAWA T., KIHARA S.I., FUNATSU K., "3-D non-isothermal flow field analysis and mixing performance evaluation of kneading blocks in a corotating twin screw extruder", *Polymer Engineering & Science*, vol. 41, pp. 840–849, 2001.

[JAN 78] JANSSEN L.P.B.M., *Twin Screw Extrusion*, Elsevier, Amsterdam, 1978.

[JON 95] JONGBLOED H.A., KIEWIET J.A., VANDIJK J.H., *et al.*, "The self-wiping co-rotating twin-screw extruder as a polymerization reactor for methacrylates", *Polymer Engineering & Science*, vol. 35, pp. 1569–1579, 1995.

[KAJ 96] KAJIWARA T., NAGASHIMA Y., NAKANO Y., *et al.*, "Numerical study of twin-screw extruders by three-dimensional flow analysis – development of analysis technique and evaluation of mixing performances for full flight screws", *Polymer Engineering & Science*, vol. 36, pp. 2142–2152, 1996.

[KAL 88] KALYON D.M., GOTSIS A.D., YILMAZER U., *et al.*, "Development of experimental techniques and simulation methods to analyze mixing in co-rotating twin screw extrusion", *Advances in Polymer Technology*, vol. 8, pp. 337–353, 1988.

[KAL 07] KALYON D.M., MALIK M., "An integrated approach for numerical analysis of coupled flow and heat transfer in co-rotating twin screw extruders", *International Polymer Processing*, vol. 22, pp. 293–302, 2007.

[KRE 94] KRELL M.J., BRANDOLIN A., VALLES E.M., "Controlled rheology polypropylenes – an improved model with experimental validation for the single-screw extruder process", *Polymer Reaction Engineering*, vol. 2, pp. 389–408, 1994.

[KRI 01] KRISHNAMOORTI R., REN J.X., SILVA A.S., "Shear response of layered silicate nanocomposites", *Journal of Chemical Physics*, vol. 114, pp. 4968–4973, 2001.

[LAR 88] LARSEN H., JONES A., "Heat transfer in twin screw extruders", *SPE Technical Papers*, vol. 34, pp. 67–70, 1988.

[LER 05] LERTWIMOLNUN W., VERGNES B., "Influence of compatibilizer and processing conditions on the dispersion of nanoclay in a polypropylene matrix", *Polymer*, vol. 46, pp. 3462–3471, 2005.

[LER 06a] LERTWIMOLNUN W., Réalisation de nanocomposites polypropylène/argile par extrusion bivis, PhD Thesis, Paris School of Mines, 2006.

[LER 06b] LERTWIMOLNUN W., VERGNES B., "Effect of processing conditions on the formation of polypropylene/organoclay nanocomposites in a twin screw extruder", *Polymer Engineering & Science*, vol. 46, pp. 314–323, 2006.

[LER 07] LERTWIMOLNUN W., VERGNES B., "Influence of screw profile and extrusion conditions on the microstructure of polypropylene/organoclay nanocomposites", *Polymer Engineering & Science*, vol. 47, pp. 2100–2109, 2007.

[LIU 01] LIU T., WONG A.C.Y., ZHU F., "Prediction of screw length required for polymer melting and melting characteristics", *International Polymer Processing*, vol. 16, pp. 113–123, 2001.

[LOR 92] LORD R., "On the instability of cylindrical fluid surfaces", *Philosophical Magazine Series 5*, vol. 34, pp. 177–180, 1892.

[LOZ 02] LOZANO T., LAFLEUR P.G., GRMELA M., "A chemical model for the dispersion of fillers in a polymeric matrix", *Canadian Journal of Chemical Engineering*, vol. 80, pp. 1135–1142, 2002.

[LOZ 03] LOZANO T., LAFLEUR P.G., GRMELA M., *et al.*, "Modeling filler dispersion along a twin screw extruder", *International Polymer Processing*, vol. 18, pp. 12–19, 2003.

[MAN 98] MANAS-ZLOCZOWER I., CHENG H.F., "Influence of design on mixing efficiency in the kneading disc region of co-rotating twin screw extruders", *Journal of Reinforced Plastics and Composites*, vol. 17, pp. 1076–1086, 1998.

[MAN 09] MANAS-ZLOCZOWER I., *Mixing and Compounding of Polymers: Theory and Practice*, 2nd ed., Hanser, Munich, 2009.

[MEI 88a] MEIJER H.E.H., ELEMANS P.H.M., "The modeling of continuous mixers. Part 1: the corotating twin-screw extruder", *Polymer Engineering & Science*, vol. 28, pp. 275–290, 1988.

[MEI 88b] MEIJER H.E.H., LEMSTRA P.J., ELEMANS P.H.M., "Structured polymer blends", *Makromolekular Chemie, Macromolecular Symposium.*, vol. 16, pp. 113–135, 1988.

[MOU 00] MOURS M., REINELT D., WAGNER H.G., *et al.*, "Melt conveying in co-rotating twin screw extruders – experiment and numerical simulation", *International Polymer Processing*, vol. 15, pp. 124–132, 2000.

[MOU 11] MOUAZEN M., POULESQUEN A., VERGNES B., "Influence of extrusion conditions on the rheological behavior of nuclear bituminized waste products (BWP)", *International Polymer Processing*, vol. 26, pp. 569–579, 2011.

[NOÉ 92] NOÉ I., Etude des écoulements de polymères dans une extrudeuse bivis co-rotative, PhD Thesis, Paris School of Mines, 1992.

[OBE 94] OBERLEHNER K., CASSAGNAU P., MICHEL A., "Local residence time distribution in a twin-screw extruder", *Chemical Engineering Science*, vol. 49, pp. 3897–3907, 1994.

[OKA 87] OKADA A., KAWASUMI M., KURAUCHI T., *et al.*, "Synthesis and characterization of a nylon 6-clay hybrid", *American Chemical Society, Polymer Preprints*, vol. 28, pp. 447–448, 1987.

[PAN 82] PANDYA J.D., SPIELMAN L.A., "Floc breakage in agitated suspensions: theory and data processing strategy", *Journal of Colloid and Interface Science*, vol. 90, pp. 517–531, 1982.

[POT 90] POTENTE H., ANSAHL J., WITTEMEIER R., "Throughput characteristics of tightly intermeshing co-rotating twin screw extruders", *International Polymer Processing*, vol. 5, pp. 208–216, 1990.

[POT 94] POTENTE H., ANSAHL J., KLARHOLZ B., "Design of tightly intermeshing co-rotating twin screw extruders", *International Polymer Processing*, vol. 9, pp. 11–25, 1994.

[POT 96a] POTENTE H., MELISCH U., PALLUCH K.P., "A physico-mathematical model for solids conveying in co-rotating twin screw extruders", *International Polymer Processing*, vol. 11, pp. 29–41, 1996.

[POT 96b] POTENTE H., MELISCH U., "Theoretical and experimental investigations of the melting of pellets in co-rotating twin screw extruders", *International Polymer Processing*, vol. 11, pp. 101–108, 1996.

[POU 01a] POULESQUEN A., Contribution à la modélisation de l'extrusion réactive en machine bivis, PhD Thesis, Paris School of Mines, 2001.

[POU 01b] POULESQUEN A., VERGNES B., CASSAGNAU P., *et al.*, "Polymerization of ε-caprolactone in a twin screw extruder: experimental study and modelling", *International Polymer Processing*, vol. 16, pp. 31–38, 2001.

[POU 03a] POULESQUEN A., VERGNES B., "A study of residence time distribution in co-rotating twin screw extruders. Part I: theoretical modelling", *Polymer Engineering & Science*, vol. 43, pp. 1841–1848, 2003.

[POU 03b] POULESQUEN A., VERGNES B., CASSAGNAU P., *et al.*, "A study of residence time distribution in co-rotating twin screw extruders. Part II: experimental validation", *Polymer Engineering & Science*, vol. 43, pp. 1849–1862, 2003.

[QUI 00] QIAN B.N., GOGOS C.G., "The importance of plastic energy dissipation (PED) to the heating and melting of polymer particulates in intermeshing co-rotating twin-screw extruders", *Advances in Polymer Technology*, vol. 19, pp. 287–299, 2000.

[RAU 01] RAUWENDAAL C., *Polymer Extrusion*, 4th ed., Hanser, Munich, 2001.

[RED 99] REDL A., MOREL M.H., BONICEL J., *et al.*, "Extrusion of wheat gluten plasticized with glycerol: influence of process conditions on flow behavior, rheological properties and molecular size distribution", *Cereal Chemistry*, vol. 76, pp. 361–370, 1999.

[RWE 90] RWEI S.P., MANAS-ZLOCZOWER I., FEKE D., "Observation of carbon-black agglomerate dispersion in simple shear flows", *Polymer Engineering & Science*, vol. 30, pp. 701–706, 1990.

[RWE 92] RWEI S.P., MANAS-ZLOCZOWER I., FEKE D., "Analysis of dispersion of carbon-black in polymeric melts and its effect on compound properties", *Polymer Engineering & Science*, vol. 32, pp. 130–135, 1992.

[SZY 87] SZYDLOWSKI W., BRZOSKOWSKI R., WHITE J.L., "Modeling flow in an intermeshing co-rotating twin screw extruder: flow in kneading discs", *International Polymer Processing*, vol. 1, pp. 207–214, 1987.

[SZY 88] SZYDLOWSKI W., WHITE J.L., "A non-Newtonian model of flow in a kneading disk region of a modular intermeshing corotating twin screw extruder", *Journal of Non-Newtonian Fluid Mechanics*, vol. 28, pp. 29–46, 1988.

[TAR 04] TARA A., BERZIN F., TIGHZERT L., *et al.*, "Preparation of cationic wheat starch by twin-screw reactive extrusion", *Journal of Applied Polymer Science*, vol. 93, pp. 201–208, 2004.

[TAY 32] TAYLOR G.I., "The viscosity of a fluid containing small drops of another fluid", *Proceedings of the Royal Society*, vol. A138, pp. 41–48, 1932.

[TAY 88] TAYEB J., VERGNES B., DELLA VALLE G., "Theoretical computation of the isothermal flow through the reverse screw element of a twin-screw extrusion cooker", *Journal of Food Science*, vol. 53, pp. 616–625, 1988.

[TAY 89] TAYEB J., VERGNES B., DELLA VALLE G., "A basic model for a twin-screw extruder", *Journal of Food Science*, vol. 54, pp. 1047–1056, 1989.

[TAY 92] TAYEB J., DELLA VALLE G., BARRES C., *et al.*, "Simulation of transport phenomena in twin-screw extruders", in KOKINI J.L., HO C.T., KARWE M.V. (eds), *Food Extrusion, Science and Technologies*, Marcel Dekker, New York, 1992.

[TEI 11] TEIXEIRA C., COVAS J.A., BERZIN F., *et al.*, "Applications of evolutionary algorithms for the definition of optimal twin-screw extruder configuration for starch cationisation", *Polymer Engineering & Science*, vol. 51, pp. 330–340, 2011.

[TOD 75] TODD D.B., "Residence time distribution in twin-screw extruders", *Polymer Engineering & Science*, vol. 15, pp. 437–443, 1975.

[TOD 88] TODD D.B., "Heat transfer in twin screw extruders", *SPE ANTEC Technical Papers*, vol. 34, pp. 54–58, 1988.

[TOD 93] TODD D.B., "Melting of plastics in kneading blocks", *International Polymer Processing*, vol. 8, pp. 113–118, 1993.

[TOM 35] TOMOTIKA S., "On the instability of a cylindrical thread of a viscous liquid surrounded by another viscous fluid", *Proceedings of the Royal Society*, vol. A150, pp. 322–337, 1935.

[TOM 36] TOMOTIKA S., "Breaking up of a drop of viscous liquid immersed in another viscous fluid which is extending at a uniform rate", *Proceedings of the Royal Society*, vol. A153, pp. 302–318, 1936.

[TZO 88] TZOGANAKIS C., VLACHOPOULOS J., HAMIELEC A.E., "Modeling of the peroxide degradation of polypropylene", *International Polymer Processing*, vol. 3, pp. 141–150, 1988.

[UTR 02] UTRACKI L.A., *Polymer Blends Handbook*, Kluwer, Dordrecht, 2002.

[VAL 07] VALETTE R., BRUCHON J., DIGONNET H., *et al.*, "Méthodes d'interaction fluide-structure pour la simulation multi-échelles des procédés de mélange", *Mécanique et Industrie*, vol. 8, pp. 251–258, 2007.

[VAL 09] VALETTE R., COUPEZ T., DAVID C., *et al.*, "A direct 3D numerical simulation code for extrusion and mixing processes", *International Polymer Processing*, vol. 24, pp. 141–147, 2009.

[VER 98] VERGNES B., DELLA VALLE G., DELAMARE L., "A global computer software for polymer flows in co-rotating twin screw extruders", *Polymer Engineering & Science*, vol. 38, pp. 1781–1792, 1998.

[VER 01a] VERGNES B., CHAPET M., "Procédés d'extrusion bivis", *Techniques de l'Ingénieur*, vol. AM3653, pp. 1–23, 2001.

[VER 01b] VERGNES B., SOUVETON G., DELACOUR M.L., *et al.*, "Experimental and theoretical study of polymer melting in a co-rotating twin screw extruder", *International Polymer Processing*, vol. 16, pp. 351–362, 2001.

[VER 06] VERGNES B., BERZIN F., "Modelling of reactive systems in twin screw extrusion: challenges and applications", *Comptes Rendus Chimie*, vol. 9, pp. 1409–1418, 2006.

[VER 08] VERGNES B., LERTWIMOLNUN W., "Impact of processing conditions on the morphology, structure and properties of polymer-organoclay nanocomposites", in SABU T., GENNADY Z. (eds), *Polymer Nanocomposite Research Advances*, Nova Science Publishers, New York, 2008.

[VER 11] VERGNES B., "The use of apparent yield stress to characterize exfoliation in polymer nanocomposites", *International Polymer Processing*, vol. 26, pp. 229–232, 2011.

[VIL 13] VILLE J., INCEOGLU F., GHAMRI N., *et al.*, "Influence of extrusion conditions on fiber breakage along the screw profile during twin screw compounding of glass fiber-reinforced polyamide", *International Polymer Processing*, vol. 28, pp. 49–57, 2013.

[WAN 89] WANG Y., WHITE J.L., SZYDLOWSKI W., "Flow in a modular intermeshing co-rotating twin screw extruder", *International Polymer Processing*, vol. 4, pp. 262–269, 1989.

[WAR 96] WARD N.J., EDWARDS H.G.M., JOHNSON A.F., *et al.*, "Application of Raman spectroscopy for determining residence time distribution in extruder reactors", *Applied Spectroscopy*, vol. 50, pp. 812–815, 1996.

[WER 76] WERNER H., Das Betriebsverhalten der Zweiwelligen Knetscheiben Schnecken-presse von Type ZSK bei der Verarbeitung von hochviskosen Flüssigkeiten, PhD Thesis, University of Munich, 1976.

[WHI 87] WHITE J.L., SZYDLOWSKI W., MIN K., *et al.*, "Twin screw extruders: development of technology and analysis of flow", *Advances in Polymer Technology*, vol. 7, pp. 295–332, 1987.

[WHI 94] WHITE J.L., CHEN Z., "Simulation of non-isothermal flow in modular co-rotating twin-screw extrusion", *Polymer Engineering & Science*, vol. 34, pp. 229–237, 1994.

[WHI 01] WHITE J.L., CORAN A.Y., MOET A., *Polymer Mixing: Technology and Engineering*, Hanser, Munich, 2001.

[WHI 10] WHITE J.L., KIM E.K., *Twin Screw Extrusion: Technology and Principle*, 2nd ed., Hanser, Munich, 2010.

[WOL 86] WOLF D., HOLIN N., WHITE D.H., "Residence time distribution in a commercial twin-screw extruder", *Polymer Engineering & Science*, vol. 26, pp. 640–646, 1986.

[WU 87] WU S., "Formation of dispersed phase in incompatible polymer blends: interfacial and rheological effects", *Polymer Engineering & Science*, vol. 27, pp. 335–343, 1987.

[WYM 75] WYMAN C.E., "Theoretical model for intermeshing twin screw extruders: axial velocity profile for shallow channels", *Polymer Engineering & Science*, vol. 15, pp. 606–611, 1975.

[XAN 92] XANTHOS M., *Reactive Extrusion: Principles and Practice*, Hanser, Munich, 1992.

[YAN 92] YANG H.H., MANAS-ZLOCZOWER I., "Flow field analysis of the kneading disk region in a co-rotating twin screw extruder", *Polymer Engineering & Science*, vol. 32, pp. 1411–1417, 1992.

[YOS 00] YOSHINAGA M., KATSUKI S., MIYAZAKI M., *et al.*, "Mixing mechanism of three-tip kneading block in twin screw extruders", *Polymer Engineering & Science*, vol. 40, pp. 168–178, 2000.

[ZHU 99] ZHU L., GENG X., "Visual research of melting mechanism of polymer pellets in intermeshing co-rotating twin screw extrusion", *SPE ANTEC Technical Papers*, vol. 1, pp. 370–374, 1999.

[ZHU 01] ZHU L, NAHR K.A., GENG X., "Modeling of particle-dispersed melting mechanism and its application in co-rotating twin screw extrusion", *Journal of Polymer Science, Part B: Polymer Physics*, vol. 39, pp. 2461–2468, 2001.

Chapter 5

Profile Extrusion

Profile-shaped products manufactured by extrusion are grouped into the following two main categories:

– complex profiles: intended for applications in construction and automotive industries, etc.;

– tubular products: this includes tubes and pipes.

In this chapter, we will study both types of products, by reviewing the problems encountered and the theoretical approach for calculating the flows in the corresponding production tools.

5.1. Profile extrusion

5.1.1. *Different types of profile dies*

There are essentially two types of profile dies.

Plate dies, Figure 5.1(a), are composed of a cylindrical plate fixed to the extruder outlet where the desired shape is machined. These dies are simple and can be easily manufactured and modified. The abrupt decrease in flow section, however, causes two major drawbacks. On the one hand, the abrupt change of geometry can lead to stagnation of the material in the die, limiting the use of plate dies to thermally stable polymers. On the other hand, this design limits the extrusion flow rate and causes significant dimensional variations. Plate dies are generally used with rubber or plasticized PVC. They can, also, be used in the extrusion of rigid PVC in short productions.

Streamlined dies, Figure 5.1(b), allow a gradual section evolution from the extruder outlet to the shape of the desired product. In this type of die, the polymer is accelerated toward the end of the die until it arrives at the terminal zone of a constant cross-section, establishing a steady-state flow. The absence of abrupt contraction prevents certain inherent problems such as the stagnation of the material. Despite being more expensive, streamlined dies are more economical for mass production.

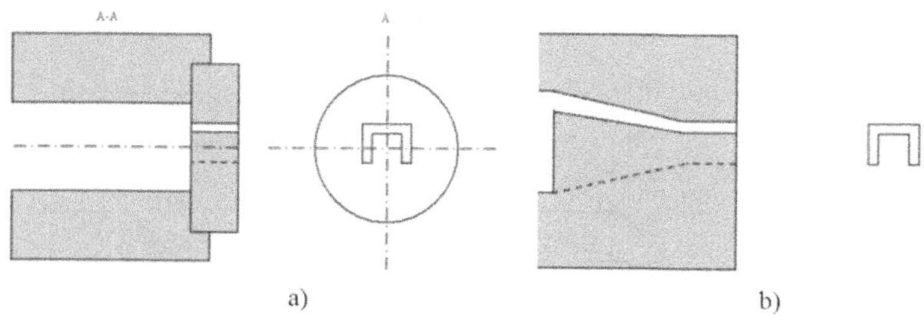

Figure 5.1. *The different types of dies: a) plate die and b) streamlined die*

In addition to these two extremes, there are many intermediate designs of dies, especially multi-plate dies, which are made up of a connecting plate, one or more transition plates and a terminal plate of a constant cross-section (Figure 5.2). The multi-plate dies are an excellent compromise between plate dies and streamlined dies, and this type of die is currently the most commonly used in the plastics industry.

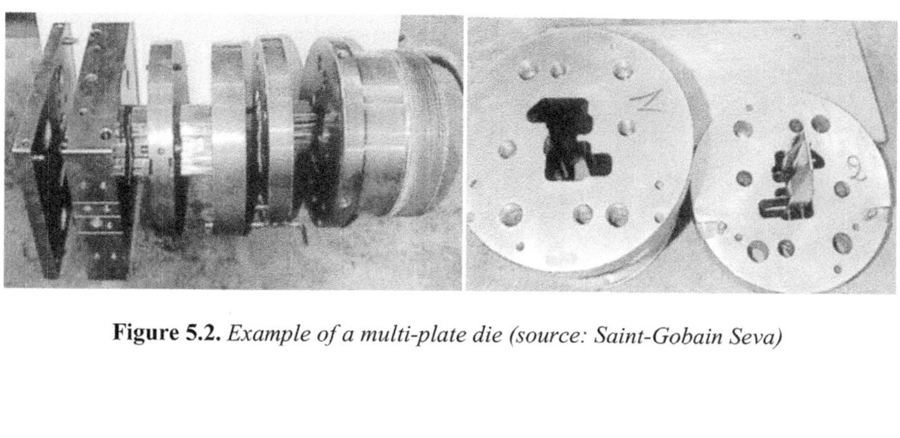

Figure 5.2. *Example of a multi-plate die (source: Saint-Gobain Seva)*

5.1.2. *Flow calculation inside profile dies*

Based on the FAN method (see Chapter 2), Huneault [HUN 92] developed a network model to predict flow inside a profile die with a varying geometry. The total width of the die is denoted by w and the length by L. This technique uses a conventional finite-difference mesh, which subdivides flow into N_x elements of dimension δ_x along the length and N_y elements of dimension δ_y along the width. Each element is identified by its coordinates (i, j) and has a depth $h(i, j)$. The pressure nodes $P(i, j)$, situated at the center of each element, are connected by resistivities $R_x(i, j)$ in the longitudinal direction and $R_y(i, j)$ in the transverse direction (Figure 5.3).

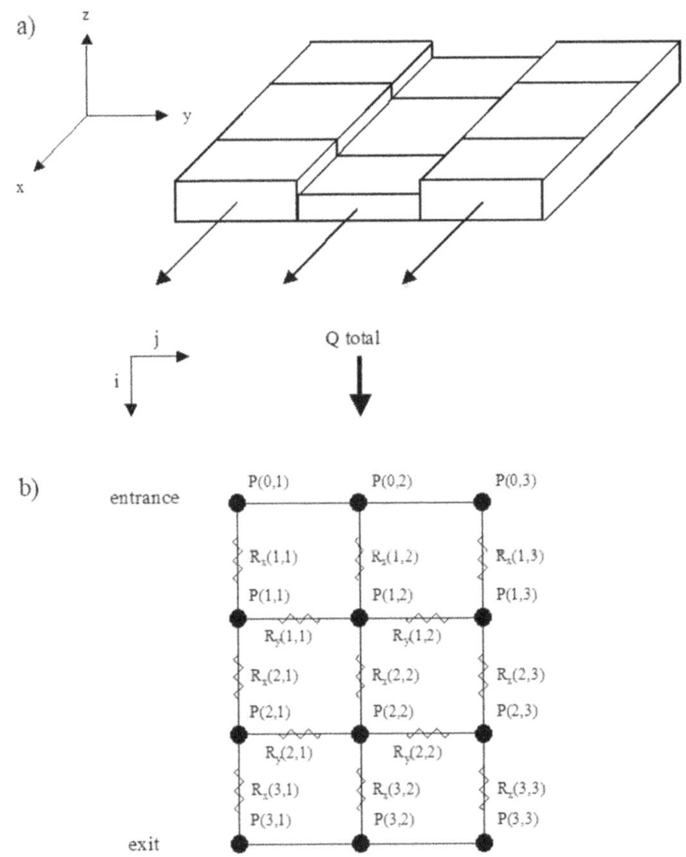

Figure 5.3. *a) Flow geometry and b) mesh representation for the network model*

In the case of a power-law fluid, these resitivities depend on the longitudinal flow:

$$R_x(i,j)=\frac{2K\delta_x}{h(i,j)}\left(\frac{2\left(2+\frac{1}{m}\right)}{\delta_y h(i,j)^2}\right)^m Q_x(i,j)^{m-1} \quad [5.1]$$

i.e.:

$$R_x(i,j)=\frac{2K\delta_x}{h(i,j)}\left(\frac{2\left(2+\frac{1}{m}\right)}{\delta_y h(i,j)^2}\right)^m \left(\frac{[P(i-1,j)-P(i,j)]}{R_x(i,j)}\right)^{m-1} \quad [5.2]$$

The transverse resistivity is estimated using the relation:

$$R_y(i,j)=\left(\frac{R_x(i,j)+R_x(i+1,j)+R_x(i,j+1)+R_x(i+1,j+1)}{4}\right)\left(\frac{\delta_y}{\delta_x}\right)^2 \quad [5.3]$$

This method allows us to calculate the pressure field, and the average temperature in the die. Subsequently, the flow distribution and the balancing of the flow in the die are calculated.

5.1.3. *Evaluation of post-extrusion phenomena*

5.1.3.1. *Post-extrusion dimensional changes*

At the die exit, many post-extrusion phenomena are capable of modifying the shape and dimensions of the extrudate, particularly extrudate swell, stretching and shrinkage, as shown in Figure 5.4.

5.1.3.1.1. Extrudate swell

Immediately upon exiting the die, the extrudate swells. This phenomenon is also known as the Barus effect and its physical interpretation is related to the balancing of the velocity profile and the viscoelasticity of the polymer.

In the case of a Newtonian fluid, there is always a swelling of 12% if extrusion occurs in a capillary die and 19% between the flat parallel plates. This phenomenon can be explained by the change from a parabolic velocity profile inside the die to a uniform one outside the die (flattening of the velocity profile).

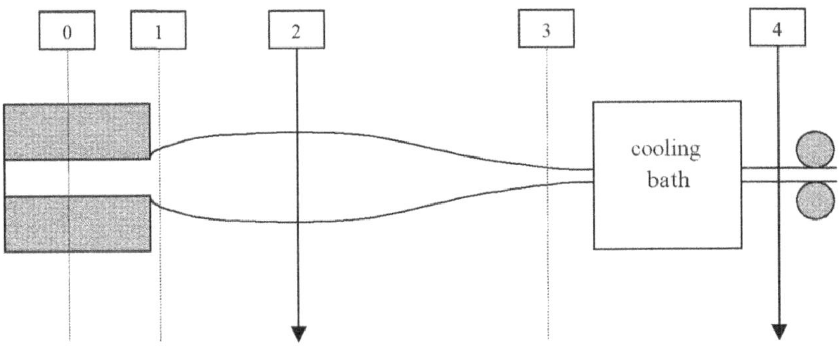

Figure 5.4. *Post-extrusion changes in extrudate shape: extrudate swell (2), stretching (3) and shrinkage (4)*

In the case of viscoelastic fluids, the velocity profile redistribution alone does not explain the significant extrudate swell observed (up to 300%). Since it was observed that the extrudate swell decreases with die length for a particular diameter (*L*/*d* ratio), and with flow rate, the concept of viscoelastic memory was introduced to account for this phenomenon. According to this, upon exiting, the polymer tends to recover the dimensions of the reservoir. In particular, in the case of short extrusion dies, most of the extrudate swell is explained by the viscoelastic memory effect.

In the case of long extrusion dies, the viscoelastic memory effect disappears. Swelling at the die exit is mainly due to deformation recovery. In fact, inside the die, the fluid is subjected to stress. At the die exit, the stresses decrease and the shear and extensional strains recover. Recovery of the shear strain does not result in a change of shape, since it involves the parallel sliding of liquid layers in the flow direction. However, recovery of the extensional strain in the flow direction causes extrudate swell in the transverse direction.

Therefore, extrudate swell at the die exit seems to be caused by a combination of the three phenomena mentioned above: Newtonian contribution, viscoelastic memory effect and strain recovery.

The quantitative prediction of extrudate swell is particularly interesting both at a fundamental level and in practice. For this reason, several models have been suggested in the literature.

Tanner [TAN 70] proposed an analytical calculation of extrudate swell caused by strain recovery induced by normal stresses for a capillary of an infinite length:

$$\frac{D}{d} = 0.12 + \left(1 + \frac{1}{2}\left(\frac{N_1}{2\tau_p}\right)^2\right)^{\frac{1}{6}} \qquad [5.4]$$

where D is the diameter of the extrudate, d is the diameter of the die, N_1 is the first normal stress difference and τ_p is the wall shear stress. The first term on the right section of the equation accounts for the Newtonian contribution.

Mendelson *et al.* [MEN 71] proposed several models based on the theory of rubber elasticity. In particular, considering that the recoverable energy is accumulated by shear in the die, and that it is recovered at the die exit by elongational contraction, they proposed the following relation:

$$N_1 = \frac{2}{3}\tau_p\left(\left(\frac{D}{d}\right)^4 + \frac{2}{\left(\frac{D}{d}\right)^2} - 3\right)^{\frac{1}{2}} \qquad [5.5]$$

In equations [5.4] and [5.5], the $N_1/2\tau_p$ ratio is referred to as the Weissenberg number, which characterizes the elastic to viscous stress ratio. We have plotted, in Figure 5.5, the evolution of extrudate swell as a function of the Weissenberg number for both proposed expressions. We find that the differences are very significant, which shows that none of these theories is actually satisfactory, especially since determining normal stress difference is often difficult at high shear rates.

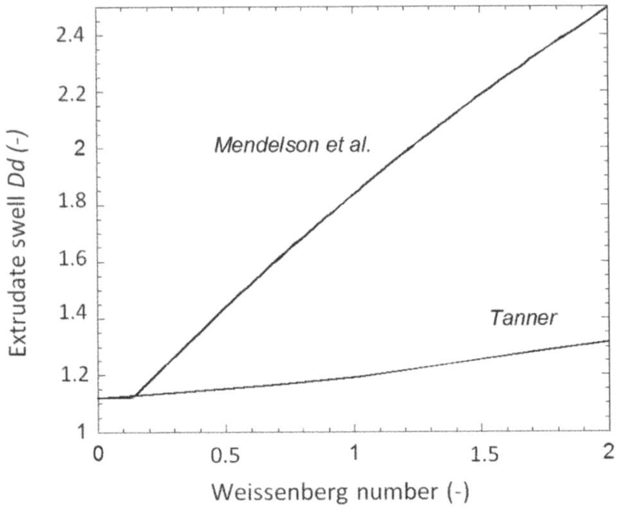

Figure 5.5. *Evolution of extrudate swell with the Weissenberg number for Tanner [TAN 70] and Mendelson et al. [MEN 71] models*

5.1.3.1.2. Stretching

To keep the extrudate horizontal as it passes through a cooling tank, and is then coiled, it is mechanically pulled at a constant velocity or stretching rate. This stretching causes a reduction in the extrudate cross-section and tends to preferentially reduce the thickness.

5.1.3.1.3. Shrinkage

The final post-extrusion phenomenon capable of affecting the extrudate dimensions is shrinkage. The cooling, and subsequent solidification of the profile, causes an isotropic reduction in the cross-section of the extrudate.

5.1.3.2. *Practical determination of post-extrusion dimensional changes*

An alternative to the theoretical modeling of post-extrusion phenomena involves evaluating the changes in shape in a semi-empirical manner. Huneault [HUN 92] proposed an experimental method to decouple the effects of the velocity profile redistribution, relaxation of normal stresses, mechanical stretching and shrinkage, upon solidification. This method is based on calculating the velocity profile inside the die, and analyzing the extrudates, obtained at points 2 and 4 in Figure 5.4. From simple mass balances, it is possible to assess the different factors affecting dimensional changes.

Flattening (thickness ratio), caused by redistribution of the velocity profile at the die exit, can be locally calculated by the following expression:

$$f_a(j) = \sqrt{\frac{V_x(i = N_x, j)}{\bar{V}(i = N_x)}} \qquad [5.6]$$

Extrudate swell, caused by deformation recovery, can be determined by analyzing the extrudates obtained at point 2. These samples are placed on a lubricated flat plate at a 45° angle and slowly cooled in air. The relaxation of stresses during slow cooling causes the extrudate to swell and then to shrink. In a capillary geometry, the swell factor can be calculated by:

$$f_g = \sqrt{\frac{A_2}{A_0}} = \left(\frac{\rho_f}{\rho_s}\right)^{\frac{1}{6}} \sqrt{\frac{M_2^*/L_2^*}{\rho_f\, A_0}} \qquad [5.7]$$

where ρ_f and ρ_s are the respective densities in the molten and solid states, and M and L are the mass and length of the sample. The index refers to the sampling points in Figure 5.4 and the asterisk indicates that the measurement was taken in the solid state after shrinkage.

The samples taken at point 4 underwent dimensional changes due to swelling, stretching and shrinkage caused by immersion in a water tray. It is possible to isolate the effect of mechanical stretching and assess the associated shape change factor by the relation:

$$f_e = \sqrt{\frac{A_3}{A_2}} = \left(\frac{\rho_s}{\rho_f}\right)^{\frac{1}{6}} \sqrt{\frac{M_4^*/L_4^*}{M_2^*/L_2^*}} \qquad [5.8]$$

Finally, shrinkage caused by solidification upon immersion of the extrudate in the water tray is isotropic and is equal to:

$$f_r = \sqrt{\frac{\rho_f}{\rho_s}} \qquad [5.9]$$

In the case of rectangular geometries, shape changes caused by swelling and stretching are not isotropic. Stevenson [STE 85] suggested using different dimensional variations for depth and width, and introduced the coefficients β_g and β_e, which are defined by:

$$\beta_g = \left(\frac{W_0}{W_2^*}\right)\left(\frac{\rho_f}{\rho_s}\right)^{\frac{1}{3}} \quad [5.10]$$

$$\beta_e = \left(\frac{W_2^*}{W_4^*}\right)\left(\frac{\rho_f}{\rho_s}\right)^{\frac{1}{6}} \quad [5.11]$$

We will later see how these shape change factors f_a, f_g, f_e, β_g, β_e and f_r can be used to calculate the final dimensions of the product.

Using the previous expressions, Beaumier [BEA 11] measured the extrudate swell of PVC for different processing conditions and different die geometries. In a capillary geometry, he showed that it was possible to precisely model the extrudate swell of PVC as a function of temperature T, apparent shear rate $\dot{\gamma}_a$ and residence time t_r by the following relationship:

$$f_g = A + BT + C\dot{\gamma}_a + D\ln t_r \quad [5.12]$$

with:

$$t_r = \frac{h * w * L}{Q} \quad [5.13]$$

where A, B, C and D are the constants specific to the polymer under consideration and are assessed by a statistical analysis of the experimental results (Table 5.1).

	Value	Uncertainty
A	+ 0.07	-
B	+ 0.0067	0.0005
C	−0.00055	0.00004
D	−0.24	0.01

Table 5.1. *Modeling the parameters for the extrudate swell of rigid PVC*

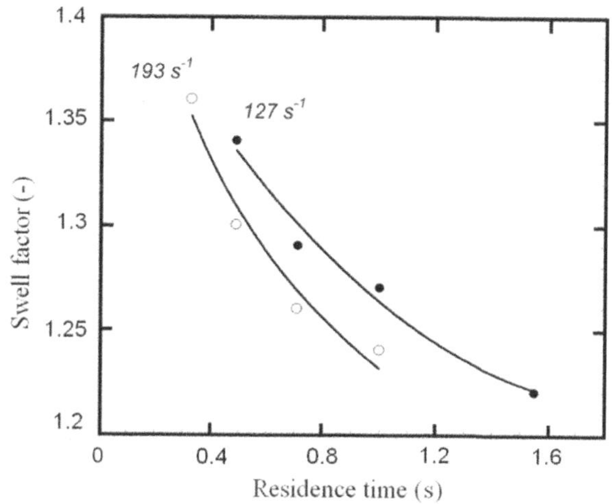

Figure 5.6. *Capillary extrudate swell of rigid PVC at 190°C as a function of residence time for two shear rates*

For rigid PVC at 190°C, Figure 5.6 shows the evolution of capillary extrudate swell as a function of the average shear rate, and residence time.

5.1.3.3. *Estimating the final dimensions of the shaped product*

The final dimensions of the shaped product can be calculated locally from the dimensions of the terminal zone in the die, the swelling, stretching and shrinkage factors. Thus, we obtain the following, for the depth and width:

$$\frac{H(j)}{h(j)} = \left[f_a \left(f_g(j) \beta_g(j) \right) \left(f_e(j) \beta_e(j) \right) f_r \right] \tag{5.14}$$

$$\frac{W(j)}{w(j)} = \left[f_a \left(\frac{f_g(j)}{\beta_g(j)} \right) \left(\frac{f_e(j)}{\beta_e(j)} \right) f_r \right] \tag{5.15}$$

where W, H are the dimensions of the profile and w, h are the dimensions of the die.

The complete flow calculation algorithm inside, and outside the die is shown in Figure 5.7.

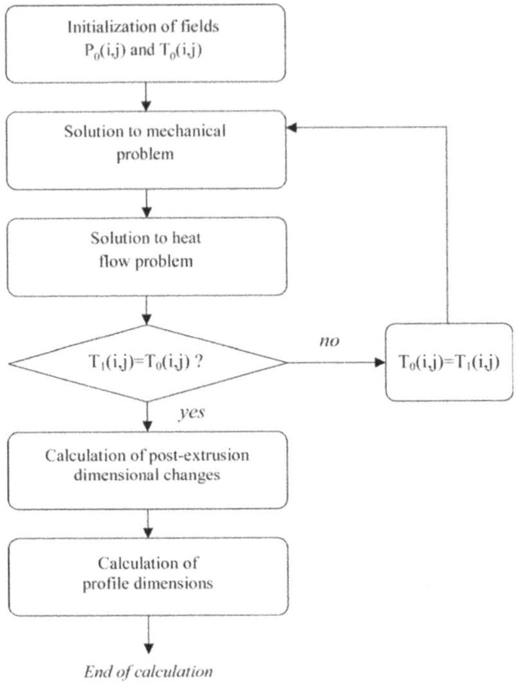

Figure 5.7. *Complete algorithm of the flow calculation*

5.1.4. *Design of extrusion dies*

Designing an extrusion die for the production of a profile of given dimensions is generally left to the experience of the designer. Considering the required dimensional accuracy and the physical phenomena occurring during the extrusion process, this task can prove to be difficult. Typically, a trial and error approach is adopted. Apart from the financial cost, such a process can significantly alter the production time. For these reasons, a reduction in trial and error cycles, even a modest one, is desirable.

5.1.4.1. *General design rules*

5.1.4.1.1. Profile design

Before an extrusion die is designed, the design of the profile itself must be assessed. Minor changes in geometry can in fact significantly affect the feasibility of the profile. In particular, the designer must respect the following considerations:

– use of the simplest design permitted by the final use;

– observation of uniformity of the wall thickness (the finest parts thickness should not be less than 20%–30% of the greatest thickness);

– use of rounded corners where possible (radius >0.5 mm).

Figure 5.8 illustrates some of the above listed suggestions for a complex product.

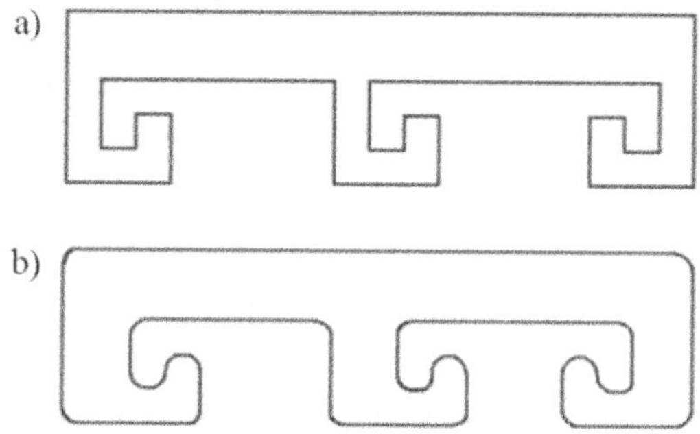

Figure 5.8. *Examples of product design: a) bad design and b) better design*

5.1.4.1.2. Design of extrusion dies

The design of extrusion dies complies with several general rules. One of the basic rules, in particular, involves avoiding abrupt changes in the geometry of the flow channel. Such changes can in fact lead to extrusion defects. For example, sudden contractions can create recirculation zones, capable of creating instabilities and degradation. Abrupt changes in the direction of flow, for example T-shaped geometries, can, also, lead to polymer degradation.

On the other hand, it is preferable that the polymer velocity increases gradually along the flow channel. This can occur by, gradually, reducing the cross-section of the die. In addition, a good design always requires a terminal zone with a constant cross-section along a length, enough to stabilize the velocity profile and to reduce the contribution of the viscoelastic memory effect to extrudate swell.

Finally, it is important for the die to be easily assembled and disassembled.

5.1.4.2. *Dimensioning of extrusion dies*

5.1.4.2.1. Dimensioning of the cross-section

The first problem associated with the design of extrusion dies involves optimizing the cross-section of the terminal plate depending on the dimensions of the profile to be produced. This problem, called the inverse problem, can be solved numerically using an iterative process [LEG 93].

In the case of the network model developed by Huneault [HUN 92], Figure 5.9 illustrates how the inverse problem can be solved. In this figure, H_p and W_p are the required dimensions of the profile, H and W are the calculated dimensions of the profile, and h and w are the dimensions of the die.

Huneault [HUN 92] and Beaumier [BEA 11] took a two-dimensional (2D) approach, which essentially applies to full profiles. In the case of more complex shapes, this method is more difficult to implement. For this reason, we, here, present a dimensioning technique that can be used for complex products. This technique is based on a 1D calculation of flow parameters and requires that the variation of extrudate swell with the shear rate be evaluated *a priori*.

Let us study the product in Figure 5.10. This product can be divided into a circular section and two simple rectangular sections. Let r be the radius of the circular section (1), h_2 and w_2 be the height and width of the rectangular section (2), and h_3 and w_3 be the height and width of the rectangular section (3). The extrusion volumetric flow rate is Q.

The method used is identical to that described in Figure 5.9. First, the dimensions of the die are assumed to be equal to those of the profile. Knowing the dimensions of each section, it is possible to calculate the local flow rate and thereby deduce the apparent shear rate. The associated extrudate swell is determined from the experimental data, and the swelling factors f_g can be calculated according to the geometry. The stretching and shrinkage factors are calculated in the same way as in section 5.1.3.2, and have the same value for the three sections.

For a rectangular geometry:

$$f_g(h) = \frac{H}{h} = \beta\left(\frac{D}{d}\right) \qquad [5.16]$$

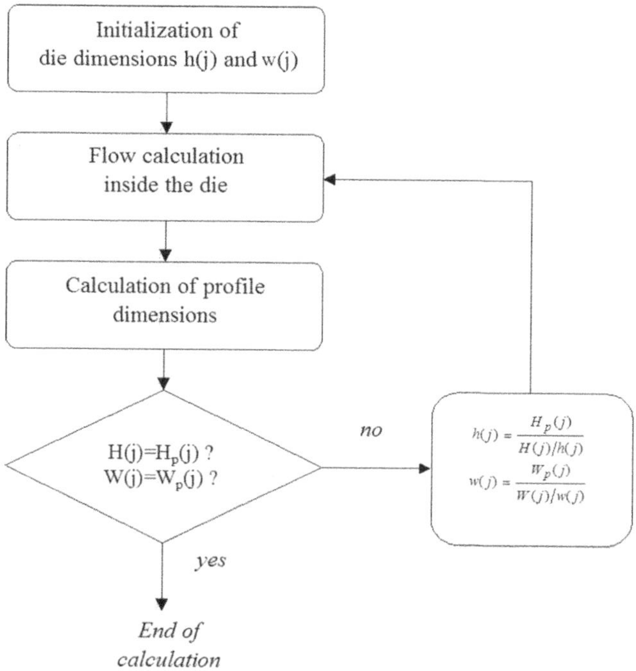

Figure 5.9. *Design process of profile dies from the network model*

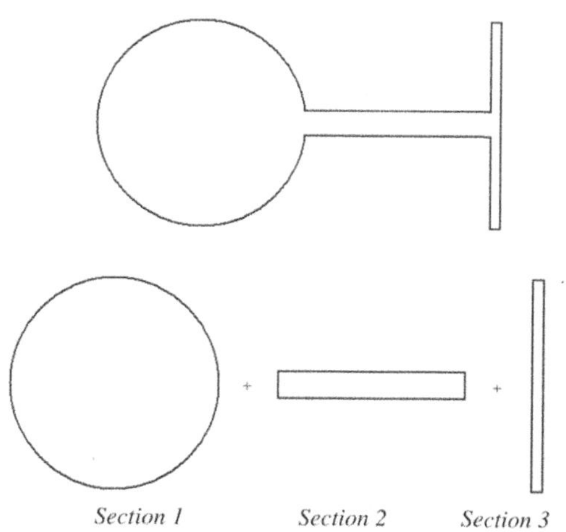

Figure 5.10. *Division of a complex product into simple sections*

$$f_g(w) = \frac{W}{w} = \frac{1}{\beta}\left(\frac{D}{d}\right) \qquad [5.17]$$

with:

$$\beta = \frac{\left(1 - \frac{w}{h}\right) + \sqrt{\left(\frac{w}{h} - 1\right)^2 + 4\left(\frac{w}{h}\right)\left(\frac{D}{d}\right)^2}}{2\left(\frac{D}{d}\right)} \qquad [5.18]$$

The dimensions of the die are recalculated using equations [5.14] and [5.15].

The equations related to the calculation of the flow parameters and shape change factors are given in Table 5.2.

	Section 1	Section 2	Section 3
Surface area (S_i)	πr^2	$w_2 h_2$	$w_3 h_3$
Volumetric flow rate (Q_i)	$\frac{Q S_1}{S_1 + S_2 + S_3}$	$\frac{Q S_2}{S_1 + S_2 + S_3}$	$\frac{Q S_3}{S_1 + S_2 + S_3}$
Apparent shear rate	$\frac{4 Q_1}{\pi r^3}$	$\frac{6 Q_2}{w_2 h_2^2}$	$\frac{6 Q_3}{w_3 h_3^2}$
Extrudate swell	$\left(\frac{D}{d}\right)_1$	$\left(\frac{D}{d}\right)_2$	$\left(\frac{D}{d}\right)_3$
Coefficient β	-	β_2	β_3
Swelling factor $f_g(h)$	-	$\beta_2\left(\frac{D}{d}\right)_2$	$\beta_3\left(\frac{D}{d}\right)_3$
Swelling factor $f_g(w)$	-	$\frac{1}{\beta_2}\left(\frac{D}{d}\right)_2$	$\frac{1}{\beta_3}\left(\frac{D}{d}\right)_3$

Table 5.2. *Calculation of local flow parameters*

According to the iterative diagram in Figure 5.9, the cross-section dimensions of the die adapted for producing a specific profile can then be estimated.

5.1.4.2.2. Flow balancing

The second problem encountered in the dimensioning of extrusion dies involves optimizing the topology of the flow pathway, that is to say, flow balancing. To solve this problem, two main strategies are used:

– minimizing the differences between local and mean velocities;

– minimizing the differences between the local mass flow rate and the intended local mass flow rate.

The method of minimizing the differences between the local velocity and the mean velocity is the most commonly used. It can be applied either by modifying the flow pathway upstream of the terminal plate [HUN 92], or by modifying the sections of the terminal plate [HUR 96] (Figure 5.11).

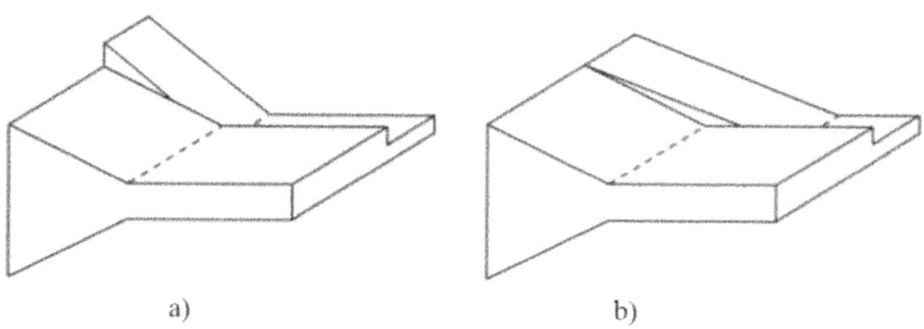

Figure 5.11. *Balancing techniques for extrusion dies: a) modification of the connecting plate situated upstream; b) modification of the terminal plate*

Let us study, for example, the die in Figure 5.12.

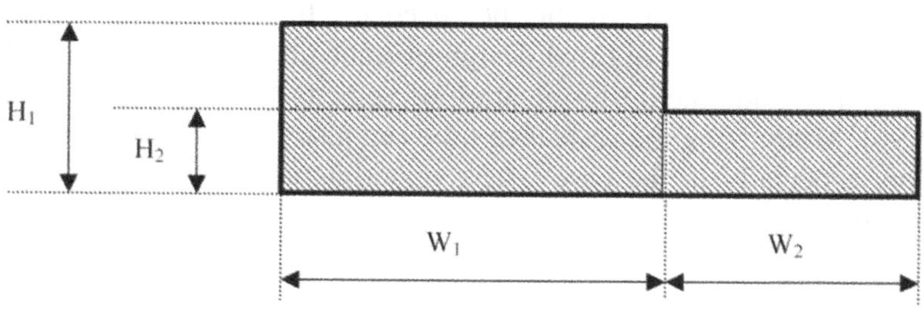

Figure 5.12. *Example of a profile cross-section*

For a 1D power-law fluid, the average velocity in section *i* is given by:

$$\bar{V}_i = \frac{\left[\dfrac{w_i h_i^2}{2(1/m+2)}\right]\left[\dfrac{h_i \Delta P_i}{2KL_i}\right]}{w_i h_i} \quad [5.19]$$

If the terminal plate is modified, then flow balancing by minimizing the differences in local velocities yields:

$$\bar{V}_1 = \bar{V}_2 \quad [5.20]$$

$$\Delta P_1 = \Delta P_2 \quad [5.21]$$

i.e.:

$$\frac{L_2}{L_1} = \left(\frac{H_2}{H_1}\right)^{1+m} \quad [5.22]$$

The process of minimizing the difference between the local mass flow rate and the intended local mass flow rate, although similar in principle to the method above, is however significantly different. First, it does not involve uniformity of the velocity profile. Therefore, a consecutive distortion of the velocity redistribution can occur at the die exit. Then, since the calculation of the local mass flow rate is an integral on a given area, only average balancing is feasible.

Note that both presented techniques provide good results only for dies with separate channels. In the absence of separate channels, the transverse flows and the fluid redistribution significantly affect overall flow. Therefore, other balancing criteria might be needed. In particular, cross flow minimization allows the optimal design of transition sections. This criterion was used, alongside the uniformization of velocity profiles by Koziey *et al.* [KOZ 96]. Minimizing the residence time can also be considered as a design criterion, especially in the case where materials are susceptible to thermal degradation.

5.1.4.3. Commercial software for the design of extrusion dies

Numerical methods for calculating the flow are a precious resource when designing extrusion dies. The conventional trial and error process is replaced by a series of numerical tests. It is, therefore, possible to design an initial geometry, such that only a small number of changes will be needed to obtain an optimal geometry [PIT 11, ETT 13].

Currently, many commercial software packages offer simulations of the flow within complex shaped dies, for example, CompuPlast© (CompuPlast Inc.), PolyDynamics© (PolyDynamics Inc.), developed in association with McMaster University (Canada), Ximex© (Sciences Computers Consultants) and Rem3D© (Transvalor), developed by CEMEF (MINES ParisTech), SPR© (Scientific Research Progress Inc.) and PolyFlow© (Fluent Inc.). These software packages use a 3D approach to the problem and the numerical solution is based on a finite element code. In Figure 5.13, we provide an example of flow balancing, in a profile die, performed using Rem3D© software. Although most codes are limited to the calculation of flow parameters inside the machine, some include free surface calculations or specialized sub-routines dedicated to the inverse problem: this is particularly the case for PROFILECAD© (PolyDynamics Inc.) and DIEDES© (Scientific Research Progress Inc.).

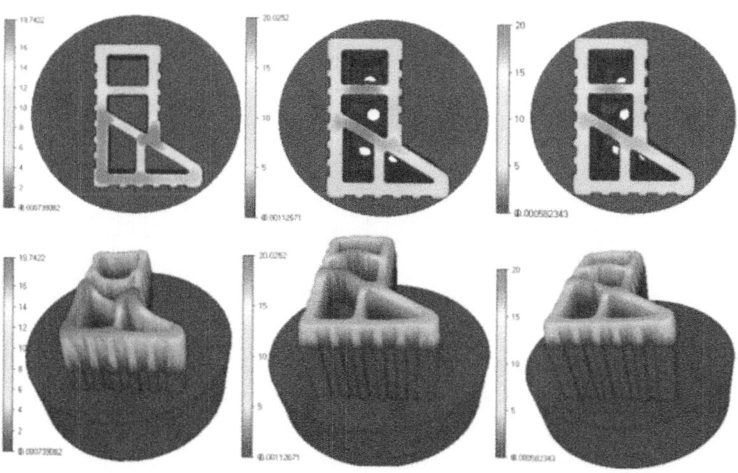

a) b) c)

Figure 5.13. *Example of a 3D calculation inside a plate die; balancing of the velocity field by geometric modification of upstream zones; a) initial geometry, b) first trial and c) second trial [ABB 03] (for a color version of this figure, see www.iste.co.uk/lafleur/polymer.zip)*

However, not many commercial software packages integrate the calculation of dimensional changes caused by extrudate swell, shrinkage and stretching. For this reason, the network model possesses a considerable advantage.

Using numerical methods to design dies requires that the physical properties of the extruded material are previously determined, including viscosity, density, thermal conductivity and specific heat capacity. Moreover, calculating post-extrusion dimensional changes requires the accurate analysis of extrudates obtained from different die geometries. Therefore, the accuracy of the results generated by the computer-aided die design software will always be limited by the accuracy of the measurements of the physical properties of the material.

5.2. Pipe extrusion

5.2.1. *Introduction*

The manufacture of pipes is usually carried out as shown in Figure 5.14: an extruder continuously feeds an axisymmetric die, which shapes the pipe. At the exit, the pipe passes through a calibrator, which cools the surface and sets the external dimensions. It then passes through several cooling tanks before being cut at the end of the line.

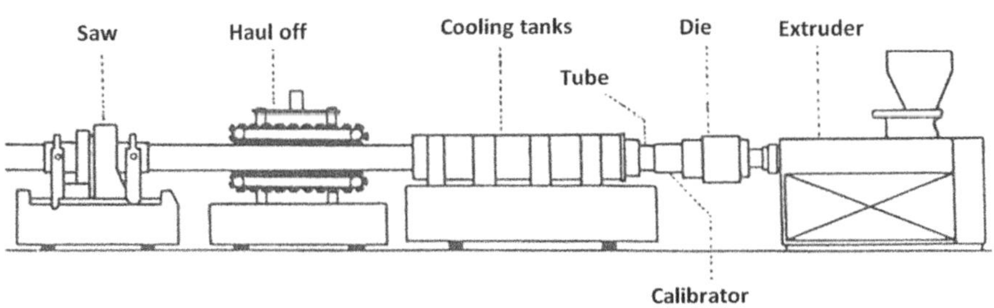

Figure 5.14. *An extrusion line for pipe manufacture*

Different types of dies can be used [RAU 01, MIC 03, CAR 12], although the most widespread is the spider die, shown in Figure 5.15.

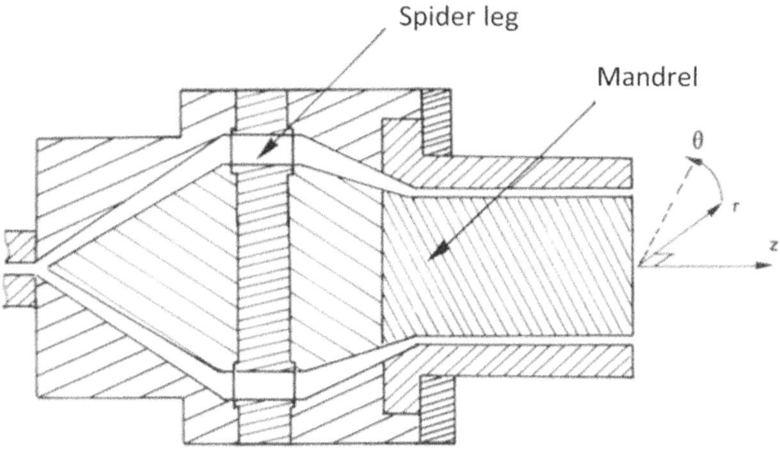

Figure 5.15. *Geometry of a spider die*

This is an axisymmetric tool composed of a mandrel fixed to the outer housing by a number of spider legs, regularly spaced around the mandrel. The molten polymer arriving from the extruder is first distributed around the mandrel and then flows around the spider legs, before entering the final zone with a constant gap that will give the dimensions to the pipe. With this type of die, a large variety of pipes can be produced, from medical catheters of several millimeters to water pipes of 2 m diameter [PUI 09].

5.2.2. *Flow calculation*

5.2.2.1. *Different approaches from 1D to 3D*

Although the geometry of a pipe is 3D, different hypotheses can be used to simplify the flow calculation. First, inside the die, the flow is confined, without a free surface or an abrupt change in cross-section (Figure 5.15). It is dominated by shear, with elongation playing only a minor role. We can, therefore, consider the behavior to be purely viscous, described by a power-law model or a Carreau model. However, heat transfer plays an important role and thermo-mechanical coupling is required for a good description of the flow.

Because pipes are generally extruded horizontally (Figure 5.14), the effects of gravity are negligible, at least in the die. In addition, as the flow cross-section changes slowly, the lubrication approximation can be applied [AGA 14] to simplify the flow equations. Consequently, three approaches are possible to characterize the flow conditions in a pipe die:

– a very simple 1D approach, which, already, allows us to obtain orders of magnitude values of pressure and temperature along the die;

– a 2D approach, based on the Hele-Shaw approximation, using finite volume or finite element methods;

– a 3D approach, to better quantify, such flows as, the flow around the spider legs.

5.2.2.2. *1D models*

Let us study the die in Figure 5.15, used to produce PVC pipes at a flow rate of 300 kg/h. These pipes have an internal diameter of 10 cm and a thickness of 5 mm. If we do not include the zone around the spider legs, the flow is axisymmetric. Applying the lubrication approximation allows us to neglect the radial velocity component *u,* compared with the component in the flow direction *w*. Moreover, component *w* only locally depends on the radial position *r*. If we assume that the polymer obeys the power law, the momentum equations (Stokes equations [AGA 14]) are reduced to a single expression, linking the pressure gradient *dp*/*dz* to the volumetric flow rate *Q*:

$$\frac{dp}{dz}=\frac{-2K(Q/2\pi)^m}{\left(\int_{R_1}^{R^*} r\int_{R_1}^{r}\left(\frac{R^{*2}}{u}-u\right)^{1/m} du dr+\int_{R^*}^{R_2} r\int_{r}^{R_2}\left(u-\frac{R^{*2}}{u}\right)^{1/m} du dr\right)^m} \quad [5.23]$$

where K is the consistency index, m is the pseudoplasticity index, R_1 and R_2 are the respective radii of the mandrel and the housing, and R^* is the location of the maximum velocity (Figure 5.16). This equation is a local equation, which means that the radii R_1, R_2 and R^* vary as a function of z. Therefore, the pressure gradient, also, evolves with z.

The mean temperature $\overline{T}$ across the flow thickness can be calculated by solving the thermal balance equation [AGA 14]:

$$\frac{d\overline{T}}{dz}=\frac{2\pi a}{Q}Nu\frac{R_2}{R_2-R_1}\left(T_p-\overline{T}\right)-\frac{1}{\rho c_p}\frac{dp}{dz} \quad [5.24]$$

where α is the thermal diffusivity, T_p is the housing temperature (assumed to be constant), ρc_p is the specific heat capacity and Nu is a Nusselt number, which describes the heat transfer between the polymer and the housing [BIR 07].

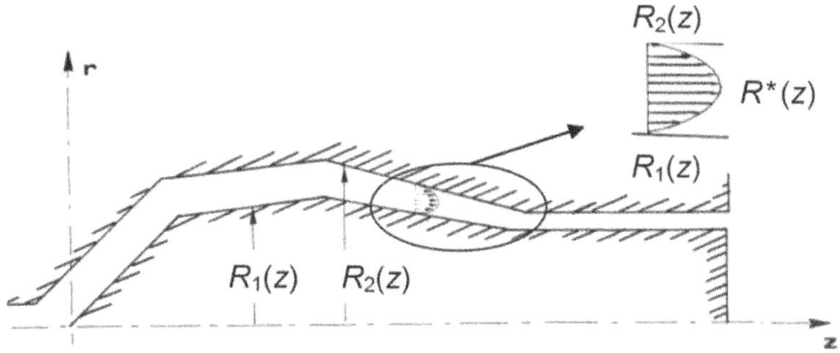

Figure 5.16. *Local geometry of a pipe die*

In this approach, we assume that the mandrel, which is, normally, not thermally regulated, is adiabatic. Equations [5.23] and [5.24] are coupled via the consistency index K. According to the Arrhenius equation, K depends on the temperature:

$$K = K_0 \exp\left[\frac{E_a}{R}\left(\frac{1}{T} - \frac{1}{T_0}\right)\right] \qquad [5.25]$$

where K_0 is the consistency index at the reference temperature T_0, E_a is the activation energy and R is the ideal gas constant.

Solving equations [5.23] and [5.24], step by step along the die using a slab method, allows the coupled evolution of pressure and temperature to be calculated (Figure 5.17). We find that, under the chosen conditions (300 kg/h), the pressure drop is 21 MPa, mainly occurring in the final section with the smallest thickness. It is, also, in this zone that we observe a significant increase in the mean temperature, from 186°C to 192°C (here the housing temperature T_p was set at 185°C).

These results clearly show that the final section of the die imparts the most effect, and it is the site of potential problems. In the case of PVC, an excessively high local temperature would lead to degradation. To better understand the flow conditions in the final section, we can, therefore, develop a 2D model for heat transfer, while keeping a 1D approach for the flow (equation [5.23]).

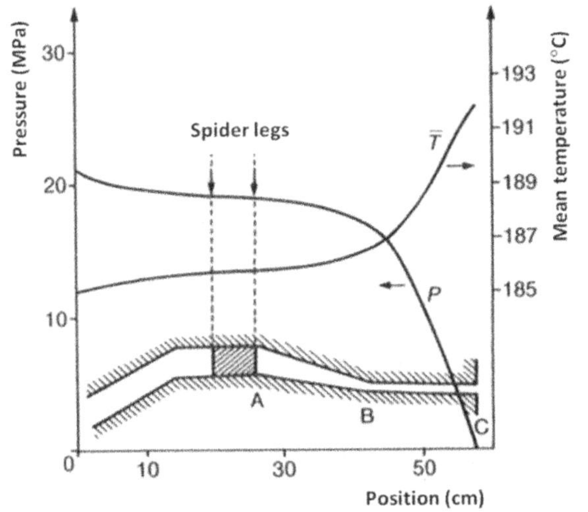

Figure 5.17. *Evolution of the pressure and mean temperature along the pipe die [VER 86]*

5.2.2.3. *Temperature profile*

If we want to calculate the temperature profile across the flow thickness and no longer the mean temperature, the following equation must be solved:

$$\rho c_p w \frac{\partial T}{\partial z} = \frac{k}{r}\frac{\partial}{\partial r}\left[r \frac{\partial T}{\partial r} \right] + K \left| \frac{dw}{dr} \right|^{n+1} \qquad [5.26]$$

where k is the thermal conductivity.

In this case, we can show that the adiabatic condition, on the mandrel, previously used, is no longer valid [SAI 82]. We are now forced to assume that the mandrel is in thermal equilibrium: on the one hand, heat is transferred from the flowing polymer, and on the other hand, some heat is lost toward the housing, by conduction, via the spider legs. After some time, the mandrel reaches an equilibrium temperature, which cannot be known *a priori*, and depends directly on the flow conditions.

Under the same conditions as above (flow rate of 300 kg/h, initial temperature of T_0 = 185°C, housing temperature of T_p = 185°C), we obtain the results shown in Figure 5.18. Three temperature profiles calculated at three different positions along the die, at points A, B and C, indicated in Figure 5.17, are shown in Figure 5.18.

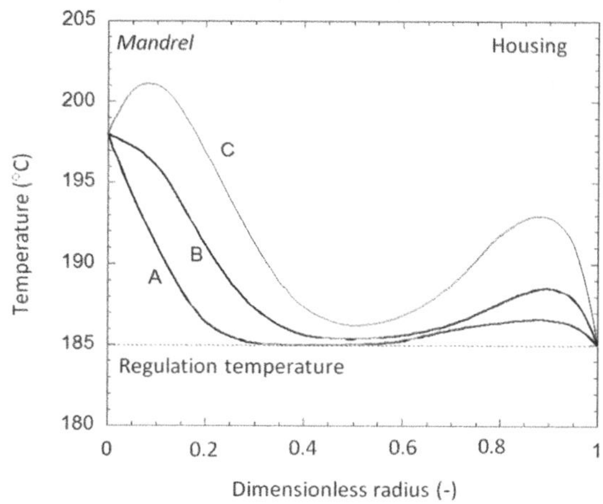

Figure 5.18. *Temperature profiles at different positions along the die (the dimensionless radius is defined by (r-R_1)/R_2-R_1)) [SAI 82]*

We find that the temperature reached at the mandrel is 198°C, which is much higher than the temperature of 185°C imposed on the housing. The profiles are heterogeneous, with hot spots near the walls, where the shear rate is very important. The longer the polymer flows, the more it heats up: the maximum temperature is attained at the die exit, close to the mandrel, with a value of 201°C, i.e. an increase of 16°C, caused by viscous heating. In the case of PVC, this can lead to a localized degradation. Moreover, a heterogeneous profile with a temperature difference of 15°C in the transverse flow direction can also cause post-extrusion problems: difficult conformation, heterogeneous cooling, creation of residual stresses, etc. [HAU 12].

Now if we want to increase the flow rate, we can see, in Figure 5.19, that the problems will only get worse (here, the housing temperature was set at 195°C). At the die exit, the profile always has the same appearance, but the temperature is much higher when the flow rate is increased. The equilibrium temperature of the mandrel increases from 204°C to 210°C and the maximum temperature increases from 207°C to 214°C when the flow rate increases from 340 to 420 kg/h. At high flow rates, the risk of degradation is significantly increased. We can show that a decrease in the regulation temperature of the housing only causes a slight decrease in temperature in the immediate vicinity of the outer wall. Therefore, the only solution to better control the temperature would be to thermally regulate the mandrel. This is feasible on large enough tools, with the cooling fluid entering and exiting the mandrel through the spider legs.

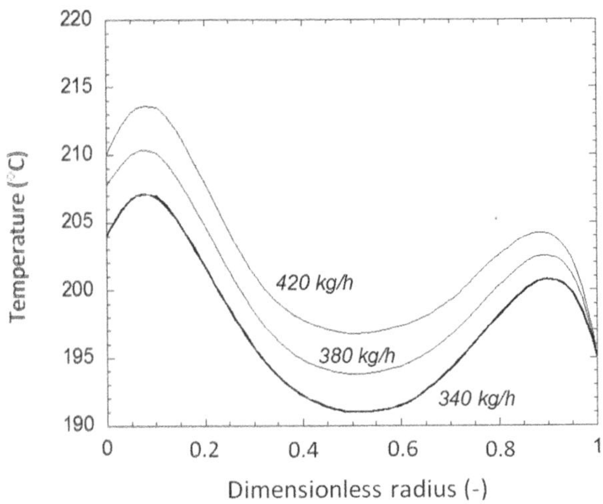

Figure 5.19. *Temperature profiles at the die exit for different flow rates (the dimensionless radius is defined by $(r-R_1)/R_2-R_1)$) [SAI 82]*

5.2.2.4. *An example of a non axisymmetric flow*

In the previous examples, the flow was considered to be completely axisymmetric. In fact, although this is true for the die, it is not necessarily always the case. To illustrate this point, we will consider a thickness defect of the pipe, encountered during industrial production, and using the previous die [SAI 84].

As seen in Figure 5.20, the defect involves the presence of two symmetrical regions where the thickness is significantly larger than the average, situated in a horizontal plane, and two regions where the thickness is significantly smaller, also symmetrical, situated in the vertical plane. Consequently, the tube has local thicknesses of 4.7 and 5.3 mm, instead of the intended 5 mm.

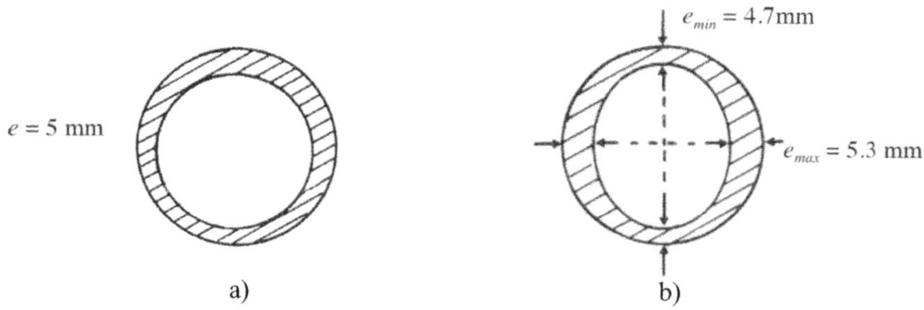

Figure 5.20. *a) Example of the correct tube cross-section b) and with a symmetrical thickness defect*

Due to its symmetry, this defect is not due to the poor centering of the mandrel with regard to the housing. In fact, it is caused by a temperature heterogeneity, similar to that we have just seen, but this time due to the flow in the extruder feeding the die. Most of the time, PVC is plasticized and pressurized in a counter-rotating twin-screw extruder. However, temperature measurements in the connecting channel between the extruder and the die reveal a very heterogeneous profile, in the flow cross-section, with two hot spots situated in a horizontal plane in the axis of each screw, Figure 5.21.

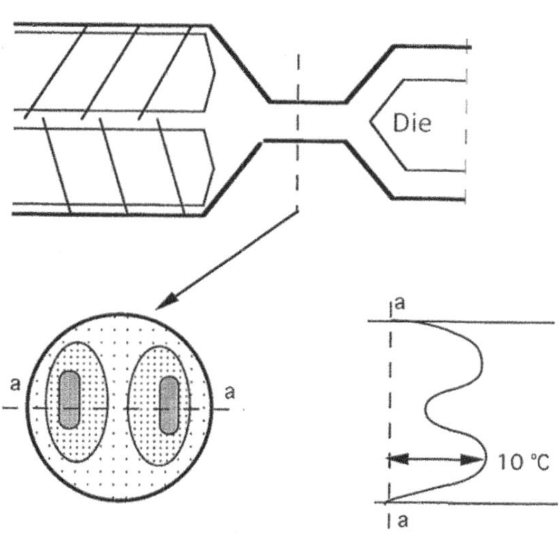

Figure 5.21. *Temperature profile in the connecting section between the extruder and the die (dark zones are the hottest)*

These zones are approximately 10°C hotter than the rest of the polymer. Therefore, we can imagine that, on reaching the mandrel, these two points will separate and flow on each side of the mandrel in the horizontal plane. Since the viscosity is lower, it leads to a greater flow rate, for the same pressure drop, and therefore an increased thickness.

To validate this hypothesis and calculate the flow, it is no longer possible to consider an axisymmetric flow. Therefore, we will perform a 2D calculation, based on the lubrication approximation.

In addition, for further simplification, we will focus on the final zone after the spider legs, which we identified above, as the one that will impart the most effect, and we will unfold and flatten it, as shown in Figure 5.22. This is possible because the flow thickness, 5 mm, is much smaller than the pipe diameter, 100 mm.

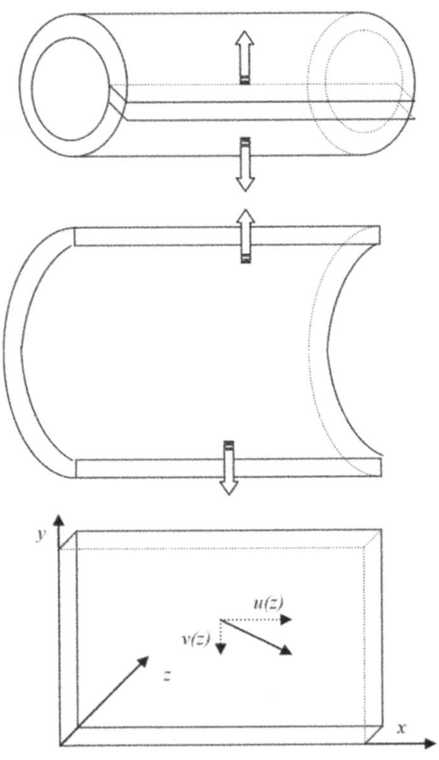

Figure 5.22. *Unrolling and flattening of the final zone of the die*

To calculate the pressure and temperature fields in the flattened geometry, we must solve the following equations [VER 86]:

$$h\Delta p+\frac{2m+1}{m}\left[\frac{\partial p}{\partial x}\frac{\partial h}{\partial x}+\frac{\partial p}{\partial y}\frac{\partial h}{\partial y}\right]-\frac{h}{Km}\left[\frac{\partial p}{\partial x}\frac{\partial K}{\partial x}+\frac{\partial p}{\partial y}\frac{\partial K}{\partial y}\right]$$
$$+h\frac{1-m}{m}\frac{\left(\frac{\partial p}{\partial x}\right)^2\frac{\partial^2 p}{\partial x^2}+2\frac{\partial p}{\partial x}\frac{\partial p}{\partial y}\frac{\partial^2 p}{\partial x\partial y}+\left(\frac{\partial p}{\partial y}\right)^2\frac{\partial^2 p}{\partial y^2}}{\left(\frac{\partial p}{\partial x}\right)^2+\left(\frac{\partial p}{\partial y}\right)^2}=0 \quad [5.27]$$

$$\bar{u}(x,y)h(x,y)\left[\frac{\partial \bar{T}}{\partial x}+\frac{1}{\rho c_p}\frac{\partial p}{\partial x}\right]$$
$$+\bar{v}(x,y)h(x,y)\left[\frac{\partial \bar{T}}{\partial y}+\frac{1}{\rho c_p}\frac{\partial p}{\partial y}\right]=\frac{2k}{\rho c_p}Nu\frac{T_p-\bar{T}}{h(x,y)} \quad [5.28]$$

where $\bar{u}$ and $\bar{v}$ are the mean velocities according to x and y, defined by:

$$\bar{u}(x,y)=-\frac{1}{h}\frac{2m}{2m+1}\left(\frac{1}{K}\right)^{1/m}\left(\frac{h}{2}\right)^{(2m+1)/m}\left[\left(\frac{\partial p}{\partial x}\right)^2+\left(\frac{\partial p}{\partial y}\right)^2\right]^{(1-m)/2m}\left(\frac{\partial p}{\partial x}\right) \quad [5.29]$$

$$\bar{v}(x,y)=-\frac{1}{h}\frac{2n}{2n+1}\left(\frac{1}{K}\right)^{1/n}\left(\frac{h}{2}\right)^{(2n+1)/n}\left[\left(\frac{\partial p}{\partial x}\right)^2+\left(\frac{\partial p}{\partial y}\right)^2\right]^{(1-n)/2n}\left(\frac{\partial p}{\partial y}\right) \quad [5.30]$$

where $p(x, y)$ is the pressure, $h(x, y)$ is the local thickness and $\bar{T}(x, y)$ is the mean temperature. The consistency index K depends on the mean temperature via the Arrhenius equation [5.25].

Equations [5.27]–[5.30] are solved using an iterative finite difference method. The results are shown in Figure 5.23. From top to bottom, we have plotted the regulation temperature profile, the mean temperature field and the flow distribution at the die exit (the ratio of the local flow rate q_x by the mean flow rate, $\bar{Q}$).

Let us consider the case where the regulation temperature is homogeneous, 185°C all around the housing (Figure 5.23(a)). At the flow entrance, we impose different polymer temperatures at different positions on the *y*-axis, corresponding to experimental observations, varying from 192°C (hot zone) to 180°C (cold zone).

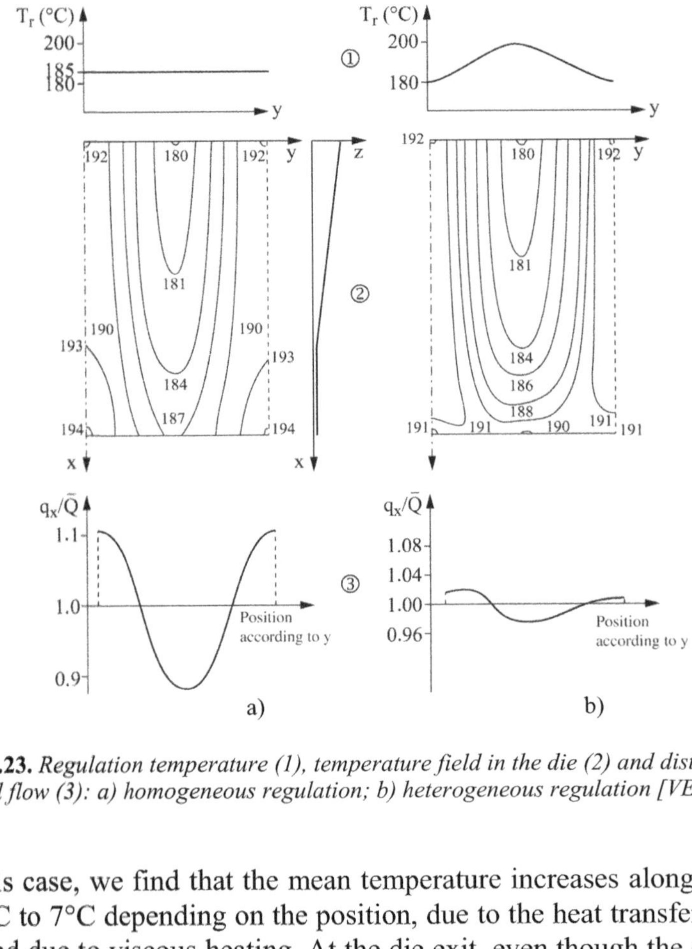

Figure 5.23. *Regulation temperature (1), temperature field in the die (2) and distribution of final flow (3): a) homogeneous regulation; b) heterogeneous regulation [VER 08]*

In this case, we find that the mean temperature increases along the flow, from 2°C to 7°C depending on the position, due to the heat transfer from the walls, and due to viscous heating. At the die exit, even though the difference is reduced, the temperature is still heterogeneous, ranging between 187°C and 194°C. Consequently, we see that the flow rate is also heterogeneous, with an overflow of approximately 10% corresponding to the hot zones, and an underflow of the same amount in the cold zones. This confirms that the temperature heterogeneity, at the die entrance, is responsible for the thickness defect.

In addition to the analysis, modeling can be used to find solutions to the above-mentioned problem. Assuming that we now control the temperature of the housing, by further heating the cold zones (T_p = 200°C), and lowering the temperature in the hot zones (T_p = 180°C) (Figure 5.23(b)). Under the same initial conditions, there is a significant modification in the temperature field, leading to a final temperature of around 191°C. At the same time, the flow distribution at the die exit is greatly improved, with variations of only ±2%, producing a tube whose thickness is now uniform. Such heterogeneous regulation systems are, nowadays, widespread in industrial facilities.

5.3. Calibrators

In the extrusion of pipes and profiles, the final adjustment of extrudate dimensions to the required specifications is often carried out using a calibrator, situated just after the die exit. The calibrator, in contact with the molten polymer, cools the extrudate. As the product leaves the calibrator, it is rigid enough to withstand the stresses imposed by the subsequent operations, while maintaining specific limits. The aim of this section is to briefly describe the different types of calibrators currently used.

5.3.1. *Friction calibrators*

The dimensions of simple rectangular profiles are generally ensured using friction calibrators. These calibrators are composed of two horizontal plates between which the profile is drawn. The plates are equipped with a water cooling system, usually with a Teflon coating, favoring a sliding contact. This coating is subjected to significant wear and must be frequently renewed. With these calibrators, the velocities reached are approximately from 1 to 4.5 m/min, depending on the thickness of the profile.

5.3.2. *External compressed air calibrators*

External compressed air calibrators are often used in the extrusion of pipes to impose limits on the external diameter. In fact, numerous tubular products are standardized by the external diameter. Compressed air is injected into the pipe through the mandrel. The generated internal pressure flattens the extrudate against the calibrator. This process is generally used for pipes with a large diameter (>90 mm in the case of polyethylene and 350 mm in the case of polyvinyl chloride).

5.3.3. *External vacuum calibrators*

In this process, small circular openings or thin slots are machined into the metal of the calibrator. Vacuum is applied, through these openings, to establish contact between the extrudate and the calibrator. This process is also used to set the dimensions of simple and complex profiles. However, for complex profiles, it is necessary to use several calibrators, connected in series, to achieve the specified dimensions.

5.3.4. *Internal calibrators*

Internal calibrators are often used in the extrusion of pipes. In this process, the die mandrel is coupled to an external mandrel to set the internal diameter dimensions of the pipe. The extruded pipe slides over the external mandrel, where it cools. Using a shaped external mandrel, this process can also be used to modify the shape of a molten pipe.

5.4. Conclusion

As for other extrusion processes, the production of pipes and profiles has greatly benefitted from flow models, which are now widely used to design tools, and to optimize processing conditions. Currently, a better understanding of the, often complex, rheological behavior of the used materials is required. In fact, PVC compounds, which form a large portion of the materials used, often manifest wall slip phenomena, which can radically modify the flow conditions. So far, the physics of these materials is still poorly understood, and, therefore, difficult to implement in numerical models.

5.5. Bibliography

[ABB 03] ABBAD H., *Optimisation de filières d'extrusion de briques,* Mémoire de mastère spécialisé, Ecole des Mines de Paris, 2003.

[AGA 14] AGASSANT J.-F., AVENAS P., SERGENT J.-P., *et al.*, *Mise en forme des matières polymères. Approche thermomécanique de la plasturgie*, Tec et Doc, Lavoisier, Paris, 2014.

[BEA 11] BEAUMIER D., *Conception des filières d'extrusion pour les profilés multiformes*, Mémoire de maîtrise, Ecole Polytechnique de Montréal, 2011.

[BIR 07] BIRD R.B., STEWART W.E., LIGHTFOOT E.N., *Transport Phenomena*, Wiley, New York, 2007.

[CAR 12] CARNEIRO O.S., NOBREGA J.M., *Design of Extrusion Forming Tools*, Smithers Rapra, Shawbury, 2012.

[ETT 13] ETTINGER H.J., PITTMAN J.F.T., SIENZ J., "Optimization-driven design of dies for profile extrusion: parameterization, strategy, and performance", *Polymer Engineering & Science*, vol. 53, pp. 189–203, 2013.

[HAU 12] HAUDIN J.M., VINCENT M., VERGNES B., "Pipe forming tools", in CARNEIRO O.S., NOBREGA J.M. (eds), *Design of Extrusion Forming Tools*, Smithers Rapra, Shawbury, 2012.

[HUN 92] HUNEAULT M.A., L'extrusion des profilés en PVC: rhéologie et conception des filières, PhD Thesis, École Polytechnique de Montréal, 1992.

[HUN 96] HUNEAULT M.A., LAFLEUR P.G., CARREAU P.J., "Evaluation of the FAN technique for profile die design", *International Polymer Processing*, vol. 11, pp. 50–57, 1996.

[HUR 96] HUREZ P., TANGUY P.A., BLOUIN D., "A new design procedure for profile extrusion dies", *Polymer Engineering & Science*, vol. 36, pp. 626–635, 1996.

[KOZ 96] KOZIEY B.L., VLACHOPOULOS J., VLCEK J., *et al.*, "Profile die design by pressure balancing and cross flow minimization", *SPE ANTEC Technical Papers*, vol. 3, pp. 247–252, 1996.

[LEG 93] LEGAT V., MARCHAL J.M., "Die design: an implicit formulation for the inverse problem", *International Journal for Numerical Methods in Fluids*, vol. 16, pp. 29–42, 1993.

[MEN 71] MENDELSON R.A., FINGER F.L., BAGLEY E.B., "Die swell and recoverable shear strain in polyethylene extrusion", *Journal of Polymer Science Part C: Polymer Symposia*, vol. 35, pp. 177–188, 1971.

[MIC 03] MICHAELI W., *Extrusion Dies for Plastics and Rubber*, Hanser, Munich, 2003.

[PIT 11] PITTMAN J.F.T., "Computer-aided design and optimization of profile extrusion dies for thermoplastics and rubber: a review", *The Journal of Process Mechanical Engineering*, vol. 225, pp. 280–321, 2011.

[PUI 06] PUISSANT S., "Lignes d'extrusion de tubes. Etapes de fabrication", *Techniques de l'Ingénieur*, vol. AM3 642, pp. 1–20, 2006,

[RAU 01] RAUWENDAAL C., *Polymer Extrusion*, Hanser, Munich, 2001.

[SAI 82] SAILLARD P., Les phénomènes thermiques dans les outillages d'extrusion des matières plastiques, PhD Thesis, Paris School of Mines, 1982.

[SAI 84] SAILLARD P., VERGNES B., AGASSANT J.F., "Study of the wall thickness defects encountered in PVC pipe extrusion", *Polymer Process Engineering*, vol. 2, pp. 53–68, 1984.

[STE 85] STEVENSON J.F., "Analysis of extrudate dimensions: die design, swell and drawdown", *Plastics and Rubber Processing and Applications*, vol. 5, pp. 325–334, 1985.

[TAN 70] TANNER R.I., "A theory of die-swell", *Journal of Polymer Science Part B: Polymer Physics*, vol. 8, pp. 2067–2078, 1970.

[VER 86] VERGNES B., AGASSANT J.F., "Die flow computations: a method to solve industrial problems in polymer processing", *Advances in Polymer Technology*, vol. 6, pp. 441–446, 1986.

[VER 08] VERGNES B., AGASSANT J.F., "Modélisation des écoulements dans les filières d'extrusion", *Techniques de l'Ingénieur*, vol. AM 3 655, pp. 1–19, 2008.

Chapter 6

Production of Films and Sheets

6.1. Introduction

Flexible packaging has been used since the beginning of mankind. Some of the first examples were sheets, used to transport food, and stomachs of sheep and goats to store water. The first flexible thermoplastic packaging, cellophane, was invented by the textile engineer Jacques Brandenberger in 1908. He was sitting in a restaurant when a client spilled wine on the tablecloth. Watching the waiter try to deal with the damage, he decided to invent a flexible and transparent film that could be applied to fabric to waterproof it. After having experimented with many materials, he found that by extruding an alkaline cellulose solution, viscose, through an acid bath, it was possible to obtain a transparent film. Originally, cellophane was used to produce gas masks. However, it found its real success in 1927, when William Charch, a researcher at DuPont de Nemours, developed a completely impermeable cellophane film to package food products.

Today, synthetic thermoplastic materials have replaced cellophane to produce protective films, self-supporting bags, agricultural films and packaging bags. Flexible packaging has many advantages. First, the ratio of transported weight to weight of the packaging is very high. For example, it is possible to store 25 kg of bulk material, such as flour or sugar, in a polyethylene sack weighing less than 140 g. Second, storing flexible packaging does not require a lot of space, and bagging operations can be carried out quickly and easily, which allows the costs to remain relatively low.

The main qualities to consider when developing flexible packaging are mechanical strength, adhesion, shine, transparency, impermeability, food compatibility and resistance to ageing. However, flexible packaging has some minor drawbacks including the lack of resistance to puncture, low modulus and breaking stress, as well as being difficult to open. Moreover, once opened, flexible packaging is almost impossible to close. This drawback has been bypassed by using a sealing system similar to a zip fastener, which increases production costs. In some cases, spouts are added to facilitate the pouring of liquids.

Until recently, flexible packaging was used less often than rigid packaging, but the innovations outlined above have greatly increased the production of flexible packaging over the past 5 years. According to Hernandez *et al.* [HER 00], flexible packaging accounts for 47% of the packaging produced worldwide, which corresponds to 19% weight of the total amount of thermoplastic resins produced. This growth has been sustained by the development of multilayered packaging, which allows different materials to be combined by coextrusion to produce a multilayer film that optimizes performance, while minimizing cost. In most cases, this type of film is made up of an inexpensive resin, acting as a mechanical structure, and an expensive layer providing, for example, barrier properties.

Initially dedicated to the extrusion of low-density polyethylene (LDPE) for the manufacture of shopping bags, different processes to manufacture flexible packaging were adapted to newly developed polyolefins. In fact, more than 80% of the produced films are now made up of high-density polyethylene (HDPE), linear low-density polyethylene (LLDPE), and polyethylene obtained using a metallocene catalyst (i.e. metallocene polyethylene (mPE)), allowing the molecular weight of the resin to be accurately controlled. Table 6.1 shows that polyolefin resins are primarily used to produce flexible packaging. LDPE, combined with many other materials, is used to produce protective films for food and agriculture as well as vapor barriers for building applications. This material is easily processed; it is flexible, has a good heat sealing capacity, has a good impact resistance and is inexpensive.

LDPE can be combined with LLDPE, HDPE and mPE to produce films with good heat sealing capacities and better mechanical properties. However, polyolefins are highly permeable to gases, which is a major disadvantage. To tackle this problem, we often combine polyolefins with materials that offer excellent gas barrier properties, such as polyethylene terephthalate (PET), polyvinylidene chloride (PVDC) and polyamides, producing multilayer films.

	Low-density polyethylene	Linear low-density polyethylene	High-density polyethylene
Injection	199	85	672
Extrusion of which	4129	1653	1764
Films	3319	1571	630
Pipes	78	35	765
Wire coating	224	35	63
Coating	508	12	30
Containers	46	12	1638
Rotomolding	5	125	12
Others	259	48	114
Total	**4638**	**1923**	**4200**

Table 6.1. *The polyethylene market in Western Europe in 2000, in thousands of tons (source: ARKEMA)*

PET, which is a fairly expensive material, has a high permeability to oxygen and carbon dioxide. PET films are transparent, form an effective barrier against odors, and are resistant to fats and oils. It is also possible to significantly improve the properties of these films by orienting them biaxially. The main downfall of PET is that, when molten, it has a low viscosity, which causes the production line start-up to be very difficult.

PVDC, used as packaging since 1946, is marketed under the name Saran, by Dow Chemical. This material is very flexible and transparent and has excellent barrier properties with regard to gas and humidity. However, it is three times more expensive than other transparent films. Its high strength, however, means that it can be used as a very thin film, which offsets the high cost. This material has a high tendency to self-adhere, which makes it desirable for the final application as packaging, but causes problems during production. It allows the taste and smell of food to be conserved very efficiently. PVDC has a high resistance to acids and alkalines, but is very sensitive to halogens.

Polyamide (nylon) was introduced onto the market in 1938, by DuPont de Nemours. Polyamide 6 (PA6) is the most frequently used polymer resin for film production. It is often considered as being three times more resistant to abrasion than polyethylene. It is resistant to weak acids and alkalines. It has a high permeability to gases, but it is a very hydrophilic material. Despite being relatively expensive, polyamide is often used since its strength and properties at high temperature are quite important in the sterile packaging market.

Among the other materials often used in the development of multilayered systems are functionalized grafted polyethylene, ethylene vinyl alcohol (EVOH), polyvinyl chloride (PVC), and more recently, polyvinylidene difluoride (PVDF). It is also possible to include a layer of a recycled material between two layers of the virgin resin to decrease losses and reduce the cost of raw materials. The main applications of thermoplastic films are shown in Figure 6.1.

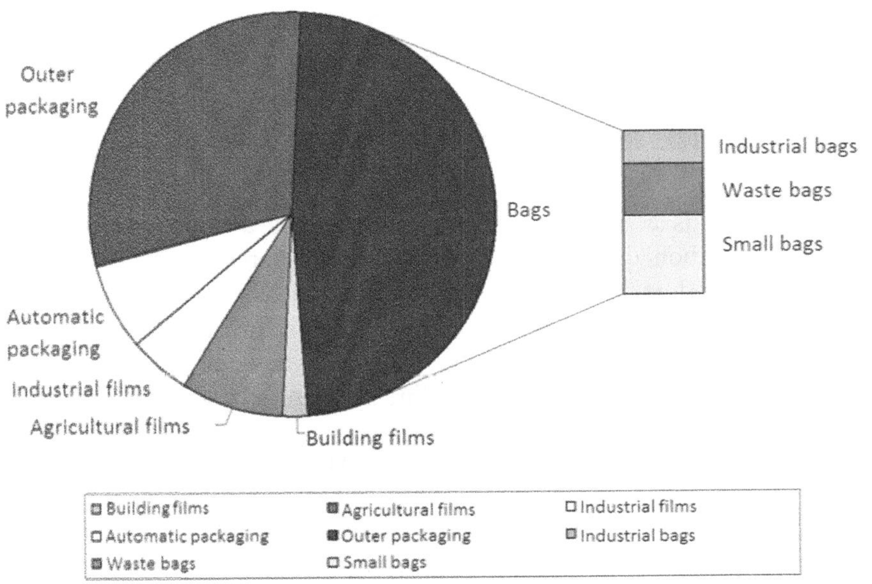

Figure 6.1. *Plastic film market in Europe in 2000 (source: Plasteurofilm)*

To produce flexible packaging, two main techniques are used: cast film extrusion and film blowing; these techniques will be the focus of the following two sections.

6.2. Cast film extrusion

6.2.1. Processing

Producing films by cast film extrusion involves using an extruder to feed a polymer through a die with a rectangular final cross-section. The size of these dies can range from several centimeters to several meters in width and from 0.2 to 25 mm in thickness. Despite thermal insulation, these dimensions result in significant heat loss that causes high energy costs. The flow channel leading to the die exit is usually coat-hanger shaped, as shown in Figure 6.2. This shape ensures a uniform distribution of the polymer and, therefore, guarantees a uniform film thickness at the die exit. In fact, because the die is very wide and the polymer is fed in the center, the polymer flowing through the center of the die is subjected to a pressure gradient different from that at the side walls, which causes variations in the local flow rates. Using coat-hanger geometry can reduce this effect. It is also necessary to mention that temperature control is extremely important to maintain a regular flow rate, through the whole die width. Modern systems are equipped with flexible metal lips, located at the die exit. These lips can be adjusted using servomotors, to accurately increase or decrease the die opening, thereby controlling the thickness of the molten polymer near the micrometer.

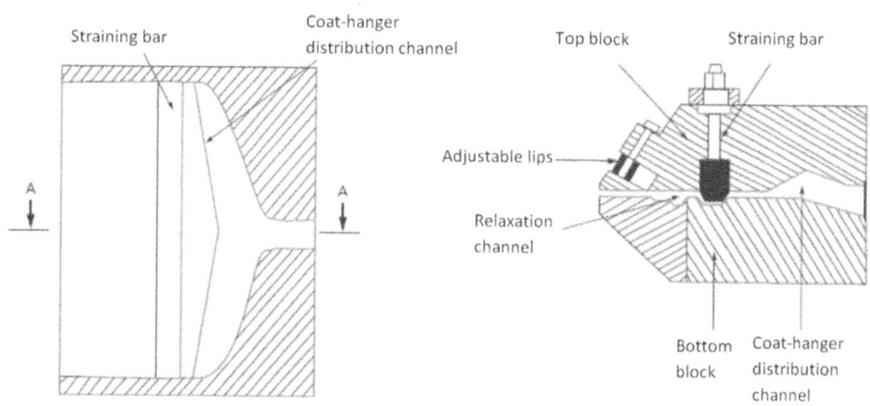

Figure 6.2. *Coat-hanger die used in cast film extrusion [VER 08]*

At the die exit, the film is axially stretched and then brought into contact with a chill roll. Typically, the distance between the die exit and the chill roll is approximately 10 cm. Rapid cooling is required to prevent the formation of crystals which can turn the film translucent. Afterwards, the film is rolled using a winder. To minimize the surface defects, the rolls involved in the cooling and stretching are polished. A diagram of the process is shown in Figure 6.3. The main advantage of this technique is that it is possible to produce transparent hard-wearing films with low turbidity, at very high production rates. However, only the film surface is cooled. To maximize this cooling, it is possible to use a second roll. For some specific applications, which require a completely transparent film, it is possible to cool the film in a water bath. However, this greatly reduces the production rate.

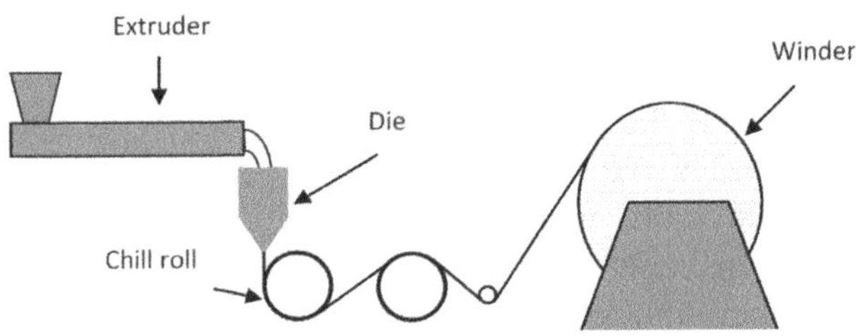

Figure 6.3. *A cast film extrusion line*

It is also necessary to cut the film edges, since they can differ greatly in thickness from the rest of the film. Typically, for a die with a width of 2.6 m, it is necessary to remove 5 cm off each side of the film. The thickness varies by approximately 3% over the width of the film. The dimensions and properties of the film are controlled by the dimensions of the die, the cooling efficiency and the stretching and extrusion rates.

6.2.2. *Designing the flat die*

The polymer exits the extruder and flows into a distribution channel with a decreasing cross-section. A straining bar followed by a relaxation channel enables a better distribution across the die width. The die design must ensure that, for a uniform die opening, the molten polymer flows at a constant temperature and local flow rates, across the width. The first flow models included analytical approaches, based on an isothermal analysis, with Newtonian or power-law behavior [TAD 79, MAT 79, VER 80]. For example, the geometry used by Tadmor and Gogos [TAD 79] is shown in Figure 6.4. It assumes that the cross-section of the distribution channel is cylindrical and that the radius varies with the l position. The curvature of the feed channel is small and forms an angle α with the horizontal. The lip opening H is constant and q is the flow rate per unit width at the die exit. If the pressure at the entry of the feed channel is constant, then the flow rate is also constant. Our aim is to uniformly distribute this flow over the final cross-section of the die. Assuming that the lip opening is constant, we can write:

$$\frac{dP}{dz} = a = \text{constant} \qquad [6.1]$$

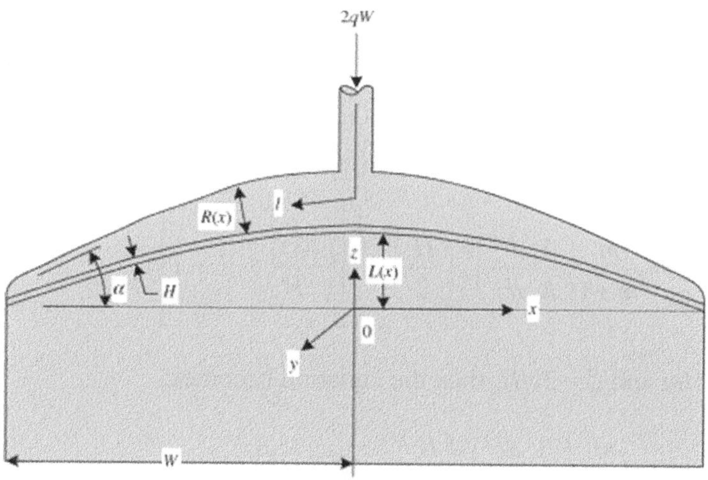

Figure 6.4. *A diagram showing the different zones of a coat-hanger die [TAD 79]*

By integrating equation [6.1], we obtain:

$$[L(0) - L(x)]a = P_0 - P'(l) \qquad [6.2]$$

where $P'(l)$ is the pressure in the distribution channel at position l and P_0 is the pressure at the entry of the distribution channel. The relationship between x and l is given by the die geometry. By differentiating equation [6.2] with respect to l, we obtain a relationship between the pressure gradient in the distribution channel, the pressure gradient in the flat section and the shape of the distribution channel.

$$\frac{dP'}{dl} = a\frac{dL}{dl} \qquad [6.3]$$

The objectives set for designing the die are represented by this equation, which applies to all types of fluids. We will differentiate the equations for a power-law fluid. Considering that in the flat region of the die, the flow passes between two parallel plates, the momentum equation is written as:

$$-\frac{dP}{dz}-\frac{d\tau_{yz}}{dy}=0 \qquad [6.4]$$

For a power-law fluid, the momentum equation can be integrated by using the following boundary conditions:

$$\frac{dV_z}{dy}=0 \text{ at } y=0 \text{ and } V(z)=0 \text{ at } y=\pm H/2$$

We obtain:

$$V_z(y)=\frac{m}{m+1}\left(\frac{1}{K}\frac{dP}{dz}\right)^{\frac{1}{m}}\left(\frac{H}{2}\right)^{m+\frac{1}{m}}\left[\left(\frac{2|y|}{H}\right)^{m+\frac{1}{m}}-1\right] \qquad [6.5]$$

If $s = 1/m$ and $\xi = 2y/H$, then the equation becomes:

$$V_z(y)=\frac{m}{m+s}\left(\frac{1}{K}\frac{dP}{dz}\right)^{s}\left(\frac{H}{2}\right)^{1+s}\left[\xi^{1+s}-1\right] \qquad [6.6]$$

This expression for velocity as a function of y can be integrated over the cross-section to obtain the relationship between the flow rate per unit width and the pressure drop:

$$a=\frac{dP}{dz}=2^{m+1}(2+s)^m K\frac{q^m}{H^{2m+1}} \qquad [6.7]$$

To study the flow within the distribution channel, we assume a steady-state local flow inside a tube. This hypothesis ignores the curvature of the channel as well as the flow within the die lips. After rearrangement, we obtain:

$$-\frac{dP'}{dl}=\left(\frac{3+s}{\pi}\right)^m 2K\frac{Q(l)^m}{R(x)^{3m+1}} \qquad [6.8]$$

The flow rate in the distribution channel, $Q(l)$, feeds the flat die:

$$Q(l) = Q(x) = q\,(W - x) \qquad [6.9]$$

By combining equations [6.8] and [6.9], we obtain:

$$-\frac{dP'}{dl} = \left(\frac{3+s}{\pi}\right)^m 2K \frac{\left[q(W-x)\right]^m}{R(x)^{3m+1}} \qquad [6.10]$$

Using equations [6.3] and [6.7], equation [6.10] takes the following form for a power-law fluid:

$$\frac{2^m (2+s)^m}{H^{2m+1}} \frac{dL}{dl} + \left(\frac{3+s}{\pi}\right)^m 2K \frac{\left[q(W-x)\right]^m}{R(x)^{3m+1}} = 0 \qquad [6.11]$$

Using equation [6.11], we can calculate the geometry of the distribution channel (dL/dl or $R(x)$), for a given die width, and a power-law fluid. In practice, the factor dL/dl is constant and the radius of the distribution channel is calculated to ensure a constant flow.

The model that we have just developed has several significant limitations (isothermal and unidirectional). Vergnes *et al.* [VER 80, VER 84] and then Arpin *et al.* [ARP 92, ARP 94] presented 2D models for non-isothermal flow of a power-law fluid in a flat die. In these models, flow is defined by using the lubrication theory and the equations are solved using the finite difference method. Figure 6.5 shows the flow lines and the pressure field.

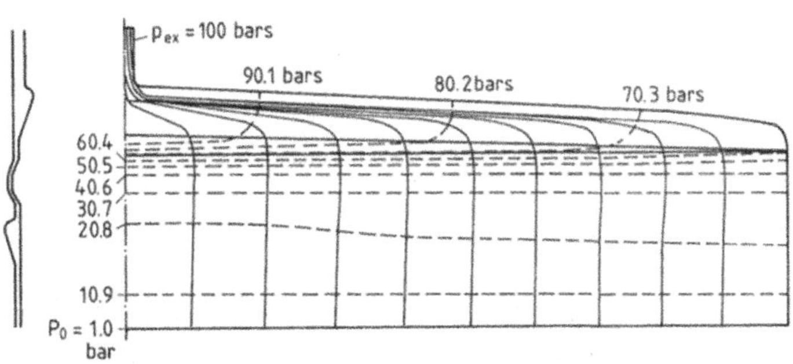

Figure 6.5. *The pressure field (--) and flow lines (——) within a flat die [VER 80]*

Figure 6.6 shows the temperature field and the flow distribution in the final section, for different die control temperature. We find that the temperature is an important factor in the flow distribution, at the die exit.

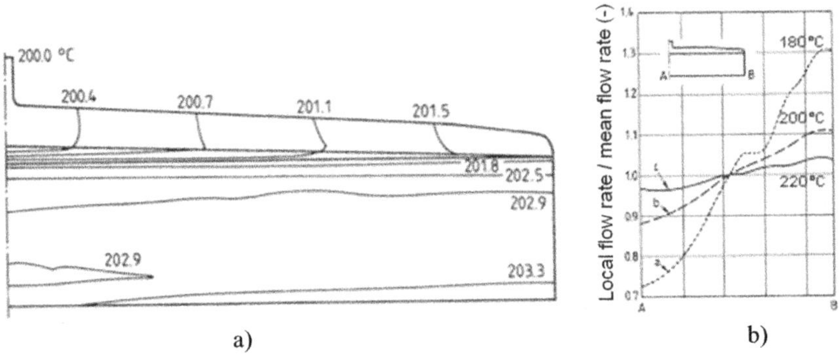

Figure 6.6. *a) Variations in temperature inside the flat die; b) effect of control temperature on flow distribution [VER 80]*

3D flow models using the finite volume or finite element methods have also been developed [WAN 91, WU 06a, WU 06b, MAH 08]. Their main purpose is to set correct boundary conditions at the die walls (no-slip boundary condition) and provide information about the temperature profiles. Figure 6.7 shows that the maximum temperature is not reached at the die lips but in the distribution channel [WU 06b]. This result is particularly important for materials sensitive to heat.

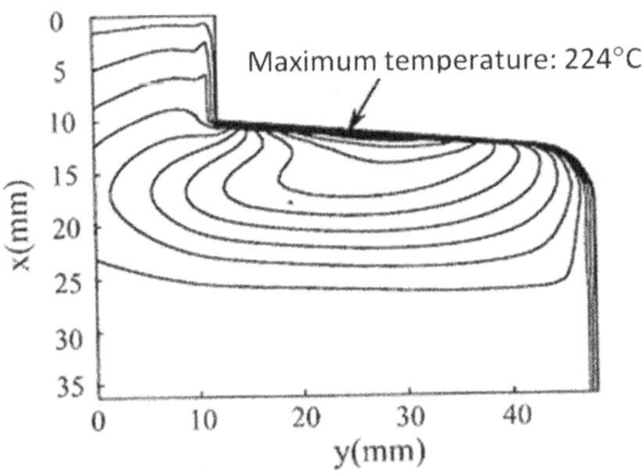

Figure 6.7. *Temperature distribution on the x-y surface of the upper quarter of the coat-hanger die at 473 K (200°C), at the entry and at the boundaries [WU 06b]*

As shown by Meng *et al.* [MEN 09] and Lebaal *et al.* [LEB 09], it is possible to couple an optimization technique with a 3D simulation algorithm employing finite elements to optimize the dimensions of the die. Meng *et al.* [MEN 09] used genetic algorithms based on two parameters: the lip gap and the angle of the flow channel. Figure 6.8 shows an optimum result obtained after 22 iterations.

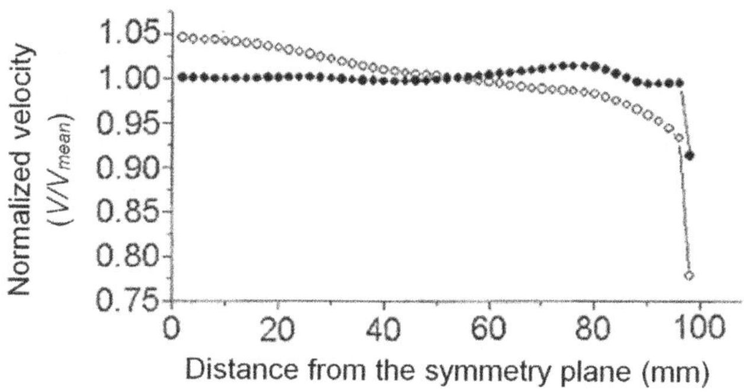

Figure 6.8. *Velocity distribution at the coat-hanger die exit. Comparison between the initial result (○) and the optimized result (●) [MEN 09]*

Using a different optimization algorithm, Lebaal *et al.* [LEB 09] optimized the flow distribution at the exit while keeping the pressure gradient to a minimum (Figure 6.9).

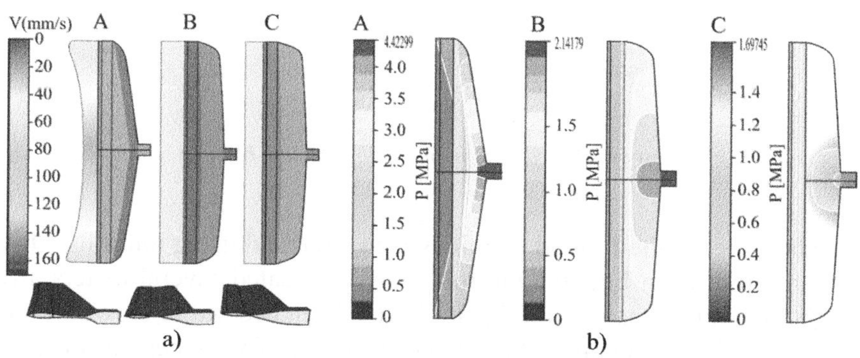

Figure 6.9. *Velocity a) and pressure b) distribution for the original die (A) and the optimized dies (B and C) [LEB 09] (for a color version of this figure, see www.iste.co.uk/lafleur/polymer.zip.)*

6.3. Film blowing

6.3.1. *Process overview*

6.3.1.1. *Introduction*

The use of the film blowing technique started at the beginning of the 1940s. The first films were made up of polyethylene to develop water-tight packaging with high durability at low temperature. Visking Corporation is known as being one of the first companies to have developed this type of process. A general diagram of the process is given in Figure 6.10: a single-screw extruder feeds molten polymer to a vertical die, which will be described in detail in section 6.3.1.2.

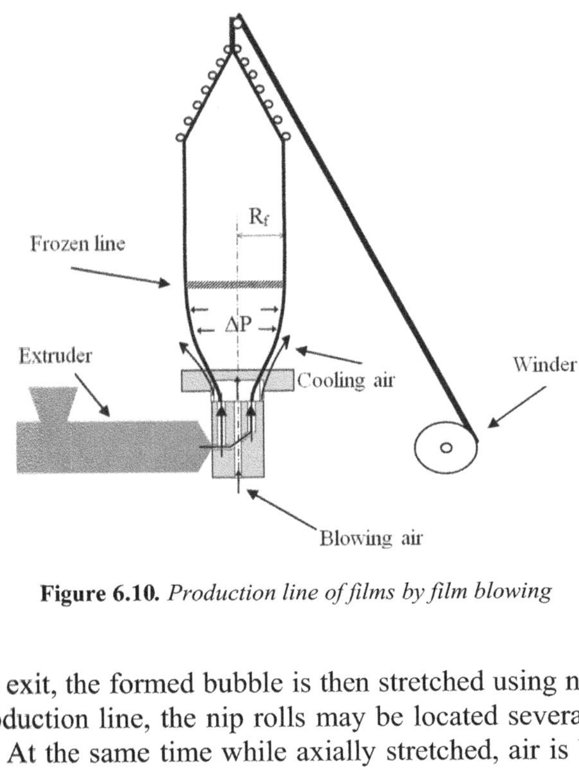

Figure 6.10. *Production line of films by film blowing*

At the die exit, the formed bubble is then stretched using nip rolls. On an industrial production line, the nip rolls may be located several meters away from the die. At the same time while axially stretched, air is blown into the bubble through an opening in the center of the die. This generates an excess of pressure that inflates the bubble. Actually, the polymer undergoes biaxial stretching. It is stretched longitudinally using the nip rollers and transversally by inflation. Alongside the stretching operations, the bubble is cooled by the system described in section 6.3.1.4.

Dimensions of film blowing lines can vary greatly according to the type of machine (laboratory or industrial), the intended products and applications. The diameter of the die may vary from 50 mm to 2 m, the gap from 0.5 to 3 mm, the flow rate from 5 to 500 kg/h. In some cases, the obtained films can have thicknesses ranging from 8 to 300 µm. The bubble diameter can vary between 10 cm and 6 m. Contemporary coextrusion dies produce films with up to 11 layers. This large number is required for high-performing barrier films. In fact, some of the used resins, such as PA and EVOH, must be protected from humidity by a protective layer. Moreover, it is often necessary to use an adhesive layer between different materials to ensure cohesion of the multilayered film.

The take-up ratio (TUR), also called draw ratio (DR), defines the longitudinal stretching of the film induced by the nip rollers:

$$TUR = \frac{V_f}{V_0} \qquad [6.12]$$

where V_f is the speed of the nip rolls and V_0 is the initial speed at the die exit.

The blow-up ratio (BUR), defines the transverse stretching of the film induced by film blowing:

$$BUR = \frac{R_f}{R_0} \qquad [6.13]$$

where R_f is the final radius of the bubble and R_0 is the initial ratio.

It is possible to evaluate the draw down ratio (DDR), by using the conservation of mass flow rate, Q_m, between the die exit and the freeze line, using the following relation:

$$DDR = \frac{e_0}{e_f} = \frac{Q_m}{\rho_0 2\pi R_0 V_0} \frac{\rho_f 2\pi R_f V_f}{Q_m} = \frac{\rho_f}{\rho_0} TUR\, BUR \qquad [6.14]$$

where ρ_0 is the density at the die exit and ρ_f is the density at the frost line.

It is, therefore, possible to control the thickness of the film by appropriately adjusting the flow rate, the TUR and the BUR.

Film blowing has many advantages. It is characterized by the fact that a bi-axial stress is imposed on the film. The impact properties of the films obtained using this technique are superior, at equal widths, to those of films obtained using cast film extrusion. The mechanical and optical properties of the film can also be controlled, to some extent, by adjusting the cooling, the TUR and the BUR. Another advantage of this manufacturing technique, compared to the previous one, is the small size of the die. In fact, to produce a film with a width of 2.5 m using cast film extrusion, a die of 2.6 m must be used. However, using the film blowing technique, with a BUR of 3, a die diameter of 0.265 m is sufficient. This process is, therefore, the only one which produces large-sized films, while using less energy, which reduces costs. Finally, it is possible to produce films of different widths using a single machine, without changing the die. This process is, therefore, highly versatile.

On the other hand, as a result of this flexibility, films produced using this technique are generally of a poorer quality compared to those produced by cast film extrusion. Usually, cooling by forced convection at ambient temperature is not so efficient and can cause significant variations in thickness (approximately 7%). In addition, the low cooling rates produce films with a high degree of crystallinity, causing the film to be less transparent. Higher crystallinity, however, improves the barrier properties and the tensile mechanical properties. It should also be noted that the start-up of this type of equipment is trickier than in cast film extrusion and results in more losses. As soon as a steady-state regime is established, however, there are very few losses, since it is not necessary to cut the edges. As a whole, this method loses little material. Film blowing is, therefore, more desirable for high volume manufacturing than for operations requiring frequent stoppages and start-ups.

6.3.1.2. *Film blowing dies*

As for cast film extrusion, film blowing requires a molten polymer to be fed through a die. In this case, an annular die. There are four types of dies used in film blowing: spider dies, dies with a distribution channel, stack dies and helical dies. The latter, due to their design, can be stationary, oscillating or rotating. Helical and stack dies can be used in simple extrusion or in coextrusion.

Using spider dies was very common several years ago. Nowadays, they are mainly used to shape heat-sensitive materials to avoid the risk of degradation, such as PVC or PVDC. The polymer is fed through the center of the housing and flows along the internal mandrel, which is composed of an inverted cone supported by spider legs, similar to that described for pipes manufacture, in section 5.2.1. The flow is disturbed by these spider legs, causing weld lines, which can, in turn, give rise to defects. This type of die is limited by the force that the spider legs are able to withstand. Usually, a diameter up to 750 mm can be reached.

Dies with a distribution channel, Figure 6.11, are similar to wire coating dies, which are described in more detail in section 7.3.1.4. A distribution channel distributes the material leaving the extruder around the central mandrel, like a kind of flat coat-hanger die rolled around itself.

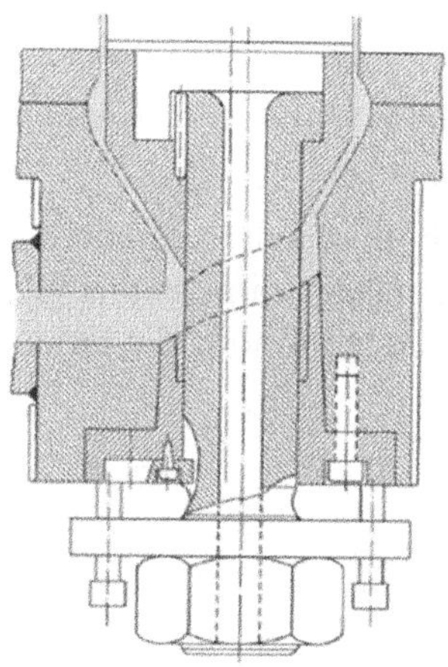

Figure 6.11. *A die with a distribution channel for film blowing [TAD 06]*

Stack dies appeared fairly recently. They are mainly used to produce multilayer films with small diameters. This type of die is composed of a series of plates fed from the sides, stacked one by one on top of each other (Figure 6.12).

In a coextrusion die, each layer possesses its own plate. An advantage of this type of equipment is that it is possible to include thermal insulation layers between the plates, thus allowing the plates to remain at different temperatures. Moreover, the system is extremely compact, compared with other types of dies. Finally, another advantage of using this die is its flexibility: in fact, it is possible to easily replace, add or remove the plates to alter the number of layers produced.

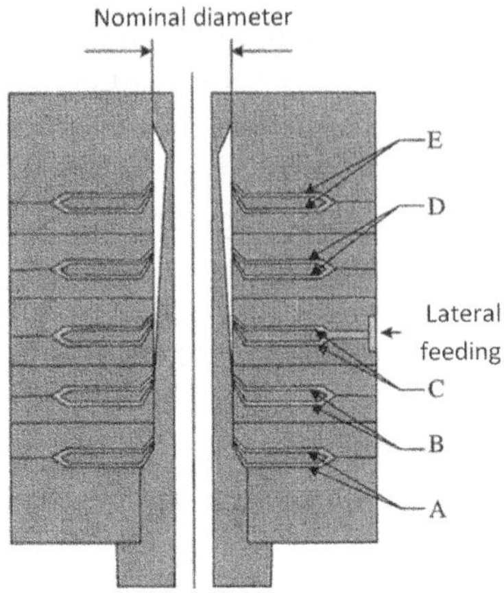

Figure 6.12. *A stack die for the coextrusion of multilayer films [TAD 06]*

Helical dies are currently the most used and they allow high flow rates to be obtained, for films of constant dimensions (Figure 6.13). We must also note that this type of die can accommodate a large variety of materials.

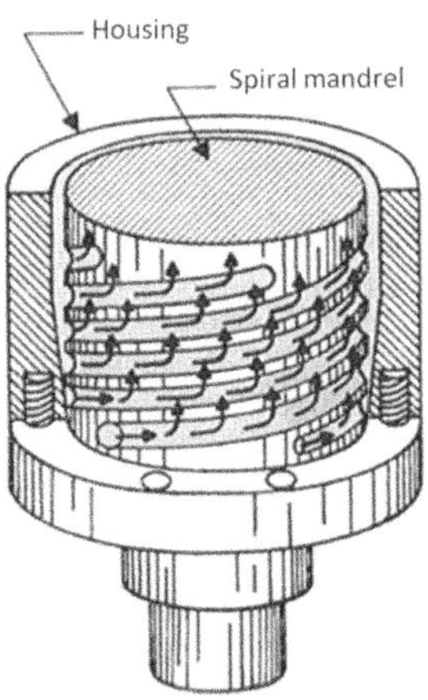

Figure 6.13. *A helical die [TAD 06]*

The molten polymer is introduced into the interconnected helical channels, which circumferentially distribute the molten polymer as uniformly as possible. The helical channels are followed by a relaxation zone which reduces the memory effect, common among viscoelastic materials.

This type of die can be used to produce multilayer films, but in this case, each resin must have its own system of helical channels and its own relaxation zone. The materials are then grouped in sections, situated above the relaxation zones, just before the die exit. As a result, the dies are extremely bulky.

6.3.1.3. *Die design and flow modeling*

Modeling flows in dies used in film blowing is complex and requires the use of numerical models. However, in all types of dies, an annular zone can be considered to represent the final portion of the flow.

For a Newtonian fluid in this type of flow, the relationship between the flow rate and the pressure drop is as follows:

$$Q = \frac{\pi R^4 \Delta P}{8\eta L}\left[1-\kappa^4 - \frac{(1-\kappa^2)^2}{2\ln\frac{1}{\kappa}}\right] \qquad [6.15]$$

where κ is the inner to outer radius ratio and is equal to $1/\beta$.

For a power-law fluid, we obtain:

$$Q = \frac{\pi R_0^3}{s+2}\left(\frac{R_0 \Delta P}{2K\,L}\right)^s \left(\frac{\beta-1}{\beta}\right)^{2+s} F(m,\beta) \qquad [6.16]$$

where R_0 is the inner radius of the die, $s = 1/m$, and $F(m, \beta)$ is a function that depends on the power-law index m and the geometry of the die [FRE 58]. For annular dies used in film blowing, the geometric factor κ is always within the range $0.5 < \kappa < 1$. For example, let us take a die with an inner diameter of 20 mm and a gap of 0.5 mm: $\kappa = 20/21 = 0.952$. For κ values in the range 0.5–1, the function $F(m, \beta)$ only depends on the geometry of the die (Figure 6.14). By substituting $F(\beta)$ by a simple function in the flow rate expression, we obtain:

$$Q = \frac{\pi R_0^3}{s+2}\left(\frac{R_0 \Delta P}{2K\,L}\right)^s \left(\frac{\beta-1}{\beta}\right)^{2+s} \frac{1+\beta}{2\beta} \qquad [6.17]$$

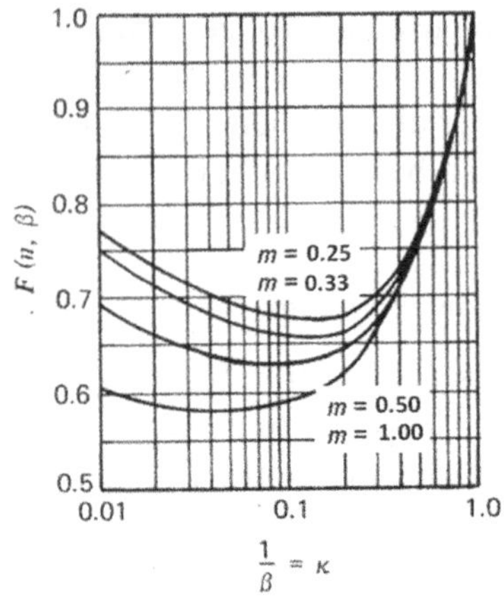

Figure 6.14. *Function F(m, β) for the flow rate of power-law fluids in the annular region [FRE 58]*

Alongside these basic approaches for calculating the flow in the helical die, we also have 2D finite volume models [SAI 84, VLC 88, HUA 98] and, of course, 3D finite element models [COY 91]. For example, in Figure 6.15 we present an example of the results obtained using a 2D model, based on the lubrication approximation [SAI 84]. It shows how altering the geometric parameters affect the circumferential flow distribution at the die exit. By systematically using this type of analysis for different geometric parameters, we can define the best possible compromise, thereby obtaining a satisfactory distribution of the material.

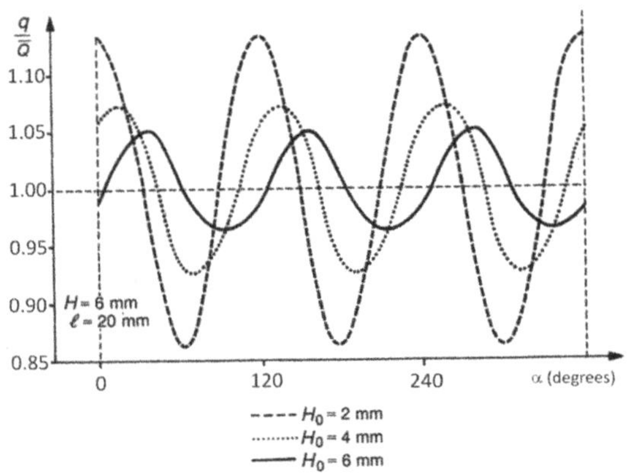

Figure 6.15. *Influence of initial channel height H_0 on the distribution of the material (local/mean flow rate) at the die exit [VER 08]*

6.3.1.4. *The cooling process*

At the die exit, the bubble is cooled by a high velocity air stream, which is blown uniformly onto the external wall, by means of a cooling ring situated directly above the die. In older machines, the air is blown using a simple cooling ring, which propels the stream axially along the bubble. Modern production lines use a double cooling ring, which feeds 20% of the total airflow across a lower lip, to propel the air in the flow direction of the molten polymer. Sidiroupoulos *et al.* [SID 96, SID 99, SID 00a, SID 00b, SID 01] carried out numerical finite element simulations, which predicted the air flow pattern around a rigid bubble, at the exit of two types of cooling rings (simple and double flow). The results indicated that there were significant differences between the two types of rings. Feeding through a simple flow ring creates a local acceleration of the air between the cooling ring and the bubble, thereby generating a negative relative pressure in the cooling ring. This phenomenon is not observed with double flow rings. Double flow rings create a local overpressure. This explains why using a simple flow ring attracts the bubble towards the ring, whereas the opposite phenomenon is observed when a double flow ring is used. In the case of a simple flow ring, cooling mainly occurs at the ring level. For a double flow cooling ring, the authors state that maximum cooling occurs a dozen centimeters above the cooling ring. They attribute this phenomenon to the increased turbulence in these regions (Figure 6.16).

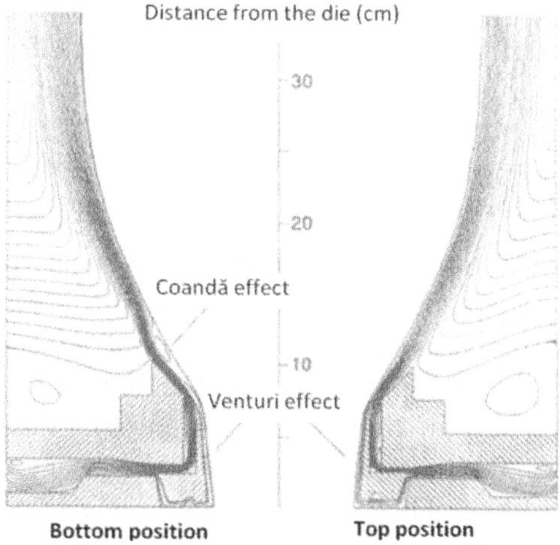

a) b)

Figure 6.16. *Modeling of the flow lines of the cooling air blown by a double flow ring around a bubble of LLDPE: a) moderate cooling and b) strong cooling [SID 00a]*

It is possible, on some coolings rings, to adjust the geometry to vary the airflow relative to the two lips. By using air cooled between 8°C and 18°C, it is possible to increase the production rate by up to 30%. Below 8°C, water vapor in the air tends to condense on the cooling ring. Other advantages of using cooled air are that there is less variability caused by climatic conditions and better optical properties due to finer crystal size. With regard to the air that is trapped inside the bubble, it rapidly heats up and its contribution to the cooling process is almost negligible. However, it is possible to use an internal bubble cooling (IBC) system, which recycles the air inside the bubble. The diameter of the bubble and the internal pressure are kept constant, by controlling the entry and exit of air inside the bubble. This increases heat transfer between the internal wall of the bubble and the inflation air. Moreover, internal cooling stabilizes the bubble, allowing a greater external airflow for cooling.

Upon cooling, the semi-crystalline polymer reaches its crystallization temperature and solidifies. The orientation of the film is, therefore, fixed. The appearance of the crystalline phase is characterized by a significant increase in the turbidity of the film. The position at which this turbidity is manifested is referred to as the frost line height (FLH). Above this position, the bubble forms a tube with a constant thickness and diameter. When the film is sufficiently cooled, the bubble is, gradually, folded back on itself through an inverted V-shaped cage, made up of guide rollers. Then, the film is coiled by a complex rolling system, which assures its uniformity and minimizes local variations in thickness. The flat tube can be directly coiled, axially cut to obtain a film twice as wide or axially cut twice to obtain two coils of equal widths.

6.3.1.5. *Instability problems*

Film blowing, just as in cast film extrusion, is particularly subject to instabilities, arising frequently due to the shaping process. In many cases, these instabilities can be lessened by using more efficient dies or calibrators. However, as in the case of fibre spinning, the final dimensions of the film are obtained by stretching in air. The molten polymer has a free surface that can move easily. It is, therefore, highly sensitive to many disturbances which can generate instabilities. These instabilities give rise to variations in the film dimensions (bubble thickness and width), which make conditioning the film very tricky, as well as significant variations in the final properties, incompatible with the market requirements. Instabilities are, therefore, responsible for reduced productivity and hence profits. Thus, it is not surprising that these instabilities were one of the first topics tackled in film blowing studies.

Han and Park [HAN 75] reported draw resonance instability, which may also be encountered in fibre spinning. This type of instability corresponds to a periodic variation in the diameter of the bubble during one-directional stretching. This instability depends on the critical DR, which depends on the polymer resin and the processing conditions. The amplitude and frequency of the resonance, progressively, increase until the bubble ruptures. Works by Minoshiwa and White [MIN 86] and Ghaneh-Fard *et al.* [GHA 96] qualitatively identified two other types of instabilities. Metastable instabilities are characterized by a periodic variation in the frost line position and helical instabilities are characterized by the rotation of the bubble around the die axis. Figure 6.17 illustrates these three types of instabilities. Although Minoshiwa and White [MIN 86] attribute helical instability to an aerodynamic effect caused by air cooling, Laffargue *et al.* [LAF 02] suggested that this instability is caused by the significant inflation of the bubble, which induces transverse resonance causing the bubble to loose axisymmetry.

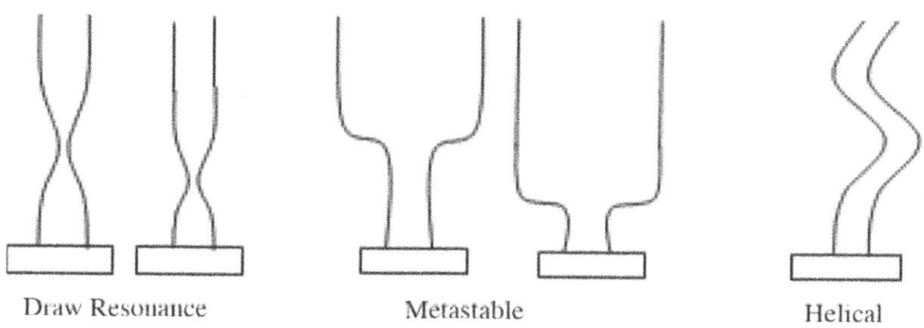

Figure 6.17. *Different types of instabilities*

Minoshiwa and White [MIN 86] and Kanai and White [KAN 84] created maps illustrating regions of stability and instability for HDPE, LDPE and LLDPE, depending on the TUR and the BUR. They suggested that a narrow molecular weight distribution would decrease the stability zone. This explains why the production of mPE films is particularly problematic. LDPE has a wider stability zone than the other two materials. This confirms a common practice in industry. In fact, HDPE and LLDPE films are frequently produced by adding a small amount of LDPE to each blend. The stabilizing effect of this type of blend was formally demonstrated by Fleissner [FLE 88], who attributed the increased stability of LDPE to strain-hardening. It appears that the resins with a marked strain-hardening behavior always have a higher stability. Finally, all the encountered studies showed that a greater cooling rate widened the stability zone.

6.3.1.6. *Post processing*

Generally speaking, despite the variety of available materials, films do not possess all of the properties required to be directly used as packaging. In fact, plastic films usually have an inert surface which does not allow adhesion of printing ink to the film surface. It is, therefore, necessary for the film surface to be treated. For packaging requiring complete impermeability to gases, humidity and light, it is necessary to laminate or even metallize the film surface by aluminum vacuum evaporation. This allows a completely hermetically sealed packaging to be obtained. Some applications require the presence of microperforations in the film, to allow air to escape from the packaging.

One of the most frequently used techniques to improve the adhesion capacity of the film is to carry out a Corona surface treatment. By applying a high voltage electrostatic discharge close to the film surface, ionized gas (plasma) forms. This causes the free radicals present in the plasma and on the film surface to interact, thereby causing local oxidation which generates fine irregularities on the film surface. This increase in surface roughness improves printability and adhesion in laminating operations. A less widespread technique involves oxidizing the film with a flame. However, this technique is hardly ever used due to the flammability of some thermoplastic materials. It is also possible to apply a very thin layer of a polar substance onto the film, which facilitates printing. The most frequently used materials are polyesters, polyurethanes and colloidal silica.

To facilitate printing, special inks were developed especially for plastic films. These inks allow attractive films to be obtained by using the appropriate printing method. Flexography is the most widespread printing technique. It involves using inexpensive flexible rubber plates fixed to a rotating roll that is supplied with fast-drying liquid ink. When the film comes into contact with the plate, the desired pattern is printed. The second technique, heliogravure, involves engraving the desired pattern onto a metal roll. This process is more expensive and better quality printing is achieved.

In both cases, the desired pattern is printed on the film in color, through the successive printing of the primary colors.

6.3.2. *Film blowing models*

6.3.2.1. *Membrane approach: Newtonian models*

The first studies aiming at mathematically describing the axisymmetric flow of a tubular liquid film with a variable radius were done by Pearson and Petrie [PEA 70a, PEA 70b]. To make solving this problem easier, the authors assume that the bubble is axisymmetric and isothermal, and that the fluid used is Newtonian. Moreover, they treat the bubble as a thin membrane, since the film is relatively thin. Therefore, the velocity profile and the stress tensor must be considered as independent from the thickness of the film. The coordinate system used by Pearson and Petrie [PEA 70a, PEA 70b] is shown in Figure 6.18. We observe that the reference point used is oriented towards the tangents and normal to the film (ξ_1, ξ_2, ξ_3). Since this reference point is located on the film surface, and given that the position and orientation of this boundary vary with the axial position, this reference point is mobile.

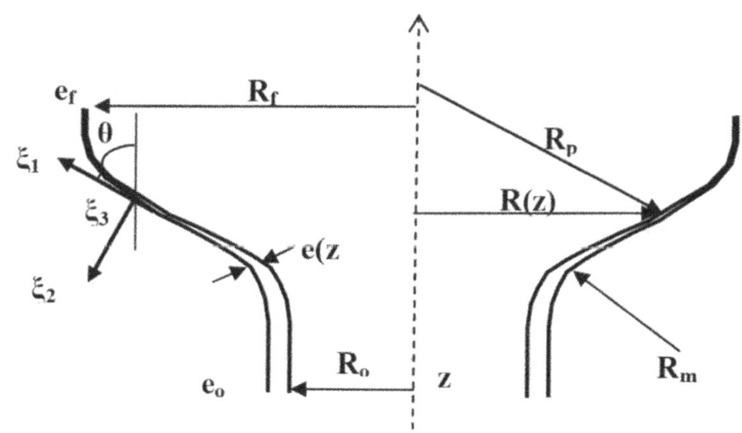

Figure 6.18. *The coordinate system used by Pearson and Petrie [PEA 70a, PEA 70b]*

It is possible to describe the evolution of the bubble along the axial position according to the evolution of the meridian curve $R(z)$. The variation in this curve representing the swelling angle θ can be described locally by the following relation:

$$\frac{dR(z)}{dz} = \tan\theta \qquad [6.18]$$

Moreover, it is possible to describe the radii of curvatures R_m and R_p according to directions ξ_1 and ξ_3 with the following relations:

$$R_m = -\left[1+\left(\frac{dR(z)}{dz}\right)^2\right]^{3/2}\left(\frac{d^2R(z)}{dz^2}\right)^{-1} \qquad [6.19]$$

$$R_p = \frac{R(z)}{\cos\theta} = R(z)\sqrt{1+\left(\frac{dR(z)}{dz}\right)^2} \qquad [6.20]$$

As the problem is axisymmetric, the flow is only in the plane (ξ_1, ξ_2). The membrane approximation allows us to disregard the velocity component in the thickness direction. The velocity vector, therefore, becomes equal to:

$$\underline{V} = V\underline{\xi_1} \qquad [6.21]$$

This membrane formula allows us to describe the problem, by expressing the different conventional equations of fluid mechanics:

– *mass conservation* is given by the following relation:

$$Q = 2\pi\, R(z)\, e(z)\, v(z) = \text{constant} \qquad [6.22]$$

– *dynamic equilibrium*: by neglecting the forces of friction with air, inertia, surface tension and mass with regard to the extensional viscosity forces and by neglecting the shear terms, we obtain a diagonal stress tensor. The membrane approximation also allows us to neglect the stress in the thickness direction. The stress tensor $\underline{\underline{\sigma}}$ becomes:

$$\underline{\underline{\sigma}} = \begin{bmatrix} \sigma_1 & 0 & 0 \\ 0 & 0 & 0 \\ 0 & 0 & \sigma_3 \end{bmatrix} \qquad [6.23]$$

The dynamic equilibrium equations projected along the directions ξ_1 and ξ_2 are written as:

– in the ξ_1 direction:

$$F_f = 2\pi R(z) e(z) \sigma_m \cos\theta + \pi\left(R_f^2 - R(z)^2\right)\Delta P \qquad [6.24]$$

– in the ξ_2 direction:

$$\frac{\Delta P}{e(z)} = \frac{\sigma_m}{R_m} + \frac{\sigma_p}{R_p} \qquad [6.25]$$

where F_f and ΔP are the pulling force at the frost line and the internal pressurization inside the bubble with regard to atmospheric pressure, respectively.

Assuming a Newtonian behavior, Pearson and Petrie [PEA 70a, PEA 70b] obtain a set of equations which can be solved by arbitrarily fixing the frost line. The numerical solution is performed from the frost line to the die exit, by employing a two-parameter shooting method. The pulling force and the pressurization inside the bubble are modified until the calculated values for the bubble radius and its thickness at the die exit are equal to those of the die. At the frost line, the swelling angle of the bubble is assumed to be zero, and the radius and velocity are known, from TUR and BUR, respectively.

An alternative solution involves applying the shooting method from the die exit to the axial position. Cain and Denn [CAI 88] and André *et al.* [AND 98, AND 99] solved the equations of Pearson and Petrie by using this approach. It is necessary to use the initial swelling angle at the die exit as the shooting parameter. However, depending on the initial angle, the calculated bubble may tend to an infinite radius or even shrink to a radius of zero. Studies by André [AND 99] also show that there are multiple solutions. In fact, the initial value chosen for the swelling angle directly affects the convergence of the equations towards a solution, with a swelling rate greater or less than one.

An example of solution multiplicity is shown in Figure 6.19. For a TUR of 10 and a dimensionless pressurization of 0.6, a bubble with a BUR of 0.5 or 4.5 can be obtained. To avoid these problems, André [AND 99] proposed to use a top-down calculation, originally proposed by Pearson and Petrie [PEA 70a, PEA 70b]. Despite the fact that heat transfer phenomena have been neglected, the results obtained appear to be qualitatively reliable and allow the effect of different processing parameters on the shape of the bubble to be identified. Nevertheless, many authors seek to improve this initial model to make it more representative of the physical reality of the process.

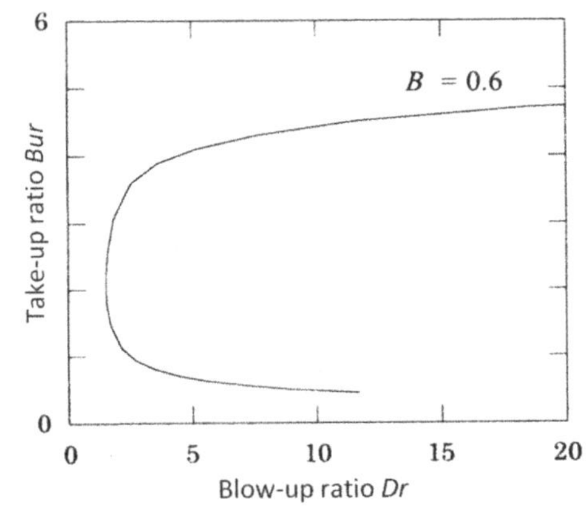

Figure 6.19. *Multiple Newtonian fluid solutions for a take-up force F and a bubble pressure B [AND 99]*

Petrie modified the model developed in 1970 to include the effects of gravity and inertia [PET 75a, PET 75b]. It still seemed evident to the author that the equations used to evaluate the heat transfer behavior would be needed to quantitatively describe the film blowing process. He therefore used experimental values for the temperature profiles obtained by Ast [AST 74] and Farber and Dealy [FAR 74] to, empirically, include the effect of temperature in the model, by linking the viscosity to temperature using the Arrhenius equation. Petrie showed that the effects of inertia and the variation in density due to the non-isothermal characteristics of the process are very weak. However, the effect of gravity, although small, is significant. The geometry and velocity profile of the bubble are affected by this phenomena. The author compared his calculated results to the experimental results of Ast [AST 74] and Farber and Dealy [FAR 74]. This revealed that the shape of the bubble is controlled by a behavior between the one corresponding to a purely elastic model and one corresponding to a purely viscous model. This is hardly surprising given the viscoelastic nature of polymers. However, the velocity profiles and the strain rate obtained with elastic and viscous calculations do not overlap with the experimental results.

Han and Park [HAN 75] presented a model that accounted for heat transfer between the cooling air and the bubble. The thermal balance equation is simplified, assuming that the temperature of the film is independent of the thickness. It is written as follows:

$$\rho C_p v \frac{\partial T}{\partial \xi_1} = \frac{\partial q}{\partial \xi_2} \quad [6.26]$$

where ρ is the density, C_p is the specific heat capacity of the polymer, v is the velocity and q is the flux term. The boundary conditions used are as follows:

– at the internal wall: $q = 0$ [6.27]

– at the external wall: $q = h_T (T - T_\infty)$ [6.28]

where h_T is the heat transfer coefficient by forced convection, T_∞ is the temperature far from the bubble and T is the temperature of the film.

The temperature and the velocity gradient are given by:

$$\frac{\partial T}{\partial \xi_1} = \frac{\partial T}{\partial z}\frac{\partial z}{\partial \xi_1} = \frac{\partial T}{\partial z}\cos\theta \quad [6.29]$$

$$v = \frac{Q}{2\pi Re} = \frac{v_z}{\cos\theta} \quad [6.30]$$

By substituting relations [6.29] and [6.30] in equation [6.26], we obtain:

$$\frac{dT}{dz} = \frac{h_T (T - T_\infty)}{\rho C_p v_z e} \quad [6.31]$$

In this first approach to calculating the heat balance, the heat transfer coefficient h_T was considered to be constant. Its value was arbitrarily set to adequately describe the experimental results. This thermal balance equation has since been used by many authors [KAN 84, LUO 85, SID 01], without the initial assumption of temperature uniformity throughout the thickness of the film being formally validated. Also, note that the convection coefficient used in all cases was adjusted accordingly in order to adequately correspond to the experimental results. This explains the strong correlation observed between the calculated and experimental results.

Using a constant convection coefficient is not always representative of the conditions occurring during the production of films. In fact, the velocity of the air blown onto the surface of the bubble using single and double flow cooling rings is much greater in the zone close to the die than in the frost line zone. To carry out applicable calculations, a realistic representation of the evolution of the convection coefficient must be obtained. It turns out that, among the publications on film blowing, some authors [ZIP 71, MEN 72, PET 74, KAN 84] were able to connect the convection coefficient to the velocity of the cooling air by coupling air velocity and bubble temperature measurements by infrared. By knowing the temperature profile and the bubble geometry, it is possible to deduce the local value of the convection coefficient, due to the heat balance we have just presented. The empirical relations presented in Table 6.2 were obtained by the authors previously mentioned. They linked the maximum velocity of the cooling air to the convection coefficient.

Authors	Relations
Petrie [PET 74]	$h_T = 4\ (V_{max})^{1.5}$
Menges and Predöhl [MEN 72]	$h_T = 3.3\ (V_{max})^{1.5}$
Zippenfeld [ZIP 71]	$h_T = 7.11\ (V_{max})^{0.78}$
Kanai and White [KAN 84]	$h_T = 0.043\ \dfrac{k_{air}}{L}\ (Re)^{0.78}$

Table 6.2. *Relations describing the evolution of the convection coefficient as a function of the cooling air velocity*

Although these relations are useful, they do not allow us to obtain the evolution of the convection coefficient as a function of the axial position, in the case where the velocity of air is unknown. Nagarajan and Campbell [NAG 95] used the air velocity distribution model, developed by Cao and

Campbell [CAO 89], to determine the evolution of the convection coefficient with axial position, using the relations obtained by the previous authors. Figure 6.20 illustrates the results obtained by Nagarajan and Campbell [NAG 95] as well as the correlations describing the evolution of the convection coefficient for each case.

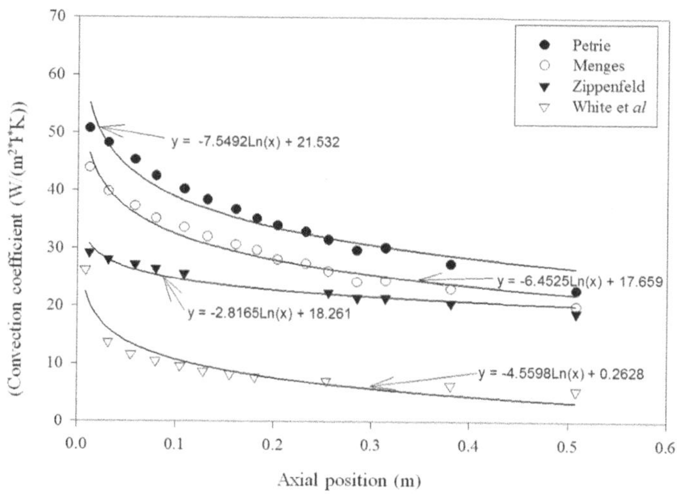

Figure 6.20. *Convection coefficients obtained by Nagarajan and Campbell with different models and fitting relations [NAG 95]*

First, in this figure, we observe that the four curves possess the same general shape. This is not surprising since all the tests were carried out using laboratory extrusion lines. The cooling rings used had a height ranging between 8 and 12 cm. This explains why the convection coefficients initially see a sharp decrease (cooling ring zone), and then decrease more gradually at a distance of approximately 15 cm from the die exit.

In Figure 6.20, we observe that the convection coefficient, resulting from the Kanai and White relation [KAN 84], is much lower than that from the data of Menges and Predöhl [MEN 72], Petrie [PET 74] and Zippenfeld [ZIP 71]. This is explained by the fact that Kanai and White [KAN 84] did not consider an overall heat transfer coefficient, including both radiation and convection, which was the case for the other three authors. In fact, Kanai and White [KAN 84, KAN 87] modified the thermal balance equation of Han and Park [HAN 75] to include the heat radiation term and the crystallization temperature. In this case, the thermal balance equation is written as:

$$2\pi R_{ext}\left[h_c(z)(\overline{T}-T_{air\_out})+\varepsilon\sigma((\overline{T}+273)^4-(T_{air\_out}+273)^4)\right]$$
$$+\rho Q\Delta H_f\cos\theta\frac{d\chi}{dz}=-\rho C_p\left(Q\cos\theta\frac{\partial\overline{T}}{\partial z}\right) \quad [6.32]$$

where ε is the emissivity of the film, σ is the Stefan-Boltzmann constant, ΔH_f is the latent heat of crystallization of the material and χ is the crystallinity.

The authors hypothesize that the flow behavior of the molten polymer is dominated by the temperature dependence of the rheological properties rather than by the rheological model used to represent it. The viscosity is estimated using a modified Arrhenius equation, which takes into account the temperature and the effect of the presence of crystals:

$$\eta=\eta_0\exp\left[\frac{E_a}{R}\left(\frac{1}{T}-\frac{1}{T_0}\right)\right]\exp G\chi \quad [6.33]$$

where η_0 is the viscosity of the non-crystallized material at the reference temperature T_0, E_a is the activation energy, R is the ideal gas constant and G is a constant deduced from experimental data.

Kanai and White [KAN 84, KAN 87] were the first to completely model the film blowing process, from the rheology of the studied materials to the final crystallinity of the films produced, by studying the evolution of internal pressure within the bubble for different TURs. The rates at which the radius and thickness of the bubble reach their final size depend, primarily, on the activation energy. These studies help illustrate the importance of taking temperature into account to adequately model the process.

Figure 6.21(a) illustrates the importance of carrying out thermomechanical calculations using the appropriate cooling coefficient. In fact, although using a constant cooling coefficient, just like in the initial method by Petrie [PET 75a], did not seem to affect the geometry of the bubble, compared to a variable cooling coefficient as in the relation by Menges and Predöhl [MEN 72], it is possible to observe that the temperature profile is different. However, in Figure 6.21(b) we find that the temperature history plays an important role in the development of stresses inside the film. The axial stresses in the bubble, calculated using a constant cooling ratio, are

20% lower than those using a variable cooling ratio. Incorporating an appropriate cooling function is, therefore, crucial if a thermomechanical calculation, representative of reality, is desired.

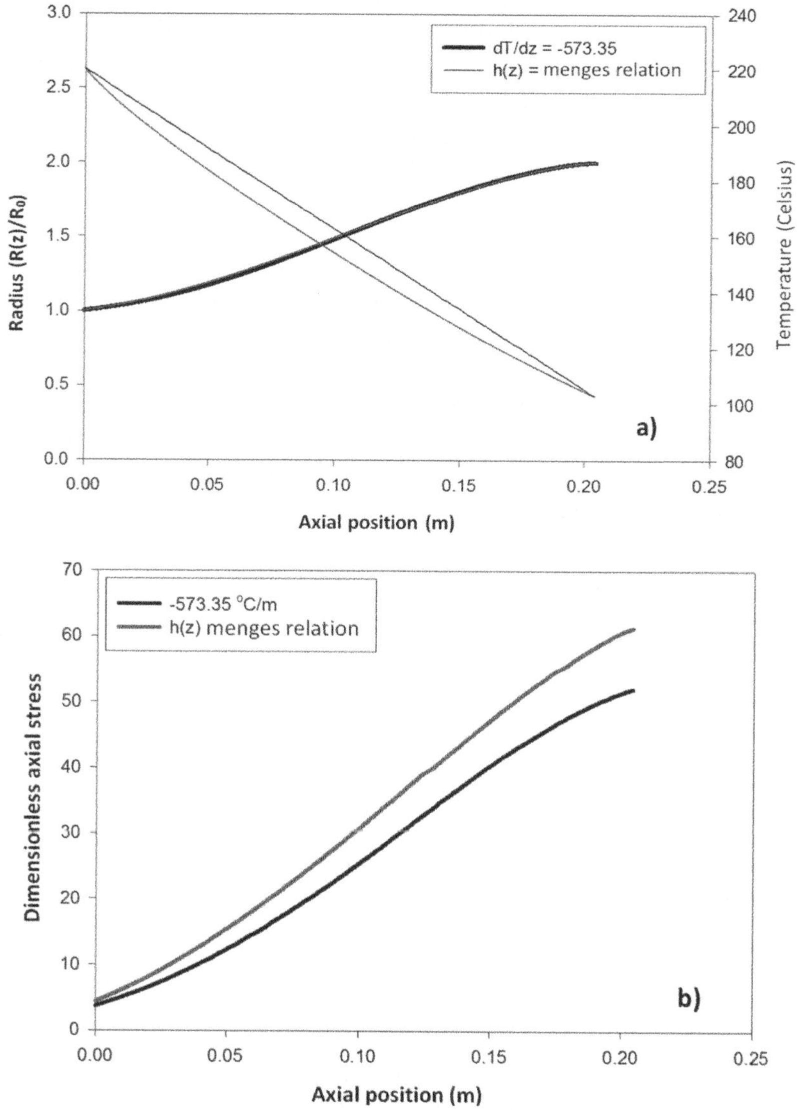

Figure 6.21. *A multilayer single material film blowing. Comparison between constant cooling rate and a calculated convection coefficient: a) radius and temperature profile and b) stress in the flow direction [GAM 04]*

By using the equations of Pearson and Petrie [PEA 70a, PEA 70b], combined with the thermal balance equation developed by Han and Park [HAN 75], it is possible to carry out calculations that reveal the effect of the material properties on the bubble temperature and geometric profiles. For example, in Figure 6.22 we find that an increase in activation energy caused the bubble to inflate very quickly. This corresponds to the experimental results. In fact, an increase in activation energy results in an increased viscosity at the frost line. It is therefore necessary to apply a greater force to deform the bubble at this position. This increase in force at the frost line is transported in the film until the die exit.

However, at this position, the viscosity is identical in all the three studied cases. The increase in force results in deformation, and therefore swelling, which is much faster when the activation energy is small. The rapid swelling results in heat loss caused by an increase in exchange surface area, and also results in a rapid decrease in temperature, which can be seen in Figure 6.22. A similar effect, although of a lesser magnitude, can be observed when the specific heat capacity is decreased. These calculations were carried out by using the data of Menges and Predöhl [MEN 72].

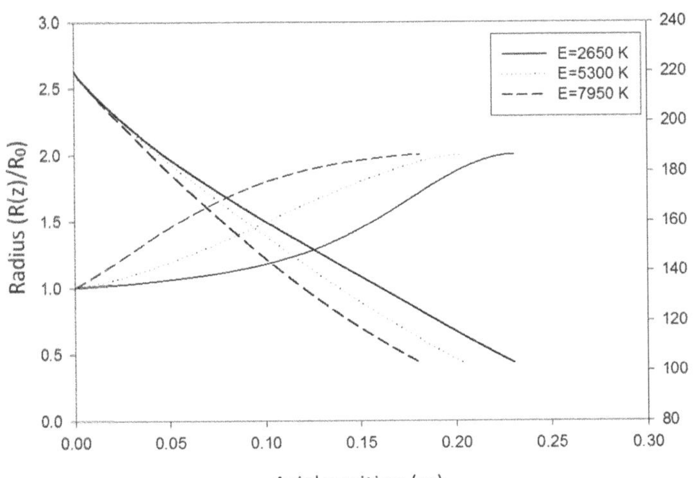

Figure 6.22. *Temperature profile and bubble shape for different activation energies (C_p = 2200 J/kg/K) [GAM 04]*

Finally, some authors have tried to take into account the effects of varying local pressure generated by the cooling ring. Cao and Campbell

[CAO 89], and then Campbell *et al.* [CAM 92], used a rigid setup, replicating several bubble shapes, to measure the distribution of pressure and velocity profiles and the axial force exerted on that rigid setup. They then carried out a macroscopic balance of the amount of motion to develop an analytical solution for the pressure exerted upon the external surface of the bubble, by the cooling air according to the axial position. The analytical solution is, then, used to alter the overpressure inside the bubble. This correction of overpressure reveals that the shape of the bubble at the die exit, the strain rate and the stress in the film can be greatly influenced by the external pressure. This confirms the experimental results. Therefore, according to these authors, it is necessary to include aerodynamic phenomena, for modeling the film blowing process, in order to accurately predict the physical properties of the films obtained. A similar approach was successfully used by André [AND 99] to illustrate the effect of single- and double-lip cooling rings on the bubble geometry.

6.3.2.2. *Membrane approach: viscoelastic models*

6.3.2.2.1. Conventional models

The first studies that introduced viscoelasticity into the isothermal film blowing calculation were carried out by Petrie [PET 73]. He studied the effect of employing the Maxwell and Oldroyd models, and experienced numerical instabilities. Petrie attributed these problems to the fact that the calculation began at the frost line, the value of which was assumed to be known, and ended at the die exit. Luo and Tanner [LUO 85] developed a non-isothermal model using the Maxwell and the Leonov models. The equations are solved from the die exit to the frost line, the initial angle of the bubble, thereby, becoming an additional shooting parameter. The authors demonstrate the importance of taking the viscoelasticity into account, which limits the swelling of the bubble through the prescribed draw force and overpressure, compared to the results obtained using the Newtonian model. The simulations tend to create slight differences in the shape of the bubble when the Maxwell and the Leonov models are used. However, it appears that the variation in the rheological properties with temperature is what, primarily, affects the shape of the bubble. Taking heat transfer into account is, therefore, more important than taking the viscoelastic effects into account when modeling the film blowing process. The results obtained using the Leonov model do not seem to describe purely elongational flows. The results obtained using the Maxwell model, with a mean relaxation time, are comparable to the experimental results obtained by Gupta and Metzner

[GUP 82]. Unfortunately, solving this problem numerically is, once again, an arduous task. The authors faced convergence problems and obtained multiple solutions under certain conditions.

The anisothermal model developed by André [AND 99] shows, in the case of the Maxwell model with a single relaxation time, that the numerical instabilities, observed by Luo and Tanner [LUO 85], actually correspond to the disappearance of a solution when the Deborah number is increased. In fact, when the Deborah number increases, we move away from a Newtonian case, where it has a value of zero. The Deborah number quantifies the significance of viscoelastic effects compared to flow time. At a critical Deborah number, >0.04, the calculated force and overpressure diverge, making the bubble calculation impossible. It is possible to calculate the Deborah number at the die exit using the following relation:

$$De_0 = \lambda_0 \frac{V_0}{R_0} = \lambda_0 \frac{Q}{2\pi H_0 R_0^2} \qquad [6.34]$$

where λ_0 is the mean relaxation time at the die exit.

Figure 6.23 illustrates the effect of viscoelasticity on the radius and velocity profiles. We find that, for a low Deborah number (0.04), the viscoelastic solutions obtained remain very close to those for a Newtonian model.

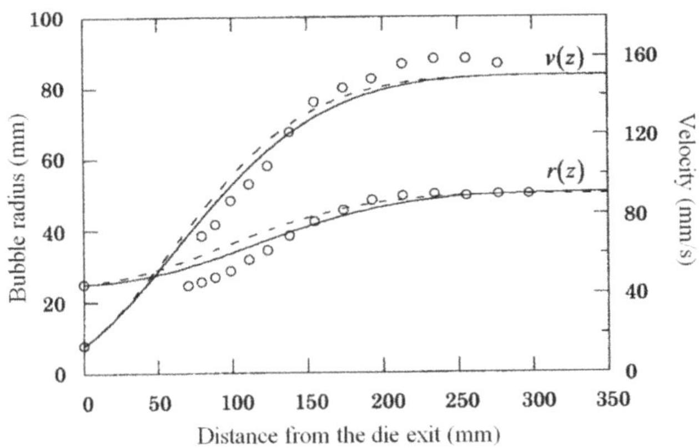

Figure 6.23. *Comparison of model and experiment; BUR = 2; DR = 10.6; FLH ≈ 300 mm; (—): Newtonian model, (- -): viscoelastic model $De_0 = 10^{-2}$, (○): experimental results: the bubble shape r(z) and velocity v(z) [GAM 04]*

6.3.2.2.2. Two-phase models

Campbell and Cao [CAM 87] and Ashok and Campbell [ASH 92] introduced a two-phase model to take crystallization into account. This hypothesis appears to be supported by the fact that the experimental results, previously obtained, reveal a temperature difference of 15–20°C between the internal and external bubble walls. It was, therefore, possible for the external wall of the film to crystallize before the internal wall. The value and evolution of the crystallization rate are determined assuming that the temperature profile and the heat transfer coefficient are known. Crystallinity along the radial direction, the bubble wall thickness, leads to an evolution in the rheological behavior. The authors, therefore, used a rheological model that takes into consideration the transition from molten to a semi-crystalline state. In this approach, the crystallization is considered to be independent of the stresses and the strain rate. Recently, Doufas *et al.* [DOU 00a, DOU 00b] and McHugh and Doufas [MCH 01] presented a much more complex thermodynamic model, which takes into account these effects on the formation of crystallization nuclei and on the evolution of the crystalline structure.

Muslet and Kamal [MUS 04] also developed a two-phase model. The constitutive equation of Phan-Thien and Tanner was used to describe the behavior during stretching of the molten film, whereas the neo-Hookean equation was used to describe the solid behavior of the film during the crystallization process. An interesting aspect of this model is that, when the swelling angle of the film reaches zero, the neo-Hookean model is used, which allows the film to become thinner, while the diameter remains unchanged. This approach is particularly interesting because it tries to include, in a single model, the effects of the temperature, the crystallization process and the stress-induced crystallization. In fact, the local convection coefficient is determined using the Sidiroupoulos and Vlachopoulos approach [SID 00a], whereas the crystallization kinematics is determined by the non-isothermal model of Nakamura, coupled with the Ziabicki equation [ZIA 67], to account for the stress-induced crystallization. The results of this calculation compare favorably to the experimental results obtained by several other authors.

6.3.2.3. *Non-orthogonal approach: Newtonian model*

So far, all the models consider the bubble to be axisymmetric. In 2003, Laffargue [LAF 03] developed a Newtonian model for single-layer film blowing, able to accommodate axisymmetric and non-axisymmetric cases. This model is based on the work by Housiadas and Tsamopoulos [HOU 00a, HOU 00b] who, unlike Pearson and Petrie [PEA 70a, PEA 70b], obtained a set of equations using a fixed cylindrical reference point rather than a mobile one. The coordinate system used is shown in Figure 6.24.

This approach has the major advantage of being able to deal with non-axisymmetric bubbles. This approach can also obtain non-stationary equations, which are no longer bound to the bubble area. It is, therefore, possible to analyze the instabilities encountered during film production. However, this approach gives rise to a system of equations that is much more difficult to process than the original equations of Pearson and Petrie [PEA 70a, PEA 70b]. The temperature decrease is assumed to be linear and is imposed from the frost line position, which is experimentally determined.

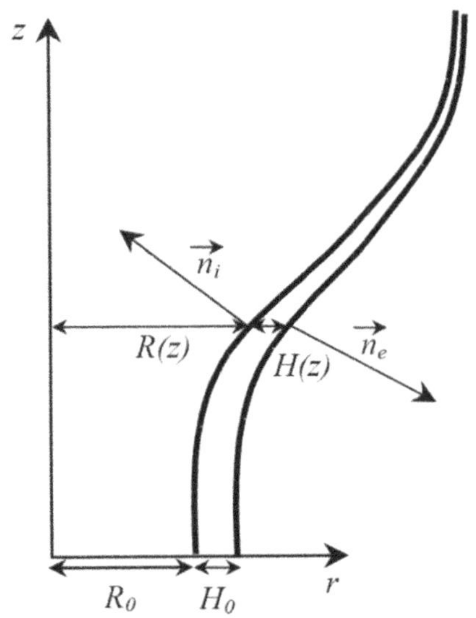

Figure 6.24. *The used coordinate system*

To illustrate the Laffargue model [LAF 03], we present the simplified axisymmetric treatment. The variables of the equations do not depend on the angle ϕ. The area occupied by the polymer is confined between:

$$0 \leq z \leq L \text{ and } R(z,t) \leq r \leq \left(R(z,t) + H(z,t)\right) \qquad [6.35]$$

The two normals at the bubble walls are defined by the following vectors:

$$\underline{n}_i(z,t) = \begin{pmatrix} -1 \\ 0 \\ \dfrac{\partial R(z,t)}{\partial z} \end{pmatrix} \quad [6.36]$$

$$\underline{n}_e(z,t) = \begin{pmatrix} 1 \\ 0 \\ -\dfrac{\partial\left(R(z,t)+H(z,t)\right)}{\partial z} \end{pmatrix} \quad [6.37]$$

6.3.2.3.1. Bubble kinematics and incompressibility

Given the assumption of axial symmetry, we can consider that at any point of the fluid, the velocity vector $\underline{v}$ can be written in the general form:

$$\underline{v} = \begin{pmatrix} v_r(r,z,t) \\ 0 \\ v_z(r,z,t) \end{pmatrix} \quad [6.38]$$

As we only study the flow of the molten polymer, we neglect the density variations of the material induced by cooling. The incompressibility equation is written as follows:

$$\frac{\partial v_r}{\partial r} + \frac{v_r}{r} + \frac{\partial v_z}{\partial z} = 0 \quad [6.39]$$

6.3.2.3.2. Behavior law and dynamic equilibrium

Laffargue [LAF 03] used a Newtonian model to describe the rheological behavior of the polymer. Under this condition, the stress tensor $\underline{\underline{\sigma}}$ is linked to the strain rate tensor by: $\underline{\underline{\sigma}} = -p\underline{\underline{I}} + 2\eta\underline{\underline{\dot{\varepsilon}}}$. Moreover, if we disregard the forces of mass, inertia, drag and surface tension, compared to the viscosity forces, the dynamic equilibrium equation is written as div $\underline{\underline{\sigma}} = 0$. All of the relations that we have just described should be solved using appropriate boundary conditions.

6.3.2.3.3. Boundary conditions on the surface

If we consider the internal and external surfaces of the bubble to be two fluid–air interfaces, we can express two types of non-stationary velocity boundary conditions using the following relations:

– at $r = R(z,t)$, $\underline{v}.\underline{n}_i = 0$, i.e.:

$$v_r\left(R(z,t),z,t\right)-\left(\frac{\partial R(z,t)}{\partial z}\right)v_z\left(R(z),z,t\right)-\left(\frac{\partial R(z,t)}{\partial t}\right)=0 \qquad [6.40]$$

– at $r = R(z,t) + H(z,t)$, $\underline{v}.\underline{n}_e = 0$, i.e.:

$$v_r\left(\left(R(z,t)+H(z,t)\right),z,t\right)$$
$$-\left(\frac{\partial\left(R(z,t)+H(z,t)\right)}{\partial z}\right)v_z\left(\left(R(z,t)+H(z,t)\right),z,t\right)-\left(\frac{\partial\left(R(z,t)+H(z,t)\right)}{\partial t}\right)=0$$

[6.41]

6.3.2.3.4. Stress conditions at the interface

The stress conditions at the interface are expressed by carrying out a balance of forces at a given axial position for each surface:

– at $r = R(z, t)$, the overpressure exerted by the air acting to inflate the bubble ΔP is applied to the internal wall of the bubble in the normal n_i direction. The balance of forces, therefore, gives:

$$\underline{\underline{\sigma}}\,\underline{n}_i = -\Delta P\,\underline{n}_i \qquad [6.42]$$

i.e.:

$$\sigma_{rr}\left(R(z),z,t\right)-\left(\frac{\partial R(z,t)}{\partial z}\right)\sigma_{rz}\left(R(z),z,t\right)=-\Delta P \qquad [6.43]$$

$$\sigma_{rz}\left(R(z),z,t\right)-\left(\frac{\partial R(z,t)}{\partial z}\right)\sigma_{zz}\left(R(z),z,t\right)=+\left(\frac{\partial R(z,t)}{\partial z}\right)\Delta P \qquad [6.44]$$

– at $r = R(z, t) + H(z, t)$, an equivalent balance to that above is also carried out on the external surface, according to the normal $\underline{n}_e$. At this boundary, the overpressure does not play a role and we simply obtain:

$$\underline{\underline{\sigma}}\,\underline{n}_e = 0 \qquad [6.45]$$

i.e.:

$$\sigma_{rr}\left(\left(R(z,t)+H(z,t)\right),z,t\right)-\left(\frac{\partial\left(R(z,t)+H(z,t)\right)}{\partial z}\right)\sigma_{rz}\left(\left(R(z)+H(z)\right),z,t\right)=0 \qquad [6.46]$$

$$\sigma_{rz}\left(\left(R(z)+H(z)\right),z,t\right)-\left(\frac{\partial\left(R(z,t)+H(z,t)\right)}{\partial z}\right)\sigma_{zz}\left(\left(R(z,t)+H(z,t)\right),z,t\right)=0 \qquad [6.47]$$

6.3.2.3.5. Non-dimensionalization

Now we have a complete system of equations and boundary conditions. To solve this system, we move onto non-dimensionalization. This allows us to express the equations according to different powers of the parameter that Laffargue [LAF 03] denoted as ε. This parameter is the ratio between the die gap and its internal diameter:

$$\varepsilon = \frac{H_0}{R_0} \qquad [6.48]$$

This parameter is very small, with the initial thickness being many orders of magnitude smaller than the radius, in this type of process. In general, this ratio has an approximate value of 10^{-2}. In the previous relation and those that follow, the index 0 is used to identify the value at the die exit. The relations that allow non-dimensionalization are:

$$R(z,t) = R_0\bar{R}(\bar{z},\bar{t}) \qquad H(z,t) = H_0\bar{H}(\bar{z},\bar{t}) \qquad \eta(z,t) = \eta_0\bar{\eta}(\bar{z},\bar{t})$$

$$r = R_0\bar{R}(\bar{z},\bar{t}) + H_0\bar{H}(\bar{z},\bar{t})\bar{r}, \quad z = R_0\bar{z}, \quad t = \frac{R_0}{V_0}\bar{t}, \quad v_r(r,z,t) = V_0 \qquad [6.49]$$

$$\bar{v}_r(\bar{r},\bar{z},\bar{t}), \quad v_z(r,z,t) = V_0\bar{v}_z(\bar{r},\bar{z},\bar{t}), \quad \sigma = \frac{\eta_0}{R_0}V_0\bar{\sigma}, \quad p = \frac{\eta_0}{R_0}V_0\bar{p}$$

With these relations, the variable r is made dimensionless using the non-orthogonal mapping technique, developed by Housiadas and Tsamopoulos [HOU 98, HOU 00c] to model the extrusion of tubes. As shown in Figure 6.25, this allows the complex geometry of the bubble to be changed to a cylindrical geometry. The bubble will, therefore, be confined between 0 and L/R_0 in the dimensionless axial position $\bar{z}$, and between 0 and 1 in the dimensionless radial position $\bar{r}$.

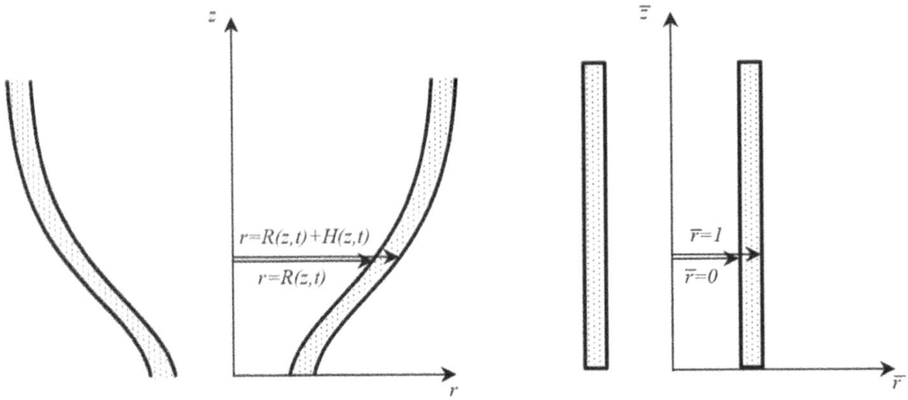

Figure 6.25. *Non-orthogonal mapping [HOU 00c]*

Rewriting dimensionless equations reveals the terms in increasing powers of ε, thereby defining each equation as the sum of several equations at different orders of magnitude. For example, here we illustrate the incompressibility equation and the velocity boundary conditions after the non-dimensionalization and non-orthogonalization stages:

$$\begin{aligned}&\frac{\partial \bar{v}_r(\bar{r},\bar{z},\bar{t})}{\partial \bar{r}}-\frac{\partial \bar{R}(\bar{z},\bar{t})}{\partial \bar{z}}\frac{\partial \bar{v}_z(\bar{r},\bar{z},\bar{t})}{\partial \bar{r}}\\&+\varepsilon\left[\frac{\bar{H}(\bar{z},\bar{t})}{\bar{R}(\bar{z},\bar{t})}\bar{v}_r(\bar{r},\bar{z},\bar{t})+\bar{H}(\bar{z},\bar{t})\frac{\partial \bar{v}_z(\bar{r},\bar{z},\bar{t})}{\partial \bar{z}}-\frac{\partial \bar{H}(\bar{z},\bar{t})}{\partial \bar{z}}\frac{\partial \bar{v}_z(\bar{r},\bar{z},\bar{t})}{\partial \bar{r}}\bar{r}\right]\\&-\varepsilon^2\left[\frac{\bar{H}(\bar{z},\bar{t})^2}{\bar{R}(\bar{z},\bar{t})}\bar{r}\,\bar{v}_r(\bar{r},\bar{z},\bar{t})\right]=0\end{aligned}\quad [6.50]$$

when $\bar{r}=0$:

$$\bar{v}_r(0,\bar{z},\bar{t})-\frac{\partial \bar{R}(\bar{z},\bar{t})}{\partial \bar{z}}\bar{v}_z(0,\bar{z},\bar{t})-\frac{\partial \bar{R}(\bar{z},\bar{t})}{\partial \bar{t}}=0 \qquad [6.51]$$

when $\bar{r}=1$:

$$\bar{v}_r(1,\bar{z},\bar{t})-\frac{\partial \bar{R}(\bar{z},\bar{t})}{\partial \bar{z}}\bar{v}_z(1,\bar{z},\bar{t})-\frac{\partial \bar{R}(\bar{z},\bar{t})}{\partial \bar{t}}-\varepsilon\left[\bar{v}_z(1,\bar{z})\frac{\partial \bar{H}(\bar{z})}{\partial \bar{z}}+\frac{\partial \bar{H}(\bar{z},\bar{t})}{\partial \bar{t}}\right]=0 \qquad [6.52]$$

6.3.2.3.6. Solving the set of equations

To solve the set of equations, we aim to approximate the value of different non-geometric dimensionless variables by series expansion with regard to the small geometric parameter ε. Thus, the strain and velocity field variables can be written in the following general form:

$$\begin{aligned}
&\bar{v}_r(\bar{r},\bar{z},\bar{t})=\bar{v}_r^0(\bar{r},\bar{z},\bar{t})+\varepsilon\bar{v}_r^1(\bar{r},\bar{z},\bar{t})+\varepsilon^2\bar{v}_r^2(\bar{r},\bar{z},\bar{t})+\dots\\
&\bar{v}_z(\bar{r},\bar{z},\bar{t})=\bar{v}_z^0(\bar{r},\bar{z},\bar{t})+\varepsilon\bar{v}_z^1(\bar{r},\bar{z},\bar{t})+\varepsilon^2\bar{v}_z^2(\bar{r},\bar{z},\bar{t})+\dots\\
&\bar{\sigma}_i(\bar{r},\bar{z},\bar{t})=\bar{\sigma}_i^0(\bar{r},\bar{z},\bar{t})+\varepsilon\bar{\sigma}_i^1(\bar{r},\bar{z},\bar{t})+\varepsilon^2\bar{\sigma}_i^2(\bar{r},\bar{z},\bar{t})+\dots\\
&\bar{p}(\bar{r},\bar{z},\bar{t})=\bar{p}^0(\bar{r},\bar{z},\bar{t})+\varepsilon\bar{p}^1(\bar{r},\bar{z},\bar{t})+\varepsilon^2\bar{p}^2(\bar{r},\bar{z},\bar{t})+\dots
\end{aligned} \qquad [6.53]$$

By introducing these expressions into different equations, it is possible to obtain equations at several distinct orders of magnitude, thereby connecting the different variables.

Analyzing these equations is the main challenge of thc Laffarguc approach [LAF 03]. In fact, it is necessary to express different equations at different orders of magnitude, and the associated surface conditions in the appropriate order, in order to deduce a complete set of equations, which will allow the stationary or non-stationary calculation to be, subsequently, carried out. For example, by analyzing the incompressibility equation at the minimal order (0), it is possible to obtain a relation that can be used to analyze the incompressibility equation at a higher order (1):

– at order 0, the incompressibility equation is:

$$\frac{\partial}{\partial \bar{r}}\left(\bar{v}_r^0(\bar{r},\bar{z},\bar{t})-\frac{\partial \bar{R}(\bar{z},\bar{t})}{\partial \bar{z}}\bar{v}_z^0(\bar{r},\bar{z},\bar{t})\right)=0 \qquad [6.54]$$

and the free surface velocity conditions are equal to:

$$\bar{v}_r^0(0,\bar{z},\bar{t}) - \frac{\partial \bar{R}(\bar{z},\bar{t})}{\partial \bar{z}}\bar{v}_z^0(0,\bar{z},\bar{t}) - \frac{\partial \bar{R}(\bar{z},\bar{t})}{\partial \bar{t}} = 0 \quad [6.55]$$

$$\bar{v}_r^0(1,\bar{z},\bar{t}) - \frac{\partial \bar{R}(\bar{z},\bar{t})}{\partial \bar{z}}\bar{v}_z^0(1,\bar{z},\bar{t}) - \frac{\partial \bar{R}(\bar{z},\bar{t})}{\partial \bar{t}} = 0 \quad [6.56]$$

It is, therefore, possible to conclude that:

$$\bar{v}_r^0(\bar{r},\bar{z},\bar{t}) - \frac{\partial \bar{R}(\bar{z},\bar{t})}{\partial \bar{z}}\bar{v}_z^0(\bar{r},\bar{z},\bar{t}) = \frac{\partial \bar{R}(\bar{z},\bar{t})}{\partial \bar{t}} \quad [6.57]$$

In doing so, Laffargue [LAF 03] obtained a complete set of equations describing the evolution of the bubble shape in the axial position and as a function of time. It is now possible to carry out a stability analysis, which identifies the processing conditions that will lead to the formation of unstable bubbles. The Laffargue model can also be applied to the stationary case, by considering that the equations are independent of time. Under these conditions, the system of equations obtained, despite being more complex, is equivalent to the model by Pearson and Petrie [PEA 70a, PEA 70b]. In fact, one of the results of the Laffargue model is that, in a first approximation (i.e. at the order of magnitude of 0), the mechanical variables can be considered to be constant in the thickness direction. This is the main assumption of the local membrane model. The dimensional analysis can, therefore, be used to justify such a hypothesis. For a steady-state regime, Laffargue obtained a differential system involving 11 unknowns that depend only on the axial position $\bar{z}$:

$$\bar{v}_r^0(\bar{z}), \bar{v}_z^0(\bar{z}), \frac{\partial \bar{v}_r^1}{\partial \bar{r}}(\bar{z}), \frac{\partial \bar{v}_z^1}{\partial \bar{r}}(\bar{z}), \bar{R}(\bar{z}), \bar{H}(\bar{z}), \bar{p}^0(\bar{z}), \bar{\sigma}_{rr}^0(\bar{z}), \bar{\sigma}_{\varphi\varphi}^0(\bar{z}), \bar{\sigma}_{zz}^0(\bar{z}), \bar{\sigma}_{rz}^0(\bar{z})$$

These 11 unknowns are linked by a system of 11 differential relations, which can be broken down into subsets:

– the kinematic equations resulting from the incompressibility relation:

- at order 0:

$$\bar{v}_r^0(\bar{z}) = \frac{d\bar{R}(\bar{z})}{d\bar{z}}\bar{v}_z^0(\bar{z}) \quad [6.58]$$

and:

$$\frac{d}{d\overline{z}}\left(\overline{H}(\overline{z})\overline{R}(\overline{z})\overline{v}_z^0(\overline{z})\right)=0 \quad [6.59]$$

- at order 1:

$$\frac{\partial \overline{v}_r^1}{\partial \overline{r}}(\overline{z})-\frac{d\overline{R}(\overline{z})}{d\overline{z}}\frac{\partial \overline{v}_z^1}{\partial \overline{r}}(\overline{z})+\overline{v}_r^0(\overline{z})\frac{\overline{H}(\overline{z})}{\overline{R}(\overline{z})}+\frac{d\overline{v}_z^0(\overline{z})}{d\overline{z}}\overline{H}(\overline{z})=0 \quad [6.60]$$

– the stress equations resulting from the dynamic equilibrium:

- at order 0:

$$\overline{\sigma}_{rr}^0(\overline{z})-\frac{d\overline{R}(\overline{z})}{d\overline{z}}\overline{\sigma}_{rz}^0(\overline{z})=0 \quad [6.61]$$

and:

$$\overline{\sigma}_{rz}^0(\overline{z})-\frac{d\overline{R}(\overline{z})}{d\overline{z}}\overline{\sigma}_{zz}^0(\overline{z})=0 \quad [6.62]$$

- at order 1:

$$\frac{d\overline{H}(\overline{z})}{d\overline{z}}\overline{\sigma}_{zz}^0(\overline{z})+\overline{H}(\overline{z})\frac{d\overline{\sigma}_{zz}^0(\overline{z})}{d\overline{z}}+\frac{\overline{H}(\overline{z})}{\overline{R}(\overline{z})}\overline{\sigma}_{rz}^0(\overline{z})-\frac{d\overline{R}(\overline{z})}{d\overline{z}}\overline{\Delta P}=0 \quad [6.63]$$

and:

$$\frac{d\overline{H}(\overline{z})}{d\overline{z}}\overline{\sigma}_{rz}^0(\overline{z})+\overline{H}(\overline{z})\frac{d\overline{\sigma}_{rz}^0(\overline{z})}{d\overline{z}}+\frac{\overline{H}(\overline{z})}{\overline{R}(\overline{z})}\left(\overline{\sigma}_{rr}^0(\overline{z})-\overline{\sigma}_{\varphi\varphi}^0(\overline{z})\right)+\overline{\Delta P}=0 \quad [6.64]$$

– the equations resulting from the rheological behavior:

- at order 0:

$$\overline{\sigma}_{rr}^0(\overline{z})=2\frac{\overline{\eta}(\overline{z})}{\overline{H}(\overline{z})}\frac{\partial \overline{v}_r^1}{\partial \overline{r}}(\overline{z})-\overline{p}^0(\overline{z}) \quad [6.65]$$

$$\overline{\sigma}_{\varphi\varphi}^0(\overline{z})=-\overline{p}^0(\overline{z})+2\frac{\overline{\eta}(\overline{z})}{\overline{H}(\overline{z})}\overline{v}_r^0(\overline{z}) \quad [6.66]$$

$$\bar{\sigma}_{zz}^{0}(\bar{z}) = -\bar{p}^{0}(\bar{z}) + 2\bar{\eta}(\bar{z})\frac{d\bar{v}_{z}^{0}(\bar{z})}{d\bar{z}} - 2\frac{\bar{\eta}(\bar{z})}{\bar{H}(\bar{z})}\frac{d\bar{R}(\bar{z})}{d\bar{z}}\frac{\partial \bar{v}_{z}^{1}}{\partial \bar{r}}(\bar{z}) \qquad [6.67]$$

$$\bar{\sigma}_{rz}^{0}(\bar{z}) = \eta(\bar{z})\frac{d\bar{v}_{r}^{0}(\bar{z})}{d\bar{z}} + \eta(\bar{z})\left(\frac{1}{\bar{H}(\bar{z})}\frac{\partial \bar{v}_{z}^{1}}{\partial r}(\bar{z}) - \frac{1}{\bar{H}(\bar{z})}\frac{d\bar{R}(\bar{z})}{d\bar{z}}\frac{\partial \bar{v}_{r}^{1}}{\partial \bar{r}}(\bar{z})\right) \qquad [6.68]$$

To take into account the variation in temperature in the axial direction, Laffargue [LAF 03] arbitrarily fixed the evolution of the mean temperature in the thickness $T(z)$, along the stretching length, from experimental data. The variation in the viscosity of the material as a function of this mean temperature is considered to obey the Arrhenius equation. However, it is possible to use relation [6.31], which includes heat transfer. To solve the system of mechanical equations, the boundary conditions must be known, in the dimensionless form, at the die exit and at the frost line. These conditions can be written as follows:

– at the die exit:

$$\bar{z} = 0, \quad \bar{R}(0) = 1, \quad \bar{H}(0) = 1, \quad \left\|\bar{V}\right\| = \sqrt{\bar{v}_{z}^{02}(0) + \bar{v}_{r}^{02}(0)} = 1 \qquad [6.69]$$

– at the frost line:

$$\bar{z} = L/R_{0}, \quad \bar{R}(L/R_{0}) = BUR, \quad \frac{d\bar{R}(L/R_{0})}{d\bar{z}} = 0, \quad \left\|\bar{V}\right\| = \bar{v}_{z}^{0}(L/R_{0}) = DR \qquad [6.70]$$

It then comes down to calculating the solution of a semi-implicit differential system, from the boundary conditions at the frost line ($\bar{z} = L/R_0$). The two parameters $\bar{F}_{pulling}$ (the pulling force) and ΔP (the overpressure) are adjusted to find out the boundary conditions at the die exit, at the end of the calculation. Initially, these two values are arbitrarily fixed. This allows the different variables to be calculated at $\bar{z} = L/R_0$. Then, the differential system at each discretization step of $\bar{z}$ position must be solved, in order to obtain the increments in the different variables. It is then possible to calculate the values of the different variables at each step until we reach $\bar{z} = 0$. If the values at this position are not equal to the boundary conditions at the die exit, we must modify the $\bar{F}_{pulling}$ and ΔP values using an optimization method. The solution is reached when the values calculated at the frost line are equal to the boundary conditions at this position.

According to the boundary conditions, at time $t = 0$ we obtain:

$$\bar{R}(L/R_0) = BUR, \quad \frac{\partial \bar{R}(L/R_0)}{\partial \bar{z}} = 0, \quad \left\| \underline{\bar{V}}(L/R_0) \right\| = \bar{v}_z^0(L/R_0) = DR \quad [6.71]$$

Using relation [6.59] (conservation of flow rate), it is possible to calculate the final thickness of the film:

$$\bar{H}(L/R_0) = \frac{1}{\bar{R}(L/R_0)\bar{v}_z^0(L/R_0)} = \frac{1}{BUR\,DR} \quad [6.72]$$

In addition, the force $\bar{F}_{pulling}$, which is initially chosen at random, allows us to calculate the stress at the frost line, using the following relation:

$$\bar{\sigma}_{zz}^0(L/R_0) = \frac{\bar{F}_{pulling}}{\bar{R}(L/R_0)\,\bar{H}(L/R_0)} = \bar{F}_{pulling}\;DR \quad [6.73]$$

Knowing the values of these five variables at time $t = 0$ allowed Laffargue [LAF 03] to completely solve the system.

Equations [6.58], [6.61] and [6.62] at the frost line give:

$$\bar{v}_r^0(L/R_0) = \frac{d\bar{R}(L/R_0)}{d\bar{z}}\bar{v}_z^0(L/R_0) \quad [6.74]$$

$$\bar{\sigma}_{rz}^0(L/R_0) - \frac{\partial \bar{R}(L/R_0)}{\partial \bar{z}}\bar{\sigma}_{zz}^0(L/R_0) = 0 \quad [6.75]$$

$$\bar{\sigma}_{rr}^0(L/R_0) - \frac{\partial \bar{R}(L/R_0)}{\partial \bar{z}}\bar{\sigma}_{rz}^0(L/R_0) = 0 \quad [6.76]$$

The second stage involves considering the differential terms in the remaining equations as unknowns in $\bar{z}$. If we count the unknowns still to be calculated under these conditions, we find that there are six derived from the variables and four additional variables:

$$\frac{\partial \bar{H}(L/R_0)}{\partial \bar{z}}, \frac{\partial \bar{v}_z^0(L/R_0)}{\partial \bar{z}}, \frac{\partial \bar{v}_r^0(L/R_0)}{\partial \bar{z}}, \frac{\partial^2 \bar{R}(L/R_0)}{\partial z^2}, \frac{\partial \bar{\sigma}_{zz}^0(L/R_0)}{\partial \bar{z}}, \frac{\partial \bar{\sigma}_{rz}^0(L/R_0)}{\partial \bar{z}};$$

$$\frac{\partial \bar{v}_r^1}{\partial \bar{r}}(L/R_0), \frac{\partial \bar{v}_z^1}{\partial \bar{r}}(L/R_0), \bar{p}^0(L/R_0), \bar{\sigma}_{\varphi\varphi}^0(L/R_0)$$

To solve this step by step, we simply use equations [6.59], [6.60] and [6.63]–[6.68], and the derivatives of equations [6.58] and [6.61]. Then, a linear system of 10 equations must be formed which, taking into account the variables that have already been calculated, becomes a linear system of the 10 unknowns mentioned above.

Figure 6.26 shows the evolution of the radius of LDPE bubbles obtained by keeping the cooling air flow constant. When the BUR increases, the frost line decreases. This is explained by the fact that an increase in the bubble diameter causes greater heat transfer, which directly translates as a decrease in the FLH.

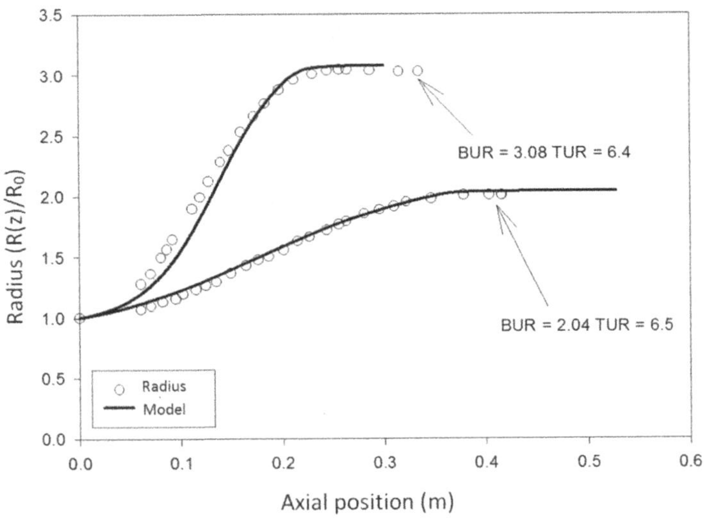

Figure 6.26. *Comparison of calculations (—) and experimental results (○) for LDPE bubbles [LAF 03]*

The simulations, carried out using the Laffargue methodology [LAF 03], adequately represent the shape of the bubble. They also revealed (not shown here) that the increase in BUR leads to an increase in axial and transverse stresses generated in the film during processing. This illustrates once more how useful these calculations are, since, as we have mentioned earlier, many studies have helped to identify links between the film properties and the stresses, developed during processing. Therefore, this calculation could, potentially, enable the design of films that meet specific customer specifications.

6.3.2.3.7. Non-stationary results

We are now going to present some of Laffargue's experimental results [LAF 03], which will allow us to better understand the parameters affecting the stability of the process.

Many LLDPE bubbles have been made with different dies at different cooling rates, by gradually increasing the TUR, for a given BUR, to obtain the stability map illustrated in Figure 6.27. This figure shows that a greater cooling delays the onset of instabilities. Under these conditions, the range of the TUR that produces a stable bubble is greatly increased when bubbles have a BUR between 2.5 and 3. A further increase in the TUR causes draw resonance instabilities to appear, which are typical in processes involving longitudinal stretching. When the BUR is greater than 3, an increase in the TUR causes helical instabilities to appear. The critical TUR required to cause this type of instability to appear is much lower when the BUR is high. According to Laffargue [LAF 03], the helical instability also appears to be enhanced by non-axisymmetric disturbances in the air blown around the bubble, during the helical movement. The flow rate of the blown air, and the type of cooling ring used, can affect the behavior of the bubble. Finally, Figure 6.27 reveals that a low cooling rate causes a flattening and reduction in the stability zone.

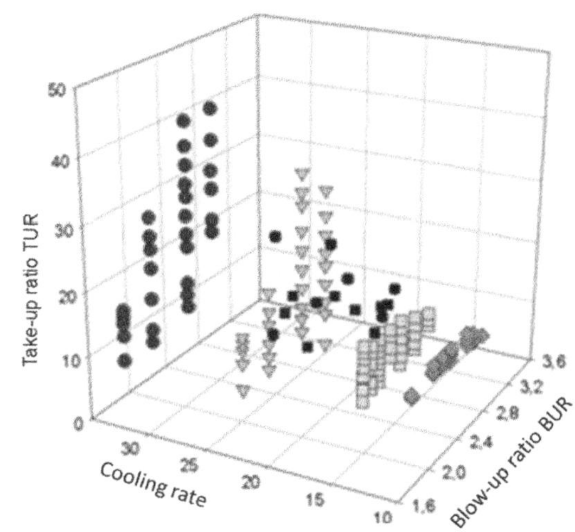

Figure 6.27. *A comparison of stable zones observed for LLDPE using different tools: (◊) CRASP-1, $dT/d\bar{z} \approx -12.6°C$; (□) CRASP-2, $dT/d\bar{z} \approx -15.8°C$; (∇) KAUFMAN-1, $dT/d\bar{z} \approx -23.4°C$ [LAS 99]; (○) KAUFMAN-2, $dT/d\bar{z} \approx -33.5°C$ [LAS 99]; (●) COLLIN 1; (■) COLLIN 2*

The most interesting part of Laffargue's work [LAF 03] was that, using the system of non-stationary equations, developed above, the onset of instabilities can be qualitatively predicted through calculation. In fact, Figure 6.28 illustrates the stability map obtained via calculation.

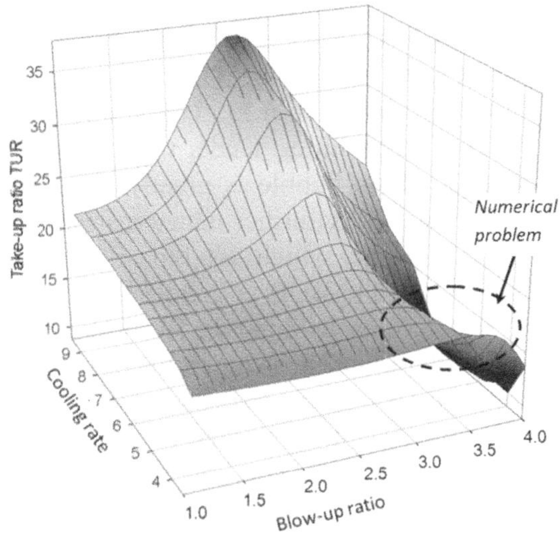

Figure 6.28. *Evolution of the critical take-up ratio as a function of the blow-up ratio and the cooling rate [LAF 03]*

Generally speaking, similar to the experimental observations, the critical TURs increase with the cooling rate. Here, it is also possible to see that the stability window becomes smaller when the cooling rate decreases. The calculated stability map is very similar to the experimental stability map, with the exception of cases with very low cooling rates. Under these conditions, it is possible to observe an increase in the critical TUR that Laffargue [LAF 03] associates with convergence difficulties, which probably leads to the wrong solutions.

6.3.3. *Multilayer films*

The production of multilayer films revealed new problems, when compared to the production of monolayer films. For example, the processing temperature and the rheological properties can vary greatly from one material to another. Moreover, it is sometimes necessary to use tie layers with certain combinations of materials, to ensure that they adhere to each other. Combining several resins also has a significant effect on film properties. In fact, Beagan *et al.* [BEA 99] studied the effect of mixing rather than coextrusion, on mPE with a LDPE. They clearly showed that multilayer films have better rupture properties than films produced by mixing. In addition, multilayer films are, usually, more stable than monolayer mPE films. This shows how important the coextrusion process, applied to film blowing, will be in the future. Moreover, for economical reasons, it is important to find the optimal quantity of each material used, in order to obtain the desired properties.

One problem encountered during the production of multilayer films is the existence of interfacial instabilities between the different materials. Perdikoulias and Tzoganakis [PER 95, PER 96] showed that the onset of interfacial instabilities was strongly delayed when the resins used had a narrow molecular weight distribution. Adhesion between the different layers is vital for highly incompatible systems, such as packaging made up of PA6 and LLDPE. To improve the structural cohesion of films, manufacturers frequently add a small amount of PE-g-MAH (maleic anhydride grafted polyethylene) to LLDPE, to favor the polymer/polymer diffusion at the interface, which, according to Sanchez-Valdes *et al.* [SAN 97] forms covalent bonds at the PA6/PE-g-MAH interface. This ensures a better adhesion between the two materials. Another problem, encountered when using multilayer packaging, is when this packaging is reused or recycled. Mendez-Prieto and Sanchez-Valdes [MEN 98] studied the effect of using recycled films on the physical properties of the packaging. They evaluated the rheological properties of a film made up of the following five layers: LLDPE/tie layer/PA6/tie layer/LLDPE, which is melted and mixed. Then, they used this mixture to produce films made up of recycled mixture/PA6 and LLDPE/recycled mixture/LLDPE, and they also evaluated the adhesion properties and permeability to oxygen and water. They concluded that using the recycled mixture in two-layer films did not significantly improve adhesion. However, increasing the amount of the recycled mixture in the three-layer films worsened the mechanical properties and the permeability to oxygen. These studies illustrate the problems associated with recycling multilayer films and highlight the importance of tie layers in the production of multilayer films. It is interesting to note that studies concerned with recycling, are relatively recent, and still seldom applied to multilayer film production.

There are, still, few studies that model the film blowing process of multilayer films. Yoon and Park [YOO 91] tackled the problem by using the membrane approach, by considering a Newtonian isothermal behavior for the internal layer and a Maxwell isothermal one for the external layer. The authors focus on the profile and evolution of the bubble thickness. According to the developed model, if the relaxation time of the viscoelastic layer is significant, it will dominate the flow in both layers, even though the flow rate is much lower than that in the Newtonian layer. The shape of the bubble is, therefore, described by the equilibrium between the viscous and viscoelastic forces. Unfortunately, the simulation was not compared to experimental results. Recently, Gamache [GAM 04] developed a non-isothermal Newtonian mechanical model that described the film blowing of a multilayer structure, which was based on the non-orthogonal approach by Laffargue [LAF 03]. The model was validated using experimental tests carried out by producing monolayer and multilayer films, using two different extrusion lines and four materials. The die dimensions and the properties of the materials used are presented in Tables 6.3 and 6.4, respectively.

Die	External diameter of the die (mm)	Clearance (mm)	Total flow rate (kg/h)	Shaping temperature (°C)
1	101.6	1.1	7.2	230
2	51.6	1.2	5	220

Table 6.3. *Dimensions of the dies*

Material		Crystallization temperature (°C)	Specific heat capacity ($J.kg^{-1}$°C^{-1})	Activation energy (kJ/mol)
Lacqtene 1003FE23	LDPE	103	2,300	50
Finacene M3410EP	mMDPE	114	1,800	33
Dow 609a	LDPE	102	2,300	44
Dow 4201	ULDPE	103	2,300	26

Table 6.4. *Properties of the used materials*

In Figure 6.29, we find that, depending on the materials and extrusion lines used, two different behaviors can be identified. First, Figure 6.29(a), illustrating the bubbles produced using die 1 by combining LDPE and ultra-low-density polyethylene (ULDPE), shows that the multilayer bubble is located in the center of the envelope, in-between the two monolayer bubbles. On the other hand, in Figure 6.29(b), which illustrates the bubbles produced using die 1 by combining LDPE and mMDPE, we find that the multilayer film blowing is faster than that for monolayers, made up of these two materials. Multilayer bubbles are, therefore, not necessarily the mean of the monolayer bubbles. Unfortunately, these experimental results did not identify whether this divergence was caused by the dies, the cooling ring or the different combinations of the materials used. The calculations carried out by Gamache [GAM 04] explain this difference in behavior.

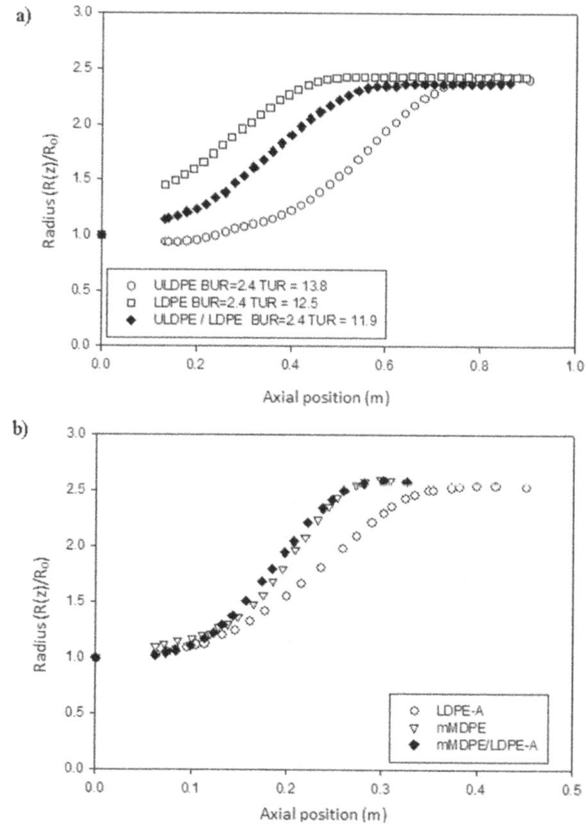

Figure 6.29. *Evolution of the bubble radius with die 1: a) LDPE and ULDPE; b) LDPE and mHDPE [GAM 04]*

Gamache [GAM 04] used a two-material model, with activation energies of 22,000 and 44,000 J/mol and an identical specific heat capacity of 2,200 J kg^{-1} K^{-1}. By comparing the monolayer and multilayer bubbles made up of these materials, he found that the position of the materials, internal or external layer, does not affect the calculated temperature and geometry profiles (Figure 6.30). He also found that the multilayer films are located in the envelope formed by both monolayer films. The obtained curve shows good agreement with the experimental results, as illustrated in Figure 6.29(a), which were obtained using materials with different activation energies and an identical specific heat capacity (Table 6.4).

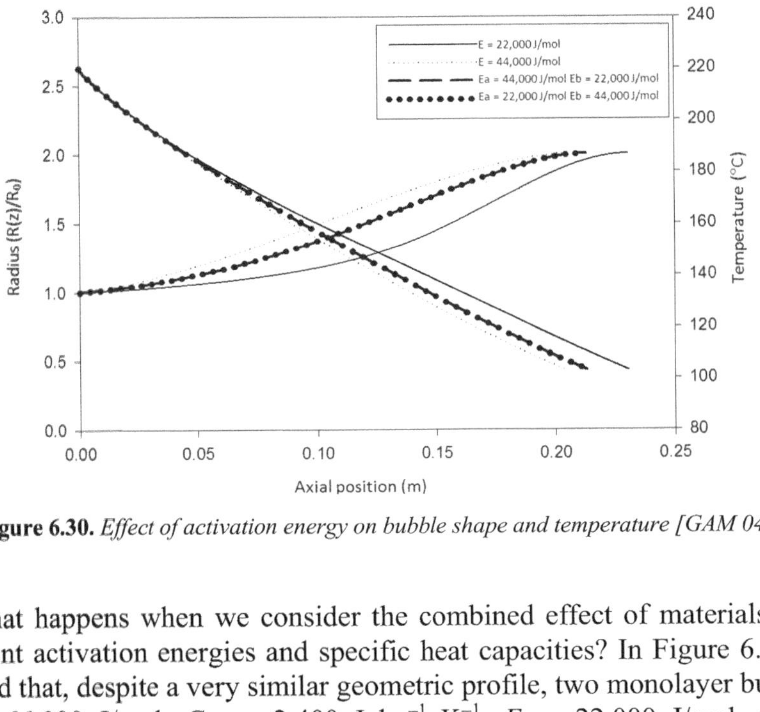

Figure 6.30. *Effect of activation energy on bubble shape and temperature [GAM 04]*

What happens when we consider the combined effect of materials with different activation energies and specific heat capacities? In Figure 6.31(a), we find that, despite a very similar geometric profile, two monolayer bubbles (E_a = 66,000 J/mol, Cp_a = 2,400 J kg^{-1} K^{-1}; E_b = 22,000 J/mol, Cp_b = 2,000 J kg^{-1} K^{-1}) have the same FLH. This is caused by the opposing effects of the activation energy and the specific heat capacity.

In fact, as we have illustrated above, a high activation energy will cause the bubble to swell rapidly, whereas, a high specific heat capacity will have the opposite effect. The multilayer bubble made up of these two materials has a geometric profile similar to the monolayer bubble, obtained with a high activation energy, and it reaches its solidification temperature

1.5 cm below monolayer bubbles. By combining the two materials, the mean specific heat capacity of the multilayer film is 2,200 J kg^{-1} K^{-1}, which means that the amount of heat lost from this film is lower than the amount of heat lost from the monolayer film with an activation energy of 66,000 J/mol. This, therefore, causes rapid swelling. A similar phenomenon, although more pronounced, was observed during the production of multilayer bubbles of LDPE/mMDPE.

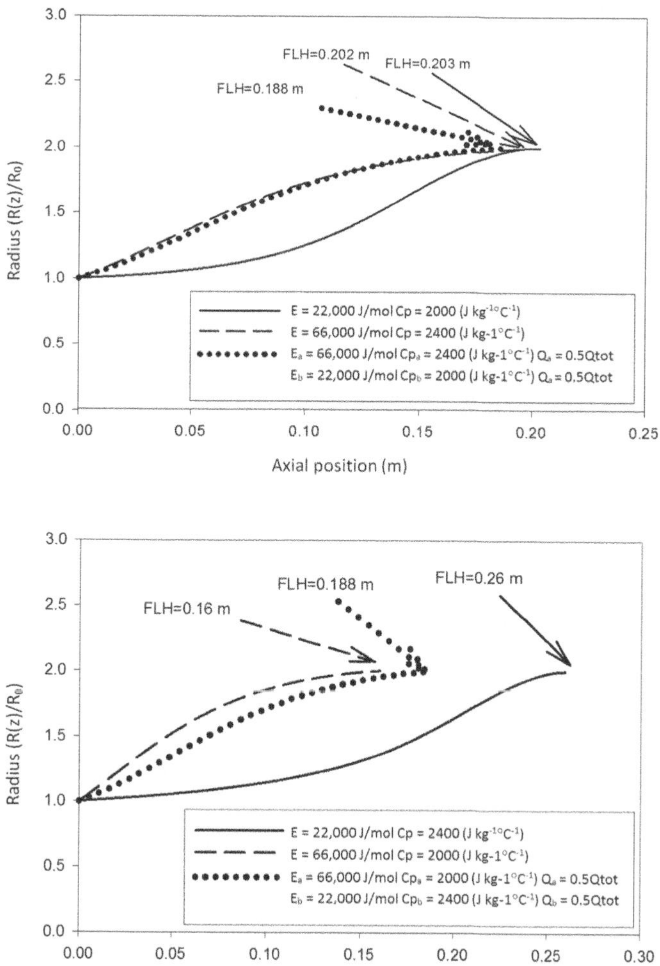

a)

b)

Figure 6.31. *Effect of activation energy and specific heat capacity on bubble shape: a) case 1 and b) case 2 [GAM 04]*

Note that Figure 6.31(b) assumes that, in all cases, the crystallization temperature is 103°C. We can imagine that, if the crystallization temperature of the fluid, with an activation energy of 22,000 J/mol, was 110°C, the multilayer films would reach this temperature much faster and the difference between monolayered and multilayer films would be much greater. Finally, in Figure 6.31(b) we see that, when the activation energies and specific heat capacities are different (E_a = 22,000 J/mol, Cp_a = 2,000 J kg^{-1} K^{-1}, E_b = 66,000 J/mol, Cp_b = 2,400 J kg^{-1} K^{-1}), the multilayer film is situated between the two monolayers.

6.4. Conclusion

The production of films or sheets, either by cast film extrusion or by film blowing, in monolayer or multilayer form, is currently well controlled. As was presented, efficient theoretical approaches allow us to better understand and optimize the different processes, both in terms of tool design and processing conditions. Further important work needs to be done to calculate and predict the orientation and structures during biaxial stretching with cooling and crystallization in order to ultimately predict the final properties of the extruded films.

6.5. Bibliography

[AND 98] ANDRÉ J.M., AGASSANT J.-F., DEMAY Y., *et al.*, "Numerical modeling of the polymer film blowing process", *International Journal of Forming Processes*, vol. 1, pp. 197–210, 1998.

[AND 99] ANDRÉ J.M., Modélisation thermomécanique et structurale du soufflage de gaine de polyéthylènes, Doctoral Thesis, Paris School of Mines, 1999.

[ARP 92] ARPIN B., LAFLEUR P. G., VERGNES B., "Simulation of polymer flow through a coat-hanger die: a comparison between two numerical approaches", *Polymer Engineering & Science*, vol. 32, pp. 206–212, 1992.

[ARP 94] ARPIN B., LAFLEUR P.G., SANSCHAGRIN B., "A personal computer software program for coat-hanger die simulation", *Polymer Engineering & Science*, vol. 34, pp. 657–664, 1994.

[ASH 92] ASHOK B.K., CAMPBELL G.A., "Two-phase simulation of tubular film blowing of crystalline polymers", *International Polymer Processing*, vol. 7, pp. 240–247, 1992.

[AST 74] AST W., "Air cooling of blown film", *Kunststoffe*, vol. 64, pp. 146–152, 1974.

[BEA 99] BEAGAN C.M., MC NALLY G.M., MURPHY W.R., "The use of metallocene catalysed polyethylene as a blend material in blown film extrusion", *Developments in Chemical Engineering and Mineral Processes*, vol. 7, pp. 201–218, 1999.

[CAI 88] CAIN J.J., DENN M.M., "Multiplicities and instabilities in film blowing", *Polymer Engineering & Science*, vol. 28, pp. 1527–1541, 1988.

[CAM 87] CAMPBELL G.A., CAO B., "The interaction of crystallinity, elastoplasticity, and a two-phase model on blown film bubble shape", *Journal of Plastics Film Sheeting*, vol. 3, pp. 158–170, 1987.

[CAM 92] CAMPBELL G.A., OBOT N.T., CAO B., "Aerodynamics in the blown film process", *Polymer Engineering & Science*, vol. 32, pp. 751–759, 1992.

[CAO 89] CAO B., CAMPBELL G.A., "Air ring effect on blown film dynamics", *International Polymer Processing*, vol. 4, pp. 114–118, 1989.

[COY 91] COYLE D.J., PERDIKOULIAS J., "Flow simulation and visualization in spiral mandrel dies", *SPE ANTEC Technical Papers*, pp. 2445–2447, 1991.

[DOU 00a] DOUFAS A.K., MCHUGH A.J., MILLER C., "Simulation of melt spinning including flow-induced crystallization. Part I. Model development and predictions", *Journal of Non-Newtonian Fluid Mechanics*, vol. 92, pp. 27–66, 2000.

[DOU 00b] DOUFAS A.K., MCHUGH A.J., MILLER C., *et al.*, "Simulation of melt spinning including flow-induced crystallization. Part II. Quantitative comparisons with industrial spinline data", *Journal of Non-Newtonian Fluid Mechanics*, vol. 92, pp. 81–103, 2000.

[FAR 74] FARBER R., DEALY J., "Strain history of the melt in film blowing", *Polymer Engineering & Science*, vol. 14, pp. 435–440, 1974.

[FLE 88] FLEISSNER M., "Elongational flow of HDPE samples and bubble instability in film blowing", *International Polymer Processing*, vol. 2, pp. 229–233, 1988.

[FRE 58] FREDRICKSON A.G., BIRD R.B., "Non-Newtonian flow in annuli", *Journal of Industrial and Engineering Chemistry*, vol. 50, pp. 347–352, 1958.

[GAM 04] GAMACHE E., Evaluation de la biréfringence en ligne et modélisation du procédé de soufflage de gaine multicouche, Doctoral Thesis, Ecole Polytechnique de Montréal, 2004.

[GHA 96] GHANEH-FARD A., CARREAU P.J., LAFLEUR P.G., “Study of instabilities in film blowing”, *AIChE Journal*, vol. 42, pp. 1388–1396, 1996.

[GUP 82] GUPTA R.K., METZNER A.B., “Modelling of non isothermal polymer processes”, *Journal of Rheology*, vol. 26, pp. 181–198, 1982.

[HAN 75] HAN C.D., PARK J.Y., “Studies on blown film extrusion. III. Bubble instability”, *Journal of Applied Polymer Science*, vol. 19, pp. 3291–3297, 1975.

[HER 00] HERNANDEZ R., SELKE S.E.M., CUTLER J., *Plastics Packaging: Properties, Processing, Applications, Regulations*, Hanser, Munich, 2000.

[HOU 98] HOUSIADAS K., TSAMOPOULOS J., “Unsteady flow of an axisymmetric annular film under gravity”, *Physics of Fluids*, vol. 10, pp. 2500–2516, 1998.

[HOU 00a] HOUSIADAS K., TSAMOPOULOS J., “Unsteady extrusion of a viscoelastic annular film I. General model and its numerical solution”, *Journal of Non-Newtonian Fluid Mechanics*, vol. 88, pp. 229–259, 2000.

[HOU 00b] HOUSIADAS K., TSAMOPOULOS J., “Unsteady extrusion of a viscoelastic annular film. II. Linearized model and its analytical solution”, *Journal of Non-Newtonian Fluid Mechanics*, vol. 88, pp. 303–325, 2000.

[HOU 00c] HOUSIADAS K., TSAMOPOULOS J., “Cooling of a viscoelastic film during unsteady extrusion from an annular die”, *Rheologica Acta*, vol. 39, pp. 44–61, 2000.

[HUA 98] HUANG C.C., “A systematic approach for the design of a spiral mandrel die”, *Polymer Engineering & Science*, vol. 38, pp. 573–582, 1998.

[KAN 84] KANAI T., WHITE J.L., “Kinematics, dynamics and stability of the tubular film extrusion of various polyethylenes”, *Polymer Engineering & Science*, vol. 24, pp. 1185–1201, 1984.

[KAN 87] KANAI T., “Theoretical analysis of tubular film extrusion and its applications for HMW-HDPE”, *International Polymer Processing*, vol. 1, pp. 137–143, 1987.

[LAF 02] LAFFARGUE J., PARENT L., LAFLEUR P.G., *et al.*, “Investigation of bubble instabilities in film blowing process”, *International Polymer Processing*, vol. 17, pp. 347–353, 2002.

[LAF 03] LAFFARGUE J., Etude et modélisation des instabilités du procédé de soufflage de gaine, Doctoral Thesis, Paris School of Mines, 2003.

[LAS 99] LASAUNIÈRE N., Etude des instabilités de bulle dans le procédé de soufflage de gaine de polyéthylène, Specialized Masters, Ecole des Mines de Paris, 1999.

[LEB 09] LEBAAL N., SCHMIDT F., PUISSANT S., "Design and optimization of three-dimensional extrusion dies, using constraint optimization algorithm", *Finite Elements Analyse and Design*, vol. 45, pp. 333–340, 2009.

[LUO 85] LUO X.L., TANNER R.I., "A computer study of film blowing", *Polymer Engineering & Science*, vol. 25, pp. 620–629, 1985.

[MAH 08] MAHDAOUI O., Etude numérique des instabilités dans le procédé de coextrusion des polyesters, Doctoral Thesis, Ecole des Mines de Paris, 2008.

[MAT 79] MATSUBARA Y., "Geometry design of a coat-hanger die with uniform flow rate and residence time across the die width", *Polymer Engineering & Science*, vol. 19, pp. 169–172, 1979.

[MCH 01] MCHUGH A.J., DOUFAS A.K., "Modeling flow-induced crystallization in fiber spinning", *Composites Part A; Applied Science and Manufacturing*, vol. 32A, pp. 1059–1066, 2001.

[MEN 72] MENGES G., PREDOHL W.O., "Film blowing process of LDPE", *SPE ANTEC Technical Papers*, vol. 18, pp. 717–721, 1972.

[MEN 98] MENDEZ-PRIETO A., SANCHEZ-VALDES S., "Multilayer structures using reprocessed coextruded LLDPE/TIE/PA-6 films", *Journal of Polymer Engineering*, vol. 18, pp. 221–234, 1998.

[MEN 09] MENG K., WANG X., HUANG X., "Optimal design of the coat-hanger die used for producing melt-blown fabrics by finite element method and evolution strategies", *Polymer Engineering & Science*, vol. 49, pp. 354–358, 2009.

[MIN 86] MINOSHIMA W., WHITE J.L., "Instability phenomena in tubular film, and melt spinning of rheologically characterized high-density, low-density and linear low-density polyethylenes", *Journal of Non-Newtonian Fluid Mechanics*, vol. 19, pp. 275–302, 1986.

[MUS 04] MUSLET I.A., KAMAL M.R., "Computer simulation of the film blowing process incorporating crystallization and viscoelasticity", *Journal of Rheology*, vol. 48, pp. 525–550, 2004.

[NAG 95] NAGARAJAN G., CAMPBELL G.A., "Experimental determination of heat transfer coefficient in blown film process", *SPE ANTEC Technical Papers*, pp. 162–166, 1995.

[PEA 70a] PEARSON J.R.A., PETRIE C.J.S., "The flow of a tubular film. Part 1. Formal mathematical representation", *Journal of Fluid Mechanics*, vol. 40, pp. 1–19, 1970.

[PEA 70b] PEARSON J.R.A., PETRIE C.J.S., "The flow of a tubular film. Part 2. Interpretation of the model and discussion of solutions", *Journal of Fluid Mechanics*, vol. 42, pp. 609–625, 1970.

[PER 95] PERDIKOULIAS J., TZOGANAKIS C., "Interfacial instability phenomena in blown film coextrusion of polyethylene resins", *SPE ANTEC Technical Papers*, vol. 53, pp. 176–180, 1995.

[PER 96] PERDIKOULIAS J., TZOGANAKIS C., "Interfacial instability in blown-film coextrusion of polyethylenes", *Plastics Engineering*, vol. 52, pp. 41–43, 1996.

[PET 73] PETRIE C.J.S., "Memory effects in a non-uniform flow. A study of the behavior of a tubular film of viscoelastic fluid", *Rheologica Acta,* vol. 12, pp. 92–99, 1973.

[PET 74] PETRIE C.J.S., "Mathematical modeling of heat transfer in film blowing", *Plastics and Polymers*, vol. 42, p. 259, 1974.

[PET 75a] PETRIE C.J.S., "Comparison of theoretical predictions with published experimental measurements on the blown film process", *AIChE J.*, vol. 21, pp. 275–282, 1975.

[PET 75b] PETRIE C.J.S., "The analysis of polymer processing operations involving free surface elongational flows", *Proceeding Conference Polymer Rheology and Plastics Processing*, pp. 307–319, 1975.

[SAI 84] SAILLARD P., AGASSANT J.F., "Polymer flow in a spiral mandrel die", *Polymer Process Engineering*, vol. 2, pp. 37–52, 1984.

[SAN 97] SANCHEZ-VALDES S., YANEZ-FLORES I., GUERRERO-SALAZAR C. *et al.*, "Characterization of LLDPE-LLDPEgMA blends by contact angle and FTIR-ATR studies", *Journal of Polymer Engineering*, vol. 17, pp. 257–268, 1997.

[SID 96] SIDIROPOULOS V., TIAN J.J., VLACHOPOULOS J., "Computer simulation of film blowing", *Journal of Plastics Film Sheeting,* vol. 2, pp. 107–129, 1996.

[SID 99] SIDIROPOULOS V., WOOD P.E., VLACHOPOULOS J., "The aerodynamics of cooling of blown film bubbles", *Journal of Reinforced Plastics and Composites,* vol. 18, pp. 529–538, 1999.

[SID 00a] SIDIROPOULOS V., VLACHOPOULOS J., "The effects of dual-orifice air-ring design on blown film cooling", *Polymer Engineering & Science*, vol. 40, pp. 1611–1618, 2000.

[SID 00b] SIDIROPOULOS V., VLACHOPOULOS J., "An investigation of Venturi and Coanda effects in blown film cooling", *International Polymer Processing*, vol. 15, pp. 40–45, 2000.

[SID 01] SIDIROPOULOS V., VLACHOPOULOS J., "Numerical study of internal bubble cooling (IBC) in film blowing", *International Polymer Processing*, vol. 16, pp. 48–53, 2001.

[TAD 79] TADMOR Z., GOGOS C.G., *Principles of Polymer Processing*, Wiley, New York, 1979.

[TAD 06] TADMOR Z., GOGOS C.G., *Principles of Polymer Processing*, 2nd ed., Wiley, New York, 2006.

[VER 80] VERGNES B., SAILLARD P., PLANTAMURA B., "Berechnungsmethoden für Breitschlitz-Extrusionwerkzeuge", *Kunststoffe*, vol. 70, pp. 751–753, 1980.

[VER 84] VERGNES B., SAILLARD P., AGASSANT J.F., "Non-isothermal flow of a molten polymer in a coat-hanger die", *Polymer Engineering & Science*, vol. 24, pp. 980–987, 1984.

[VER 08] VERGNES B., AGASSANT J.F., "Modélisation des écoulements dans les filières d'extrusion", AM 3655, *Techniques de l'Ingénieur*, Paris, 2008.

[VLC 88] VLCEK J., PERDIKOULIAS J., VLACHOPOULOS J., "Determination of output uniformity from spiral mandrel dies", *International Polymer Processing*, vol. 2, pp. 174–181, 1988.

[WAN 91] WANG Y., "The flow distribution of molten polymers in slit dies and coathanger dies through three-dimensional finite element analysis", *Polymer Engineering & Science*, vol. 31, pp. 204–2012, 1991.

[WU 06a] WU T., JIANG B., XU S., *et al.*, "Three-dimensional nonisothermal simulation of a coat hanger die", *Journal of Applied Polymer Science*, vol. 101, pp. 2911–2918, 2006.

[WU 06b] WU T., JIANG B., XU S., *et al.*, "Nonisothermal comprehensive 3D analysis of polymer melt flow in a coat-hanger die", *Polymer Engineering & Science*, vol. 46, pp. 406–415, 2006.

[YOO 91] YOON K.S., PARK C.W., "A study on two-layer blown-film coextrusion", *SPE ANTEC Technical Papers*, pp. 2315–2317, 1991.

[ZIP 71] ZIPPENFELD R., "Possibilities for increasing production from air cooled film blowing equipment", *Kunststoffe*, vol. 61, pp. 147–151, 1971.

[ZIA 67] ZIABICKI A., "Kinetics of polymer crystallization and molecular orientation in the course of melt spinning", *Applied Polymer Symposium*, vol. 6, pp. 1–18, 1967.

Chapter 7

Wire Coating and Cable Insulation

Some of the main applications of the extrusion process include metal wire coating and cable insulation. The term "wire" refers to a single strand of metal, normally copper, varying between 0.1 and 10 mm in diameter. A cable is usually two or more wires, normally coated, wrapped together and then insulated. Wire coating and cable insulation are used in the production of telephone, electrical and submarine communication cables, etc. The diversity of the polymers available for coating and insulating makes it possible for a wide range of products with different properties to be produced by extrusion. In Figure 7.1, we can see a telephone wire (Figure 7.1(a) copper conductor of 0.4 mm diameter, insulated with a polymer layer of a few tenths of a mm, extruded at up to 3,000 m/min) and energy transfer cables (Figure 7.1(b): metallic conductor of a few centimeters in diameter, surrounded by insulating layers of several centimeters, extruded at several m/min [PUI 06]; Figure 7.1(c): energy cable with three conductors).

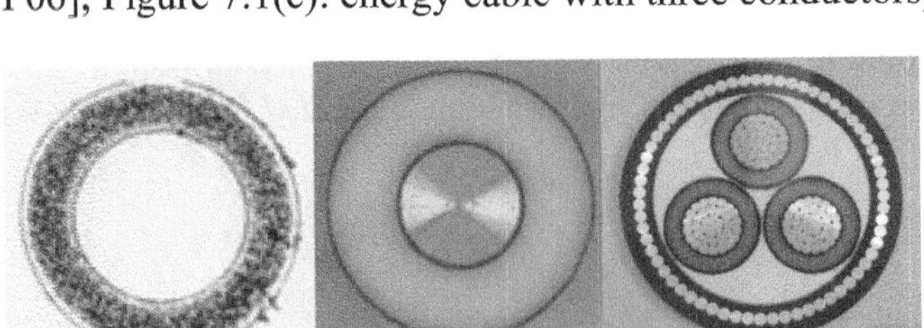

a) b) c)

Figure 7.1. *Examples of cables produced by extrusion: a) telephone wire, b) energy transfer cables and c) an energy cable with three conductors [PUI 06]*

7.1. General process

7.1.1. *Production line*

A typical wire coating production line is presented in Figure 7.2. First, the reel unrolls the conductor and feeds the cross-head die (i.e. perpendicular to the extruder) via a winch, whose role is to tighten the wire. The wire must be preheated before entering the die in order to better adhere to the polymer and also to remove any moisture from the conductor. Once inside the die, the wire is coated by the molten polymer coming from a connected extruder.

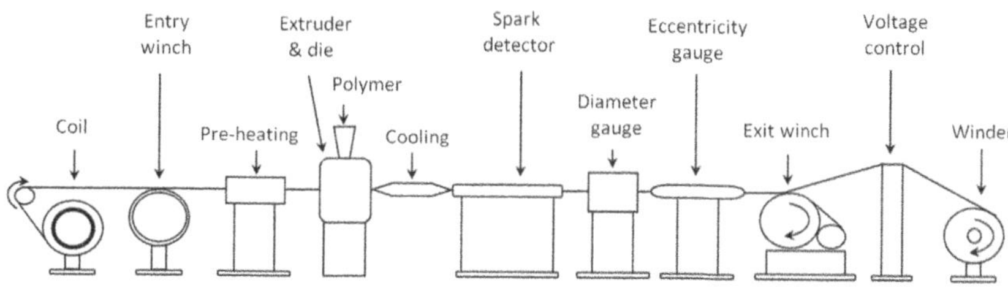

Figure 7.2. *Wire coating production line*

In wire production, extrusion screws have a diameter that varies between 35 and 250 mm, with a flow rate of between 100 and 160 kg/h. The length to diameter ratio is typically 20:1, with the molten material in the die having a pressure of approximately 55 MPa [POD 88]. At the die exit, the *coated conductor*, a term used interchangeably with wire here, enters the cooling tanks, whose length is determined by the diameter of the wire, the thickness of the polymer layer and the process speed. In some cases, the length can reach up to 100 m. It is important to detect any defects (lack of polymer, weak points) along the coated surface. For this reason, the conductor passes through a spark detector, where a high voltage is applied. If there is a small defect or if the coating (insulation) is inadequate, a discharge is observed and recorded. Generally, different parameters such as wire diameter, eccentricity of the conductor and the presence of air bubbles are measured and quantified using a variety of instruments installed on the production line. Finally, a winch draws the coated wire wound around a coil, automatically substituting a full coil for an empty coil to prevent production stoppages or slowdowns.

The production line must be adapted to the desired finished product. The elements of a wire coating line and those of cable insulation differ slightly:

the feeding and winding coils must account for the size of the cables. Similarly, coating metal rods (Figure 7.3) requires a specific hopper that is capable of feeding the die, as well as an adapted cooling system.

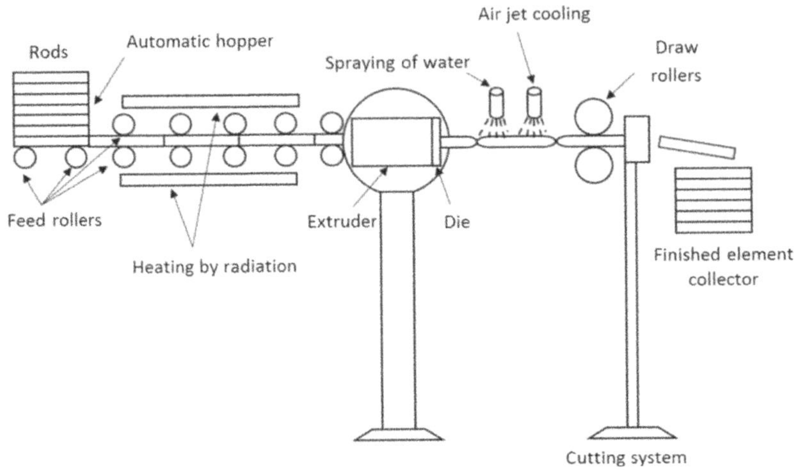

Figure 7.3. *An extrusion line for coating solid rods [LEV 78]*

It is also possible to coat a wire or insulate a cable with foam (expanded polymer). In fact, its low thermal conductivity, good ability to maintain a fixed shape and good resistance to moisture make an expanded thermoplastic a very attractive material for cable insulation. Blowing agents are then incorporated into the molten polymer and remain in the solution for the whole time spent in the extruder, as long as high pressure is applied. As soon as the material exits the die, the foam forms instantly due to the pressure drop [WIL 02]. This technique is, therefore, quite adaptable to wire coating and cable insulation.

7.1.2. *Wire coating dies*

7.1.2.1. *The different assemblies*

The cross-head die is the main element in insulation and coating by extrusion. Both pressure and tube assemblies are frequently used geometries, according to the type of the product manufactured (Figure 7.4).

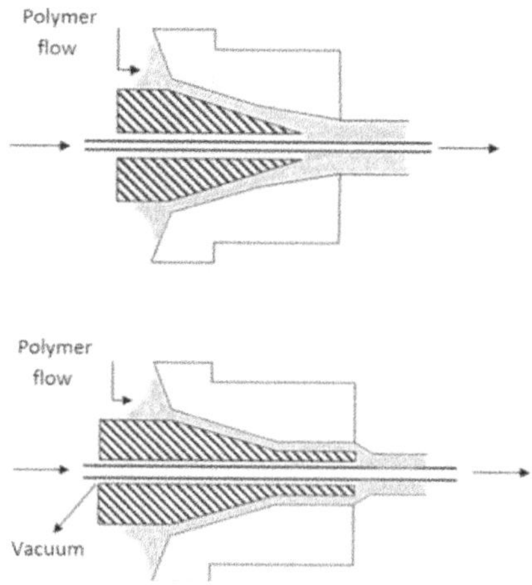

Figure 7.4. *Wire coating die: a) pressure assembly and b) tube assembly*

Typically, the wire or cable to be coated passes through the guide-wire (or mandrel) and is coated by molten polymer near the die exit. In the case of a pressure assembly, the molten polymer is pressurized before exiting the die. At the outlet, the wire is already coated, unlike in the tube assembly where a polymer tube is extruded around the conductor. In this case, the tube is pressed against the conductor outside of the die, by applying a vacuum in the guide-wire. The concentricity and thickness uniformity of the polymer covering the wire or cable can be adjusted by altering the bolts centering the die with regard to the guide-wire [RAU 01].

The pressure assembly, whose production rates can reach up to 3,000 m/min in the case of telephone wire coating, is also used for cable insulation. The tube assembly is essentially used for cable insulation, where the diameters are large (up to 10 cm) and the production rates are relatively low (approximately 5–10 m/min). The most fragile wires are also insulated using the tube assembly, since the tensile stress to draw the conductor through the molten polymer is lower than that in the pressure assembly.

The die is usually composed of two zones (Figure 7.5). The first zone is the feed zone, whose role is to distribute the molten polymer from the extruder around the conductor, with a 90° bend (hence the name cross-head). The second zone is the axisymmetric zone, which is composed of two sections: first, the Poiseuille zone where the polymer flows between two fixed boundaries, the mandrel and the housing; second, the shear zone where the polymer is only in contact with the conductor [VER 08]. In the case of a tube assembly, only the Poiseuille zone is present.

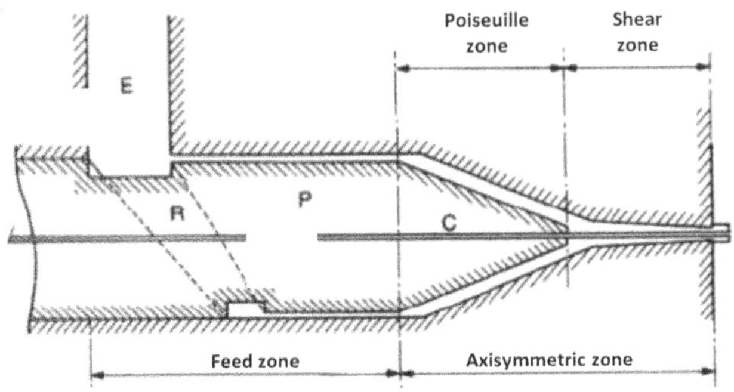

Figure 7.5. *Main zones of a wire coating die [VER 81]*

7.1.2.2. *Coextrusion*

It is also possible to adapt coextrusion to wire coating. One of the first studies on this topic was done by LeNir [LEN 74], who adapted the technique of Fenner and Williams [FEN 67] to coextrusion. This allowed two (or more) layers to be simultaneously extruded on the same support. The two molten polymers are extruded from different extruders and come into contact in the pressure-assembly die (Figure 7.6). The coextrusion line is identical to that used for simple extrusion, except for the die, and the presence of additional extruders [HAN 80, BAS 81]. The coextrusion process is often used because it is economical.

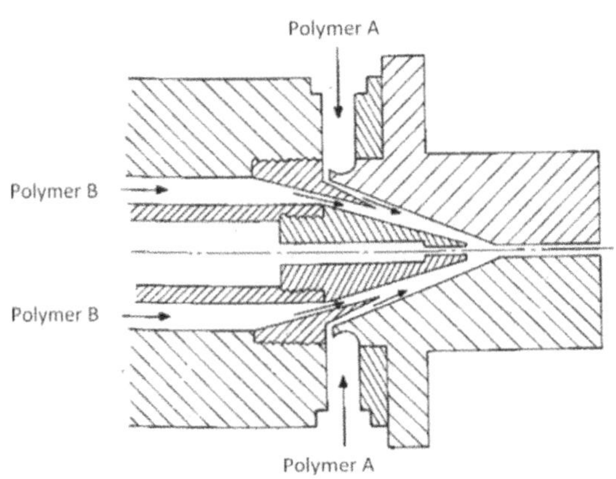

Figure 7.6. *Pressure assembly coextrusion die [HAN 80]*

7.2. Commonly encountered problems

Some problems encountered during wire coating and film production with a flat die (see Chapter 6) are quite similar. In fact, the geometry of the feed zone in the coating die looks like the one in the flat die. Therefore, problems with flow distribution, temperature, stagnant points, etc., may occur.

However, a variety of new challenges arise in the axisymmetric zone and particularly in the shear zone, mainly, in applications where the wire speed is high. They are associated with the high shear rates occurring in this zone. Damage to the conductor is a serious and recurrent problem: the polymer exerts a force on the wire, which depends on several factors including the geometry, the speed of the wire, etc. If the tensile strength on the wire increases beyond its limit, the conductor will break. Even if no breakage occurs, the conductor may undergo an irreversible deformation, which might degrade its electrical and other properties.

High shear rates can also lead to very significant local viscous heating [CAR 79], polymer degradation, as well as deterioration in the surface quality of the finished product (e.g. roughness). Flow instability can also cause surface defects when the wall stresses become excessive.

Sometimes, the converging geometry of the wire coating dies results in a significant elongational flow. This may create stresses that do not have time to relax inside the die, which leads to defects [EVE 74].

Finally, it is worth mentioning the centering of the conductor: a poorly centered conductor will have a non-uniform polymer coating. This can lead to a deterioration of the properties of the finished product.

There are also some peculiar problems associated with coextrusion, such as interfacial instability and migration, as well as poor adhesion between the polymer layers [HAN 80].

7.3. Analyses and solutions

Since the late 1960s, wire coating has been widely studied, both theoretically and experimentally.

7.3.1. *Theoretical studies*

Theoretical analyses mainly involve the axisymmetric zone of the pressure assembly die. In fact, problems are more rare in a tube assembly die, since the conductor velocity is low, and the flow is fairly standard (pressure flow only). In addition, unlike in the axisymmetric zone, there is no excessive temperature rise in the feed zone, where the polymer is evenly distributed around the conductor. Recent studies are, therefore, concentrating more on the shear zone, where the stresses are greatest.

7.3.1.1. *Analytical approach*

One of the first fundamental studies of wire coating dies was done by Fenner and Williams [FEN 67]. It involves a non-Newtonian fluid and applies the lubrication theory. It is based on the following assumptions:

– laminar flow;

– incompressible fluid;

– isothermal flow;

– no-slip at the walls;

– no elastic effects.

The die used for this analysis is shown in Figure 7.7, and it has become a reference die for many authors.

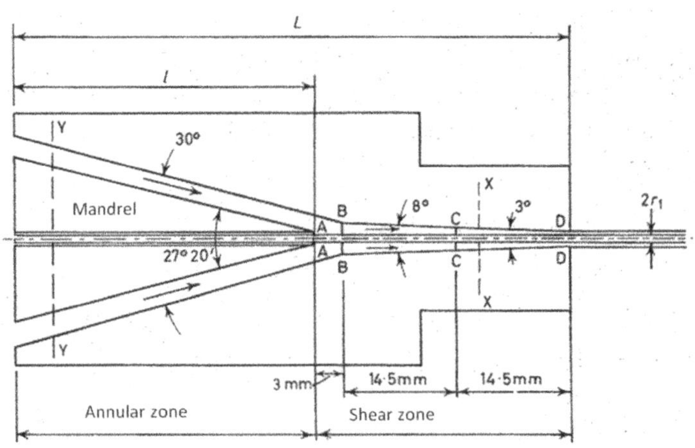

Figure 7.7. *Wire coating die of Fenner and Williams [FEN 67]*

Solving the flow equation gives the pressure gradient P_z in direction z for a given flow rate Q [FEN 67]. The flow equation, under the conditions listed above, is written in the axial direction:

$$\frac{1}{r}\frac{d}{dr}(r\tau)=\frac{dp}{dz}=P_z \qquad [7.1]$$

where p is the pressure and τ is the shear stress.

We assume that the shear stress follows the power law:

$$\tau=\eta_0\frac{dw}{dr}\left|\frac{dw}{dr}\right|^{m-1} \qquad [7.2]$$

where η_0 is the reference viscosity, m is the power-law index and $w(r)$ is the fluid velocity in the z-direction.

The mean shear stress $\bar{\tau}$, calculated for a mean shear rate V/H, is given by:

$$\bar{\tau}=\eta_0\left(\frac{V}{H}\right)^m \qquad [7.3]$$

where V is the conductor velocity and H is the radial gap between the conductor and the die wall.

Fenner and Williams [FEN 67] defined the annular flow parameter π_A as:

$$\pi_A=\frac{Q\bar{\eta}}{CH^2(-P_z)} \qquad [7.4]$$

where C is the mean circumference, such as $C=\frac{\pi}{2(D+d)}$, where D and d are the diameters of the die and the conductor, respectively. The negative sign is added for convenience, since the axial pressure gradient term P_z is usually negative. The mean viscosity of the polymer $\bar{\eta}$ is defined by:

$$\bar{\eta} = \eta_0 \left(\frac{Q}{AH} \right)^{m-1} \quad [7.5]$$

with A being the flow section: $A = \pi\ (D^2\text{-}d^2)/4$. Usually, π_A is a function of m and $\kappa = D/d$, but it becomes independent of κ when this value is small. When $\kappa = 1$, it becomes:

$$\pi_A = \left(\frac{2m}{2m+1} \right)^m \frac{1}{2^{2m+1}} \quad [7.6]$$

The mass flow rate of the polymer is constant and is written as follows:

$$Q_m = \rho V \frac{\pi}{4} \left[(d+2h)^2 - d^2 \right] \quad [7.7]$$

where h is the final thickness of the coating after cooling to room temperature and ρ is the density of the polymer at this temperature. For an incompressible isothermal flow, the volumetric flow rate is constant and it is expressed as:

$$Q = \frac{Q_m}{\rho_m} \quad [7.8]$$

with ρ_m being the density of the molten polymer at the processing temperature.

Thus, by defining $z = 0$ as the beginning of the annular zone, the initial pressure p is expressed by:

$$p = \frac{Q}{\pi_A} \int_0^l \frac{\bar{\eta}}{CH^3} dz + \int_l^L \frac{\pi p \bar{\tau}}{H} dz \quad [7.9]$$

where π_A and Q are calculated from the polymer properties and the geometry, using the previous equations, with all the other variables being known functions of z.

For example, Figure 7.8 shows the evolution of pressure, calculated by Fenner and Williams [FEN 67], for a polymer with a power-law index of $m = 0.48$. We observe a slight decrease in the Poiseuille zone, followed by a sharp increase in the shear zone, and a rapid decrease just before exiting the die.

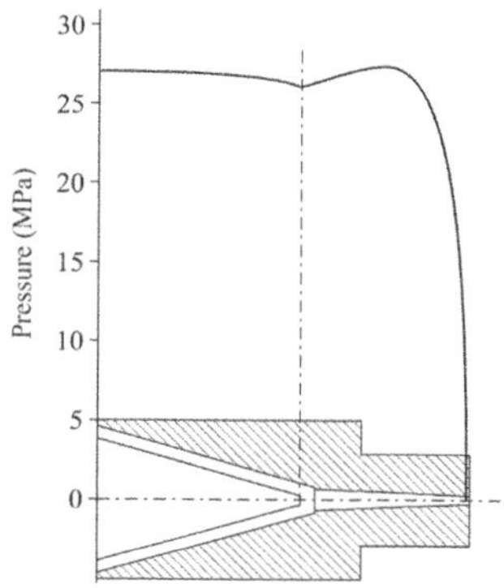

Figure 7.8. *Pressure profile in the wire coating die at a wire speed of 300 m/min [FEN 67]*

Two types of flow are in competition during wire coating: the shear flow (or drag flow), due to the velocity of the conductor passing through the die, which draws the polymer, and the flow exerted by the pressure gradient (pressure-driven flow) (Figure 7.9). Tiu [TIU 86] proposed a simplified approach for the geometry presented in Figure 7.10, based on the combination of both basic flows.

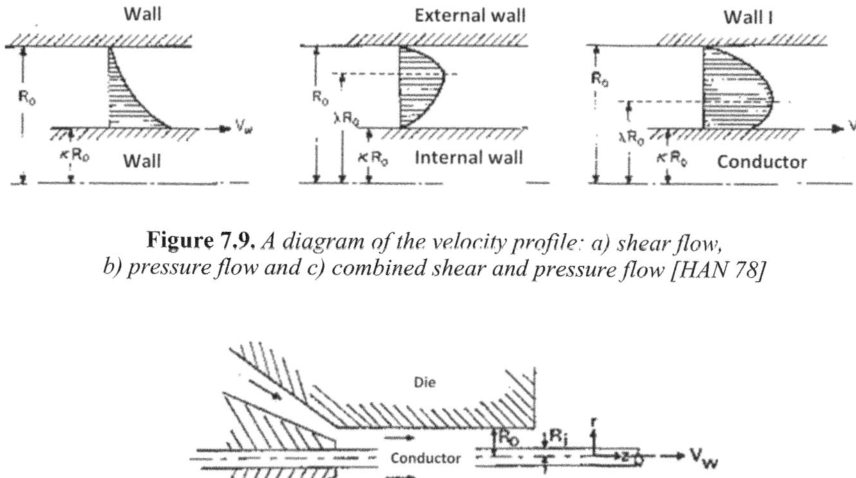

Figure 7.9. *A diagram of the velocity profile: a) shear flow, b) pressure flow and c) combined shear and pressure flow [HAN 78]*

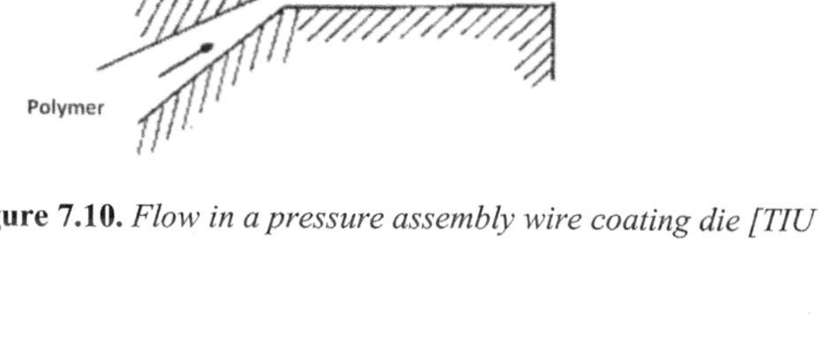

Figure 7.10. *Flow in a pressure assembly wire coating die [TIU 86]*

7.3.1.1.1. Shear flow

When a fluid flows between two coaxial cylinders in the z-direction, with the interior cylinder having a velocity V_w, the z component of the flow equations is written as:

$$\frac{d}{dr}(r\tau_{rz})=0 \qquad [7.10]$$

where τ_{rz} is the shear stress. After integration, we obtain:

$$\tau_{rz}=\frac{C_1}{r} \qquad [7.11]$$

The integration constant C_1 can be deduced using the boundary conditions:

$$r = R_i = k\,R_0,\ \tau_{rz} = \tau_i \text{ or } r = R_0,\ \tau_{rz} = \tau_0$$

where τ_0 is the shear stress at the boundary. Thus:

$$C_1 = \tau_0 R_0 = \tau_i R_i \qquad [7.12]$$

At this point, the values of τ_0 and τ_i are unknown and need to be determined. For a fluid whose properties are independent of time, the model generally used is:

$$\tau_{rz}=\eta\left(-\frac{dv_z}{dr}\right)=\eta\dot{\gamma} \qquad [7.13]$$

where the viscosity η can either depend on the shear stress τ_{rz} or the shear rate $\dot{\gamma}_{rz}$. By combining [7.11] and [7.13], and by integrating from r to R_0, the velocity due to shear is expressed by:

$$v_z=\int_r^{R_0}\frac{C_1}{\eta r}dr \qquad [7.14]$$

Using equation [7.12], the velocity as a function of τ_{rz} is as follows:

$$v_z=-\tau_0\,R_0\int_{\tau_{rz}}^{\tau_0}\frac{1}{\eta\tau}d\tau \qquad [7.15]$$

So, the volumetric flow rate Q_d due to shear is written as:

$$Q_d=\int_{R_i}^{R_0}v_z 2\pi r\,dr=2\pi R_0^2\int_k^1 v_z\xi\,d\xi \qquad [7.16]$$

where k is the R_i/R_0 ratio and ξ is the dimensionless radial position r/R_0.

Integration by parts, after changing the variable, gives:

$$Q_d' = \frac{Q_d}{2\pi R_0^2 V_w} = -\frac{k^2}{2} - \frac{\tau_0^2 R_0}{2V_w}\int_{\tau_i}^{\tau_0} \frac{1}{\eta\tau^3} d\tau \qquad [7.17]$$

Once the process parameters are known, the shear flow rate can be calculated according to the chosen rheological model.

7.3.1.1.2. Pressure flow

Similarly, the pressure flow rate can be determined for a uniform, isothermal laminar flow of an incompressible non-Newtonian fluid. The method involves obtaining two geometric parameters, *a* and *b*, which depend on the $k = R_i/R_0$ ratio and whose values were calculated by Kozicki *et al.* [KOZ 66], for an annular section (Figure 7.11).

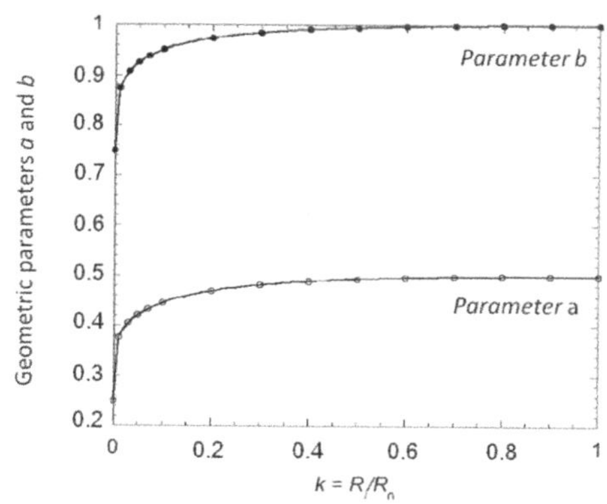

Figure 7.11. *Geometric parameters for an annular flow [KOZ 66]*

This method is based on the Rabinowitsch equation, which is written as:

$$\frac{2V}{r_H} = \frac{1}{a\tau_w^{b/a}}\int_0^{\tau_w} \frac{\tau^{b/a}}{\eta} d\tau \qquad [7.18]$$

where V is the mean velocity, r_H is the hydraulic radius and τ_w is the wall shear stress. Generally, the b/a ratio, which is equal to 3 for a circular pipe and 2 for parallel plates, is denoted by ζ.

The hydraulic radius r_H, for the annular section, is expressed by:

$$r_H = \frac{R_0(1-k)}{2} \qquad [7.19]$$

The mean wall shear stress, for a pressure drop Δp, is given by:

$$\tau_w = r_H \frac{-\Delta p}{L} \qquad [7.20]$$

If we express equation [7.18] as a function of the volumetric flow rate Q_p, due to the pressure gradient in the annular zone, we obtain:

$$Q_p' = \frac{Q_p}{2\pi R_0^2 V} = \frac{R_0(1-k)^2(1+k)}{8a\tau_w^{\varsigma}} \int_0^{\tau_w} \frac{\tau^{\varsigma}}{\eta} d\tau \qquad [7.21]$$

As with equation [7.17], equation [7.21] allows us to calculate the pressure flow rate for any type of behavior.

Table 7.1 summarizes the different expressions of Q'_p and Q'_d, according to the model used: Newtonian or power-law fluid. Note that the shear flow of the power-law fluid is undefined when $m = 1$ and $q = 1-1/m = 0$, m being the power-law index. For the pressure flow, the power-law model is reduced to the Newtonian model when $q = 0$ or $\alpha\,(\alpha = 1)$ and $\eta_0 = 2\eta$.

Behavior	Shear flow rate Q_d'	Pressure flow rate Q_p'
Newtonian	$\frac{k^2-1-2k^2\ln k}{4\ln k}$	$\frac{R_0^2(1-k)^3(1+k)}{16\eta V_w a(1+\xi)}\left(-\frac{\Delta p}{L}\right)$
Power law	$-\frac{k^2}{2}+\frac{q(1-k^{2+q})}{2(2+q)(1-k^q)}$	$\frac{(1-k^2)(1-k)^{2-q}}{a(1-q+\xi)}\frac{1}{4V_w}\left(\frac{R_0}{2}\right)^{2-q}\left(-\frac{\Delta p}{KL}\right)^{1-q}$

Table 7.1. *Non-dimensional shear and pressure flow rates for different behaviors [TIU 86]*

When the coated conductor leaves the die, the polymer is bound to the wire and it advances at the same velocity V_w. Middleman [MID 77] showed that the dimensionless thickness of the conductor coating is written as follows:

$$\delta' = \frac{\delta}{R_i} = \sqrt{\frac{\rho_m}{\rho_s}}\sqrt{\frac{\rho_s}{\rho_m} + \frac{2}{k^2}(Q'_d + Q'_p)} - 1 \qquad [7.22]$$

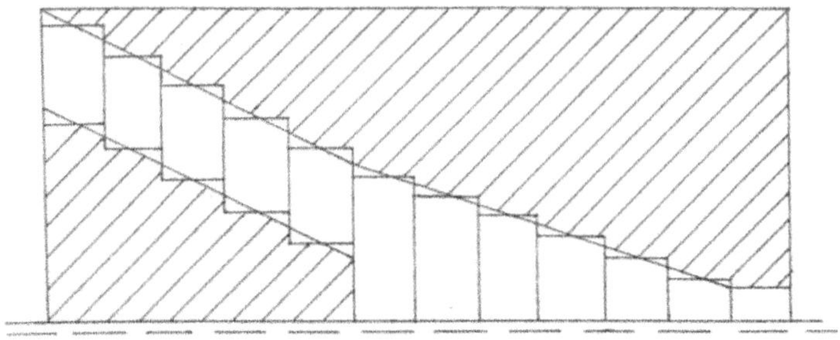

where δ is the thickness of the layer, R_i is the radius of the conductor and ρ is the density (m indicates that the material is molten and s indicates that the material is solid). This equation allows us to determine the thickness of the polymer layer on the wire, for any fluid whose behavior is independent of time, by using the previous expressions for Q_d' and Q_p'.

These methods have the main advantage of being analytical. However, their legitimacy may be questioned. In fact, we saw in Chapter 3 that the superposition of a pressure flow and a shear flow is only valid for the Newtonian behavior. For power-law fluids, we can no longer combine a shear flow and a pressure flow, which limits the applicability of such methods in practice. Nevertheless, they are still useful when simply calculating individual flows in this type of geometry.

7.3.1.2. *Semi-analytical approach*

More legitimate methods, based on the rigorous flow and heat transfer equations, were proposed during the 1980s [CAR 79, VER 81, BAS 81]. Here, we will present in detail the model proposed by Lafleur *et al.* [LAF 94]. As mentioned previously, the axisymmetric zone is composed of a Poiseuille zone and a shear zone. In the first zone, the polymer flows between two fixed boundaries whereas, in the second zone, the internal boundary moves at the velocity of the wire. Only the boundary conditions differ. The following assumptions are made:

– established laminar flow;

– permanent regime;

– non-Newtonian flow;

– non-isothermal flow;

– incompressible flow;

– application of the lubrication approximation;

– axisymmetric flow: $\partial/\partial\theta = 0$;

– non-elastic flow;

– no-slip boundary condition.

These assumptions give the following approximations:

$$\frac{\partial}{\partial z} \ll \frac{\partial}{\partial r}, \quad V_r \ll V_z, \quad V_z = V_z(r,z), \quad P = P(z) \qquad [7.23]$$

The analysis is carried out using a slab method in which each slab represents a uniform annular section as shown in Figure 7.12.

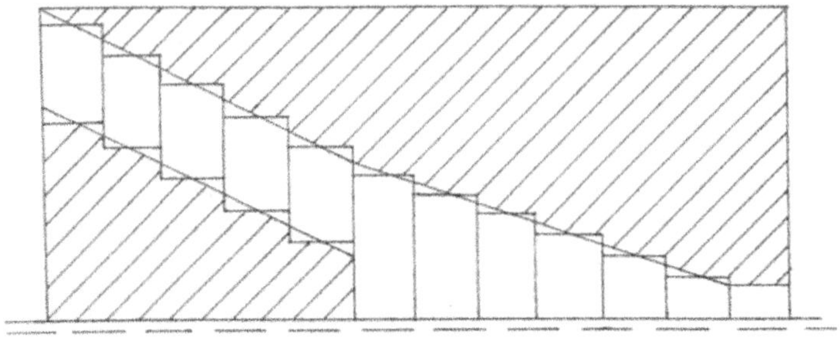

Figure 7.12. *The die divided into slabs*

The flow equation is discretized using the finite differences method: the section is divided into $N + 1$ elements. The position of the variables is shown in Figure 7.13.

The velocity boundary conditions imply no-slip conditions at all walls:

– Poiseuille zone: $V_z = 0$ at $r = R_i(z)$, $V_z = 0$ at $r = R_e(z)$;

– Shear zone: $V_z = V_c$ at $r = R_i(z)$, $V_z = 0$ at $r = R_e(z)$.

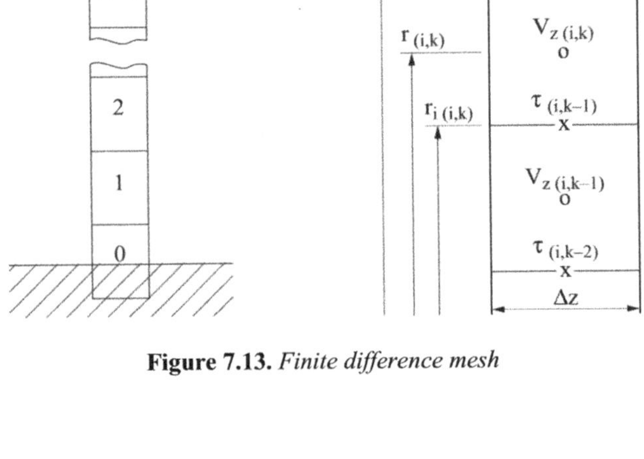

Figure 7.13. *Finite difference mesh*

7.3.1.2.1. Isothermal analysis

Taking the lubrication approximation into account, the flow equation is:

$$\frac{\partial P}{\partial z} = -\frac{1}{r}\frac{\partial}{\partial r}(r\tau_{rz}) \qquad [7.24]$$

The continuity equation is replaced by its integral form:

$$Q = 2\pi\int_{R_i(z)}^{R_e(z)} V_z(r,z)r\,dr \qquad [7.25]$$

Using difference quotient approximations, we obtain the following tridiagonal velocity matrix (AX = B):

A X B

$$\begin{bmatrix} B_2 & C_2 & 0 & \cdots & & & 0 \\ A_2 & B_2 & C_2 & 0 & \cdots & & 0 \\ 0 & A_4 & B_4 & C_4 & 0 & \cdots & 0 \\ \vdots & \vdots & \vdots & \vdots & \vdots & & 0 \\ & \vdots & \vdots & \vdots & \vdots & & 0 \\ 0 & & & A_{N-2} & B_{N-2} & C_{N-2} & 0 \\ 0 & & \cdots & 0 & A_{N-1} & B_{N-1} & C_{N-1} \\ 0 & 0 & & \cdots & 0 & A_N & B_N \end{bmatrix} \begin{bmatrix} V_z(2) \\ V_z(3) \\ V_z(4) \\ \vdots \\ \vdots \\ V_z(N-2) \\ V_z(N-1) \\ V_z(N) \end{bmatrix} = \begin{bmatrix} D - A_2V_z(1) \\ D \\ D \\ \vdots \\ \vdots \\ D \\ D \\ D \end{bmatrix} \qquad [7.26]$$

where the parameters are defined by:

$$A_k = \frac{r_i(k)\eta_{k-1}}{r_k},\; C_k = \frac{r_s(k)\eta_k}{r_k},\; B_k = -(A_k + C_k),\; D = \frac{\partial P}{\partial z}\Delta_r^2$$

The viscosities are calculated using the second invariant $II_{\dot{\gamma}}(k)$ and any rheological model:

$$II_{\dot{\gamma}}(k) = 2\left[\frac{V_z(k+1) - V_z(k)}{\Delta_r}\right]^2 \qquad [7.27]$$

To solve this system, the pressure gradient must be known. First, since this gradient is unknown, we perform a first approximation with the Newtonian equations. We assume that there is an initial gradient and the velocity profile is calculated. We then integrate the velocity profile (continuity equation) and compare it to the experimental flow rate Q. If the difference is too large, the pressure gradient is corrected according to equation [7.28] (iterative process):

$$\frac{\partial P^{(n+1)}}{\partial z} = \frac{\partial P^{(n)}}{\partial z} + \beta \frac{Q_z^{(n)} - Q}{Q} \qquad [7.28]$$

with β being the convergence factor. The choice of β is important since a too large value can cause divergence and a too small value can cause very slow convergence. $Q_z^{(n)}$ is the integrated flow rate at iteration n. The calculation algorithm of V_z is shown in Figure 7.14.

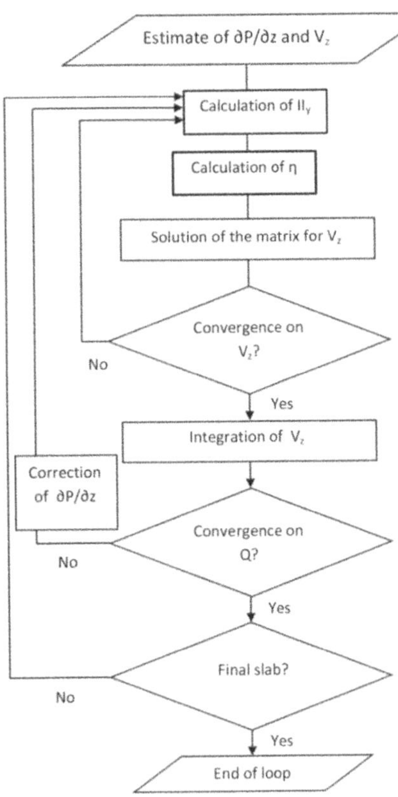

Figure 7.14. *Calculation algorithm of V_z for an isothermal analysis of wire coating*

After convergence on Q, we can move to the next slab calculation. The pressure gradient is only estimated once. For the following slabs, the gradient of the previous slab is used as a first estimation.

7.3.1.2.2. Non-isothermal analysis

Other assumptions about the non-isothermal calculation are that the physical properties are constant and the axial conduction is negligible compared to convection. The boundary conditions are expressed using the Biot number Bi, such that:

$$\frac{\partial T}{\partial z} = 0 \text{ in } z = 0 \qquad [7.29]$$

$$\frac{\partial T}{\partial z} = Bi \frac{T_r - T_p}{H} \text{ in } z = H/2 \qquad [7.30]$$

where H is the mid-height of the slab, T_r is the regulation temperature and T_p is the wall temperature. The Biot number is defined as $Bi = h_T L/k$.

The values most frequently used are the following:

– adiabatic regime: $Bi = 0$;

– isothermal regime: $Bi = \infty$;

– normal regime: $Bi = 10$ to 100.

Then, using the finite difference method, we obtain the following matrix system (AX = B):

$$\begin{bmatrix} 1+\frac{Bi}{N} & -1 & 0 & \cdots & & & 0 \\ A_2 & B_2 & C_2 & 0 & \cdots & & 0 \\ 0 & A_3 & B_3 & C_3 & 0 & \cdots & 0 \\ \vdots & \vdots & \vdots & \vdots & \vdots & & 0 \\ & \vdots & \vdots & \vdots & \vdots & & 0 \\ 0 & & & A_{N-1} & B_{N-1} & C_{N-1} & 0 \\ 0 & & \cdots & 0 & A_N & B_N & C_N \\ 0 & 0 & & \cdots & 0 & -1 & 1+\frac{Bi}{N} \end{bmatrix} \begin{bmatrix} T(i,1) \\ T(i,2) \\ T(i,3) \\ \vdots \\ \vdots \\ T(i,N-1) \\ T(i,N) \\ T(i,N+1) \end{bmatrix} = \begin{bmatrix} \frac{Bi\,Tc}{N} \\ D_2 \\ D_3 \\ \vdots \\ \vdots \\ D_{N-1} \\ D_N \\ \frac{Bi\,T_r}{N} \end{bmatrix} \qquad [7.31]$$

in which the parameters are defined by:

$$A_k = -r_{i(k)}\Phi\,,\ B_k = 2\Phi r_k + V_{z(k)} / \Delta z\,,\ C_k = -r_{s(k)}\Phi\,,$$

$$D_k = \frac{\eta_k \Pi_{\gamma(k)}}{2\rho C_p} + \frac{V_{z(k)}T(i-1,k)}{\Delta z}\,,\ \Phi = \frac{k}{\rho C_p r_k \Delta r^2}$$

The overall solution algorithm is summarized in Figure 7.15. The calculation of V_z and the pressure gradient form a part of the flow calculation.

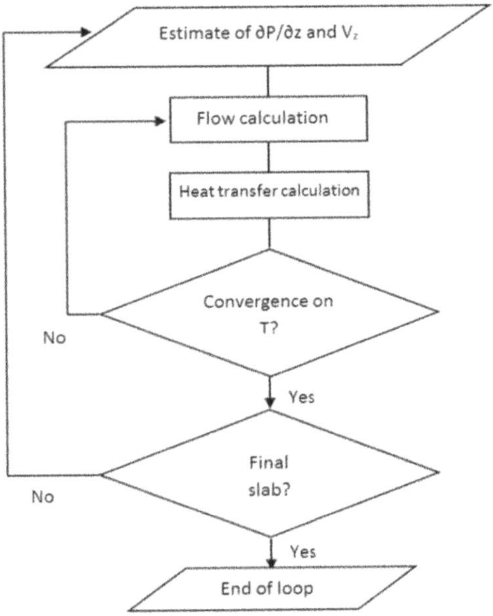

Figure 7.15. *Calculation algorithm of V_z for a non-isothermal analysis of wire coating*

For each slab, the average temperature can be calculated using the following integral:

$$T_{mel}(z) = \frac{\int_{R_i(z)}^{R_e(z)} V_z(r,z)T(r,z)\,r\,dr}{\int_{R_i(z)}^{R_e(z)} V_z(r,z)\,r\,dr} \qquad [7.32]$$

Figure 7.16 shows the pressure profile and average temperature as a function of the axial position. It is clear that the pressure drop is concentrated close to the die exit, which is a typical behavior for this process. The temperature increases continually, with a significant increase at the start of the shear zone and at the die exit.

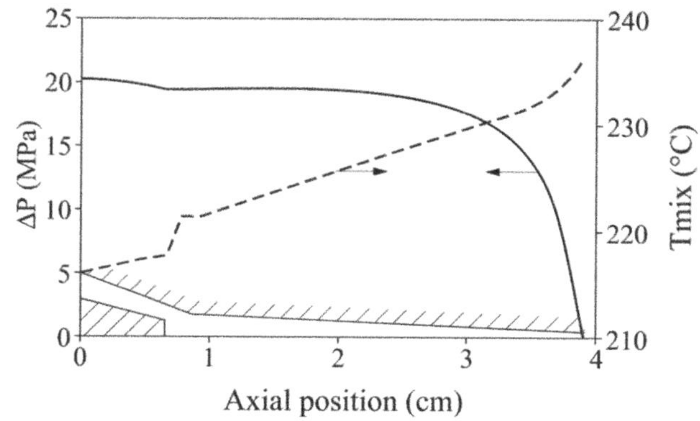

Figure 7.16. *Average temperature and pressure profiles (V_c = 100 m/min and Q = 5.09 kg/h) [LAF 94]*

7.3.1.2.3. Force of interaction

The force of interaction between the conductor and the molten polymer is obtained by integrating the shear stress on the conductor, for the whole shear zone:

$$F = \int_0^{2\pi} \int_0^{L_c} \tau_c R_c \, dz \, d\theta \qquad [7.33]$$

With the slab method, the double integral takes a discrete form:

$$F = \sum_{i=1}^{N_{zc}} \tau_c(i) 2\pi R_c \Delta z_c \qquad [7.34]$$

Figure 7.17 illustrates the influence of the conductor velocity on the maximum stresses (on both the die wall τ_f and the conductor wall τ_c) as well as on the force of interaction. The maximum stresses tend to stabilize at high velocity, whereas the force of interaction continues to increase. The former is often the limiting factor. Its maximum must be lower than the rupture force of the conductor to prevent breakage or deformation.

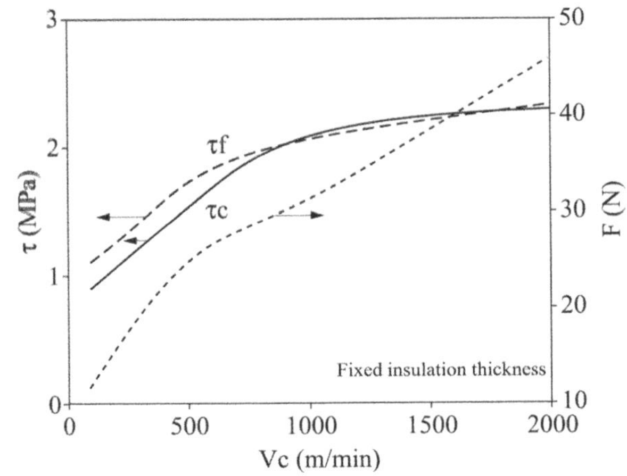

Figure 7.17. *Influence of the conductor velocity on the stresses and the force of interaction [LAF 94]*

7.3.1.3. *Finite element method*

In addition to the methods mentioned above, based on strong simplifications or on the lubrication approximation, more accurate 2D methods, which use the finite element method, have been available since the end of the 1970s [CAS 78, CAR 79, WAG 85, MIT 86].

The analysis of the axisymmetric zone, carried out by Mitsoulis [MIT 86], is one example of this. It analyzes the die of Fenner and Williams [FEN 67] and uses the Criminale–Eriksen–Filbey (CEF) rheological model. An example of a finite element mesh is shown in Figure 7.18. This involves dividing the calculation domain into elements (here triangular) for velocity/pressure calculations. Because of the axisymmetry of the problem, only a 2D domain needs to be considered. The mesh was designed to concentrate a large number of elements close to the impact zone and at the die exit, where the stresses are expected to be much greater. This mesh, which looks too coarse, was found to be adequate, at that time, since the results obtained were virtually identical to those obtained with a denser mesh (with more elements).

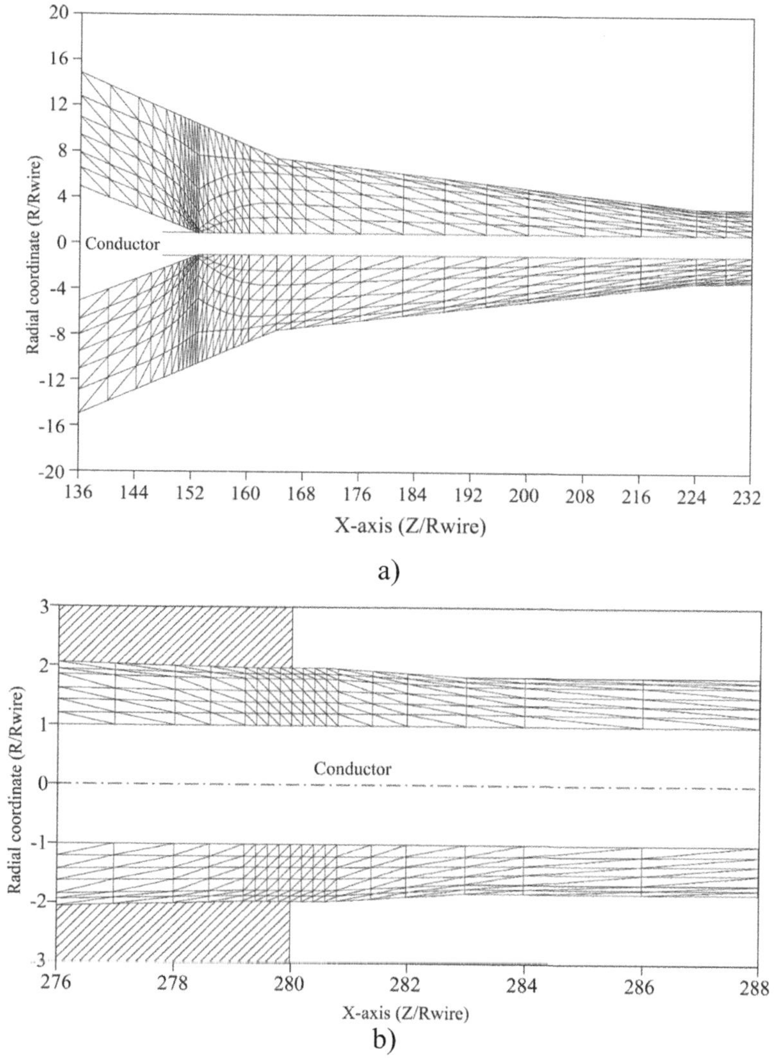

Figure 7.18. *a) Finite element mesh of the impact zone and b) finite element mesh at the die exit [WAG 85]*

Using the finite element mesh of the flow domain shown in Figure 7.19, the results for the axial pressure distribution were compared with those of Caswell and Tanner [CAS 78]. As shown in Figure 7.20, both studies are in agreement in the annular zone; however, this is no longer the case in the impact region, where the pressure is at its maximum. Numerous calculations have shown that the maximum pressure was underestimated when the mesh was less dense in this zone. Caswell and Tanner [CAS 78] used significantly fewer elements than Wagner and Mitsoulis [WAG 85]. It is, therefore, essential to optimize the discretization of the mesh to obtain a realistic prediction. The finite element method, therefore, allows us to obtain an accurate prediction of the flow conditions in the die and to modify it when necessary. It was also applied to viscoelastic materials in order to attempt to better estimate the stress field [MAT 00, MAT 02].

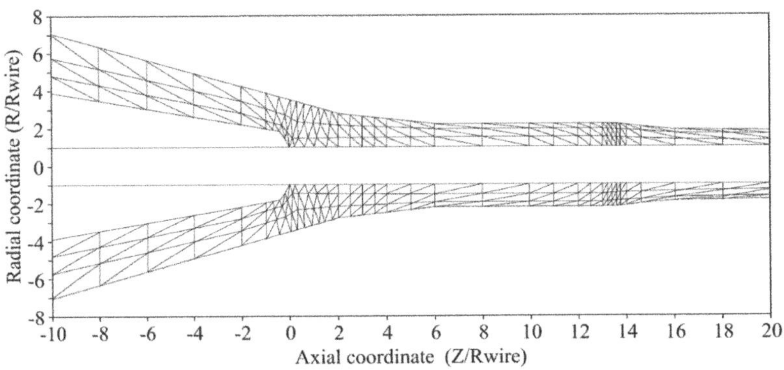

Figure 7.19. *Finite element mesh of the Caswell and Tanner die [WAG 85]*

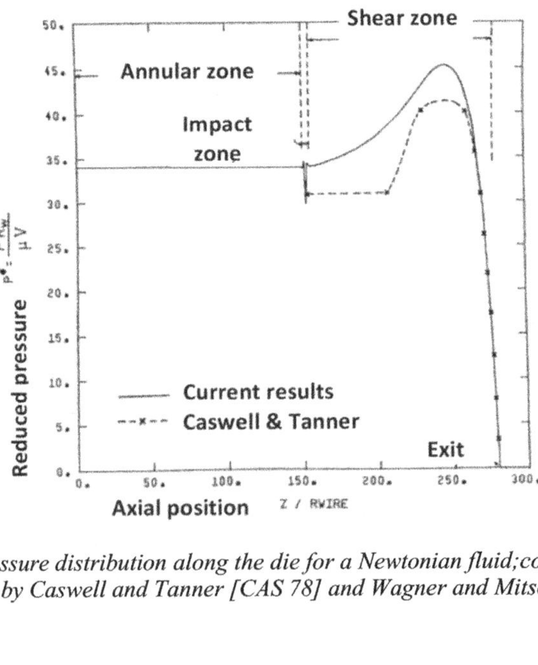

Figure 7.20. *Pressure distribution along the die for a Newtonian fluid;comparison of the results obtained by Caswell and Tanner [CAS 78] and Wagner and Mitsoulis [WAG 85]*

Interestingly, the finite element method also showed that there is a significant recirculation in the impact zone between the conductor and the polymer (Figure 7.21). When the power-law index decreases, the range and intensity of recirculation also decrease. When the index is low enough ($m < 0.35$, according to Mitsoulis [MIT 86]), this recirculation phenomenon disappears. However, this recirculation can also be revealed using one-dimensional (1D) methods based on the lubrication approximation presented above [VER 79].

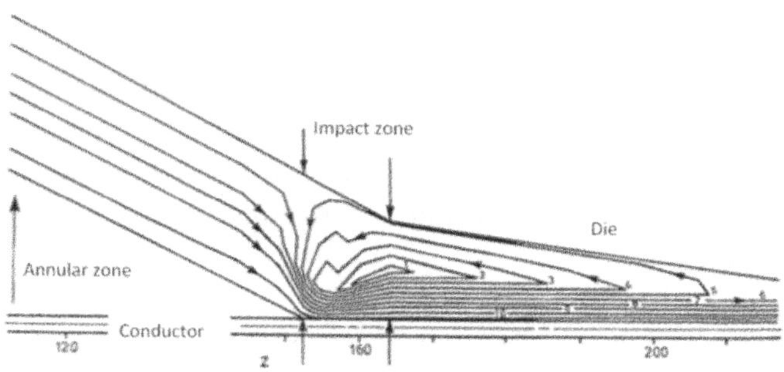

Figure 7.21. *Recirculation lines of a Newtonian fluid in the impact zone [CAS 78]*

At the die exit, some swelling of the Newtonian fluid is observed (Figure 7.22): the polymer free surface rearranges before the final coating radius is reached. It was shown that the normal stresses increase both the swelling at the die exit and the recirculation in the impact zone, where the stresses are greatest. However, they also help to reduce the pressure drop.

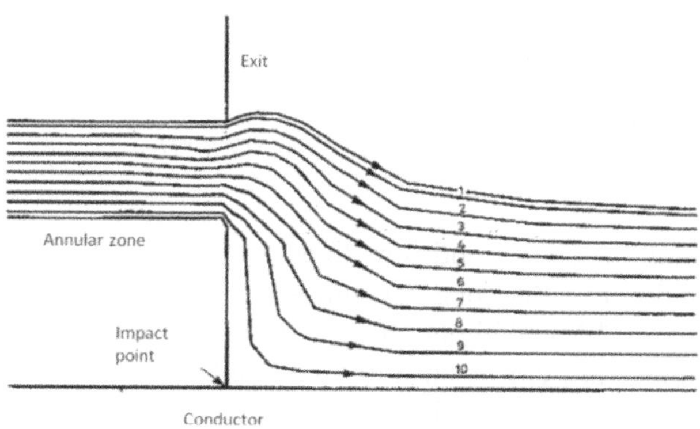

Figure 7.22. *Newtonian fluid swelling at the die exit in the case of a tube assembly [MIT 86]*

7.3.1.4. *Analysis of the feed zone*

Because of the more complex flow conditions, this section of the process has not been studied as extensively as the axisymmetric zone. However, it does play an essential role, since it must uniformly distribute the polymer, exiting the extruder, around the mandrel. As mentioned above, this flow is similar to that found in a flat die. It can, therefore, be calculated using 2D methods, based on the Hele–Shaw approximation, after having "unrolled" and flattened the geometry, as was the case for the pipe die (section 5.2.1.4) [FEN 79, VER 81, SCH 95].

For illustration purposes, Figure 7.23 shows the feed zone of a telephone wire coating die, once unrolled and flattened (due to symmetry, only half of the zone is considered). This zone includes a distribution channel with a constant thickness, and a decreasing width, followed by a flow zone of a much smaller thickness.

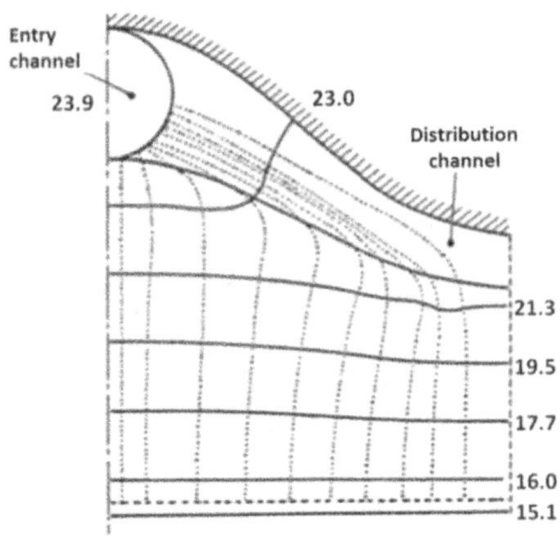

Figure 7.23. *Isobars in MPa (–) and flow lines (...) in the flattened feed zone [VER 81]*

After plotting the isobars and flow lines, we see that, under the employed processing conditions, this zone consumes a significant amount of pressure (approximately 9 MPa), without satisfactorily distributing the polymer. In fact, the flow lines are closer on the "right" side, which corresponds to the bottom part of the die. This will result in a larger flow rate in this area and, therefore, in a poor centering of the conductor. This is due to an improper design of the distribution channel, whose large cross-section results in less flow through the narrow zone and therefore causes an excess material flow in the lower part of the mandrel. As in many other cases, modeling could thus be used to define the optimal shape of this channel, so that the polymer is uniformly distributed, at the end of the distribution zone, i.e. at the start of the axisymmetric zone [LEB 12].

7.3.2. *Experimental studies*

Most experimental studies, carried out in the 1960s, tried to establish the effect of the die geometry on the quality of the finished product, more specifically its surface quality.

Hammond [HAM 60] studied the influence of the "land" length (final section of the die) as well as the angle of the conical section on surface quality. He showed that the quality improves when the angle decreases at a fixed wire speed. It, however, worsened when the wire speed was increased at a fixed angle. By using dies with different "land" lengths (L) and exit diameters (D), an optimum in surface quality was observed, at an $L/D \approx 11.4$ (fixed angle). King and Durno [KIN 64], however, believe that there is no optimum: the greater L is, the better the surface quality.

At high speeds, the two main problems are extrusion defects and breakage of the conductor, caused by an excessive force of interaction. In addition, this force of interaction, particularly in the cooling tank, limits the speed to a theoretical maximum of 3,500 m/min. One alternative would be to use a spray cooling system, where the isolated conductor is no longer immersed in water, but cooled by jets as it moves. A die with a diameter smaller than that of the coated conductor is also recommended, since a swell of 10% at the exit would decrease the force of interaction by 30%. However, this method can cause surface defects since there is also an increased pressure drop within the die.

The distance between the mandrel and the die (gum space) can also be optimized to obtain the best possible surface quality [HAA 74].

7.4. Conclusion

The wire coating and cable insulation industry is a very technically active sector, even though there are fewer theoretical studies than in other processes. Making use of the proper thermomechanical models, in this domain, has the potential to greatly improve the understanding and optimization of flow. This is particularly true in the two extreme cases: the production of telephone wires, due to high production rates, and the production of energy transport cables, due to the thick insulation and extremely sensitive cooling conditions.

7.5. Bibliography

[BAS 81] BASU S., "A theoretical analysis of non-isothermal flow in wire-coating co-extrusion dies", *Polymer Engineering & Science*, vol. 21, pp. 1128–1138, 1981.

[CAS 78] CASWELL B., TANNER R.I., "Wire coating die design using finite element methods", *Polymer Engineering & Science*, vol. 18, pp. 416–421, 1978.

[CAR 79] CARLEY J.F., ENDO T., KRANTZ W.B., "Realistic analysis of flow in wire-coating dies", *Polymer Engineering & Science*, vol. 19, pp. 1178–1187, 1979.

[CHU 86] CHUNG T.S., "The effect of melt compressibility on a high-speed wire-coating process", *Polymer Engineering & Science*, vol. 26, pp. 410–414, 1986.

[EVE 74] EVERAGE Jr. A.E., BALLMAN R.L., "A mechanism for polymer melt or solution fracture", *Journal of Applied Polymer Science*, vol. 18, pp. 933–937, 1974.

[FEN 67] FENNER R.T., WILLIAMS J.G., "Analytical methods of wire-coating die design", *Translations of the Journal of Plastics Institute*, vol. 35, pp. 701–706, 1967.

[FEN 79] FENNER R.T., NADIRI F., "Finite element analysis of polymer melt flow in cable-covering crossheads". *Polymer Engineering & Science*, vol. 19, pp. 203–209, 1979.

[HAA 74] HAAS K.U., SKEWIS F.H., "Wire coating process: die design and polymer flow characteristics", *SPE ANTEC Technical Paper*, vol. 20, pp. 8–12, 1974.

[HAM 60] HAMMOND L.R., "The design of dies for high-speed coating", *Wire and Wire Products*, vol. 35, pp. 725–783, 1960.

[HAN 78] HAN C.D, RAO D.A., "Studies on wire coating extrusion. I. The rheology of wire coating extrusion", *Polymer Engineering & Science*, vol. 18, pp. 1019–1029, 1978.

[HAN 80] HAN C.D., RAO D.A., "Studies on wire coating extrusion. II. The rheology of wire coating coextrusion", *Polymer Engineering & Science*, vol. 20, pp. 128–139, 1980.

[KIN 64] KING J.S., DURNO J.A., "The effect of die land length on the surface appearance of low melt index polyethylene extrudates", *Wire and Wire Products*, vol. 39, pp. 553–614, 1964.

[KOZ 66] KOZICKI W., CHOU C.H., TIU C., "Non-Newtonian flow in ducts of arbitrary cross-sectional shape", *Chemical Engineering and Science*, vol. 21, pp. 665–679, 1966.

[LAF 94] LAFLEUR P.G., ARPIN B., LENIR V., "Computer aided design of wire coating dies", *Advances in Polymer Technology*, vol. 13, pp. 297–304, 1994.

[LEB 12] LEBAAL N., PUISSANT S., SCHMIDT F.M. *et al.*, "An optimization method with experimental validation for the design of extrusion wire coating dies for a range of different materials and operating conditions", *Polymer Engineering & Science*, vol. 52, pp. 2675–2687, 2012.

[LEN 74] LENIR V., "Melt flow analysis in wire coating dies for dual extrudates", *Wire Journal*, vol. 7, pp. 59–64, 1974.

[LEV 78] LEVY S., "Process and equipment for in line post extrusion forming", *Plast. Mach. Equip.*, vol. 7, pp. 24–26, 1978.

[MAT 00] MATALLAH H., TOWNSEND P , WEBSTER M.F., "Viscoelastic multi-mode simulations of wire-coating", *Journal of Non-Newtonian Fluid Mechanics*, vol. 90, pp. 217–241, 2000.

[MAT 02] MATALLAH H., TOWNSEND P., WEBSTER M.F., "Viscoelastic computations of polymeric wire-coating flows", *International Journal of Numerical Methods in Heat Fluid Flow*, vol. 12, pp. 404–433, 2002.

[MID 77] MIDDLEMAN S., *Fundamentals of Polymer Processing*, McGraw-Hill, New York, 1977.

[MIT 86] MITSOULIS E., "Finite element analysis of wire coating", *Polymer Engineering & Science*, vol. 26, pp. 171–186, 1986.

[POD 88] PODOLSAK A.K., TIU C., "A review of wire coating and cable sheating extrusion processes", *Polymer and Plastics Technology and Engineering*, vol. 27, pp. 389–427, 1988.

[PUI 06] PUISSANT S., BONVIN P.Y., "Lignes d'extrusion en câblerie. Câbles, contrôle qualité, rentabilité", *Techniques de l'Ingénieur*, vol. AM3640, pp. 1–14, 2006.

[RAU 01] RAUWENDAAL C., *Polymer Extrusion*, 3rd. ed., Hanser, Munich, 2001.

[SCH 95] SCHLÄFLI D., "Analysis of polymer flow through coat hanger melt distributors", *Intern. Polym. Proc.*, vol. 10, pp. 195–199, 1995.

[TIU 86] TIU C., "Wire coating extrusion – an approximate solution", *Polymer Engineering & Science*, vol. 26, pp. 957–961, 1986.

[VER 79] VERGNES B., Modélisation de l'écoulement d'un polymère dans une filière de câblerie, PhD Thesis, Paris School of Mines, 1979.

[VER 81] VERGNES B., "Etude de l'écoulement d'un polymère fondu dans une filière de câblerie téléphonique", *Matériaux et Techniques.*, vol. 6, pp. 187–191, 1981.

[VER 08] VERGNES B., AGASSANT J.-F., "Modélisation des écoulements dans les filières d'extrusion", *Techniques de l'Ingénieur*, vol. A3655, pp. 1–20 2008.

[WAG 85] WAGNER R., MITSOULIS E., "Effect of die design on the analysis of wire coating", *Advances in Polymer Technology*, vol. 5, pp. 305–325, 1985.

[WIL 02] WILLETT J.L., SHOGREN R.L., "Processing and properties of extruded starch/ polymer foams", *Polymer*, vol. 43, pp. 5935–5947, 2002.

Index

D, E

F, G,

H, I, K

L, M, N

O, P, R